57 Springer Series in Solid-State Sciences

Edited by Manuel Cardona

Springer Series in Solid-State Sciences

Editors: M. Cardona P. Fulde H.-J. Queisser

40 **Semiconductor Physics -** An Introduction By K. Seeger

41 **The LMTO Method** Muffin-Tin Orbitals and Electronic Structure By H. L. Skriver

42 **Crystal Optics with Spatial Dispersion, and Excitons** By V. M. Agranovich and V. L. Ginzburg

43 **Resonant Nonlinear Interactions of Light with Matter** By V. S. Butylkin, A. E. Kaplan, Yu. G. Khronopulo, and E. I. Yakubovich

44 **Elastic Media with Microstructure II** Three-Dimensional Models By I. A. Kunin

45 **Electronic Properties of Doped Semiconductors** By B. I. Shklovskii and A. L. Efros

46 **Topological Disorder in Condensed Matter** Editors: F. Yonezawa and T. Ninomiya

47 **Statics and Dynamics of Nonlinear Systems** Editors: G. Benedek, H. Bilz, and R. Zeyher

48 **Magnetic Phase Transitions** Editors: M. Ausloos and R. J. Elliott

49 **Organic Molecular Aggregates,** Electronic Excitation and Interaction Processes Editors: P. Reineker, H. Haken, and H. C. Wolf

50 **Multiple Diffraction of X-Rays in Crystals** By Shih-Lin Chang

51 **Phonon Scattering in Condensed Matter** Editors: W. Eisenmenger, K. Laßmann, and S. Döttinger

52 **Superconductivity in Magnetic and Exotic Materials** Editors: T. Matsubara and A. Kotani

53 **Two-Dimensional Systems, Heterostructures, and Superlattices** Editors: G. Bauer, F. Kuchar, and H. Heinrich

54 **Magnetic Excitations and Fluctuations** Editors: S. Lovesey, U. Balucani, F. Borsa, and V. Tognetti

55 **The Theory of Magnetism II** Thermodynamics and Statistical Mechanics By D. C. Mattis

56 **Spin Fluctuations in Itinerant Electron Magnetism** By T. Moriya

57 **Polycrystalline Semiconductors,** Physical Properties and Applications Editor: G. Harbeke

58 **The Recursion Method and Its Applications** Editors: D. Pettifor and D. Weaire

59 **Dynamical Processes and Ordering on Solid Surfaces** Editors: A. Yoshimori and M. Tsukada

Volume 1-39 are listed on the back inside cover

Polycrystalline Semiconductors

Physical Properties and Applications

Proceedings of the International School
of Materials Science and Technology
at the Ettore Majorana Centre, Erice, Italy,
July 1–15, 1984

Editor: G. Harbeke

With 159 Figures

Springer-Verlag
Berlin Heidelberg New York Tokyo 1985

Dr. Günther Harbeke

Laboratories RCA Ltd., Badenerstr. 569
CH-8048 Zürich, Switzerland

Series Editors:

Professor Dr. Manuel Cardona
Professor Dr. Peter Fulde
Professor Dr. Hans-Joachim Queisser

Max-Planck-Institut für Festkörperforschung, Heisenbergstrasse 1
D-7000 Stuttgart 80, Fed. Rep. of Germany

ISBN-13: 978-3-642-82443-2 e-ISBN-13: 978-3-642-82441-8
DOI: 10.1007/978-3-642-82441-8

Softcover reprint of the hardcover 1st edition 1985

Bookbinding: J. Schäffer OHG, 6718 Grünstadt.
2153/3130-543210

Preface

In terms of structure, the field of semiconductors spans a wide range, from the perfect order of single crystals to the non-periodic, disordered amorphous state. The two extremes of this range attract a large amount of interest. On one side, glamorous novel phenomena are being found which can only occur in specially tailored ultra-perfect periodic lattices. On the other side, the exotic and challenging nature of the amorphous state has triggered a surge of activity in recent years.

Polycrystalline semiconductors are in between. They are among the workhorses in the field, useful in many applications, a handy solution to many practical problems and still - they have not received in the past the amount of research interest that they deserve. It is the aim of the present book to improve this situation. The book originated from the lectures and seminars presented at the course on "Polycrystalline Semiconductors - Physical Properties and Applications" of the International School on Materials Science and Technology, held at the Centre for Scientific Culture "Ettore Majorana" in Erice, Italy, July 1-15, 1984.

The physics of polycrystalline semiconductors is largely the physics of grain boundaries. Grain boundaries play an essential part in polycrystalline semiconductors. The charge carriers perhaps see themselves in the role of the traveller in ancient times who could fairly easily traverse wide areas of flat country or long stretches of a calm sea but had great difficulties in crossing steep mountain ridges and deep river gorges.

The present volume is divided into four parts. In the first part, the fundamental aspects of the physics of grain boundaries are laid out. The use of modern research techniques has greatly increased our knowledge about the detailed atomic structure of grain boundaries and the type of defects which may exist in them. This knowledge forms the basis for total energy calculations and facilitates the discussion of atom migration phenomena. In the second part, the present state of electronic characterization of grain boundaries is reviewed. The authors deal with the important issues of electronic states at and charge transport across grain boundaries. One of the major objectives of these studies is to establish a relationship between the electronic properties of a single boundary in a bi-crystal (ideal situation, easier to study) and those of films with a virtually infinite number of grain boundaries (real situation, of practical interest).

The remaining two parts are mainly devoted to a discussion of the physics involved in using polycrystalline semiconductors as parts of active or passive electronic and optoelectronic devices. Polycrystalline silicon offers the invaluable advantage of a high degree of interfacial compatibility with single crystal silicon and silicon dioxide in microelectronics. Besides its crucial function in integrated-circuit technology it is also a key material for polycrystalline solar cells. Polycrystalline structures of semiconducting compounds such as ZnS and various oxides have found their place in the technology of electroluminescent displays or sensors and varistors, respectively. All the chapters on device physics stress the necessity of a fruitful exchange between fundamental aspects, materials characterization and the technological application. This fruitful exchange became a major achievement of the Erice lecture course, as is reflected in this book.

I wish to thank all lecturers and participants for their stimulating contributions to the success of the course and to the present book. We all have expressed our appreciation for the fine facilities afforded us by the Ettore Majorana Centre and are indebted to its Director, Professor A. Zichichi, to Dr. A. Gabriele, Dr. Pinola Savalli and the staff of the Centre. I am grateful for the generous support, in particular for student grants, given by the sponsoring institutions European Physical Society, European Research Office, General Electric Co. Wembley, IBM Italia, Italian Ministry of Education, Italian Ministry of Scientific and Technological Research, Italian CNR Group of the Structure of Matter, Laboratories RCA Zurich, Sicilian Regional Government and Siemens AG Munich. Last but not least, I appreciate the advice of the Director of the School, Professor M. Balkanski, and of many friends and colleagues in the field of semiconductors.

Zurich, Switzerland
December 1984

Günther Harbeke

Contents

Part I

Fundamental Aspects of Grain Boundaries

Atomic Structure of Grain Boundaries

A. Bourret

Département de Recherche Fondamentale - Physique du Solide, CEN-Grenoble
F-85 X-38041 Grenoble Cedex, France

ABSTRACT The geometrical description of a general grain boundary in a crystalline structure is first summarized. This includes the definition of the coïncidence site lattice (CSL) at particular twin angles and the DSC lattice. Then the dislocation modelling is presented using primary and secondary dislocations. However it is emphasized that this description is not unique. Therefore it is necessary to introduce the structural unit model which describes the core structure of the boundary. A detailed description of these units in cubic diamond structure and their relation to dislocation core structure will be given. A review of the experimental techniques giving some information on atomic structure is made. Few selected examples particularly using the high resolution electron microscopy technique are used to illustrate the different theoretical approaches with emphasis on the structural unit model. It is concluded that our knowledge of the atomic structure of grain boundary in semiconductors is still incomplete.

1. INTRODUCTION

Grain boundaries (G.Bs) are one of the most common two dimensional extended defects in crystals. Their influence on a large variety of physical properties of materials has been largely recognized. Firstly emphasis was given to the influence of GB on mechanical properties at high temperature (G.B. creep or sliding), or on transport properties (enhanced diffusion along G.Bs). In semiconductors their influence on electrical properties is of prime importance since the growing interest in employing cheap polycrystalline silicon. It was remarked for a long time, that some G.Bs have practically no effect on electron mobility, although some others have a deleterious effect and affect the efficiency of photovoltaic solar cell : it was clear that such differences should be attributed to the local structure of the G.B. The earliest ideas on the G.B. atomistic structure were given by Rosenhaim and Humphrey [1] who proposed that an amorphous layer is present between the two crystals. Although this old model was discarded for many years it has been proved to be valid in some multicomponent ceramics and sintered powder [2]. However in all other cases and specially in metals or semiconductors it rapidly became evident that GBs were closely related to the crystalline nature of the material : for instance the orientation dependence of many macroscopic properties is well established. Hargreaves and Hills [8] first proposed models based on a periodic medium at the G.B. interface composed of one atomic plane occupying intermediate positions right between the two crystals (transition lattice theory). Then Mott [9] proposed the island model with region of good fit separated by region of poor fit. One of the first observations by etch pit or copper decoration of small angle G.B. in semiconductors [3] gave a firm basis to the idea proposed originally by Burgers [10] that a sub G.B. is composed of a regular array of dislocations. This was formalized by Brandon et al. [11] and developed further by Bollmann [4] who introduced the coincidence site lattice (CSL) model and its extension, the O-lattice model. The existence of secondary relaxations predicted by Bollmann was experimentally confirmed by Schober and Balluffi [5] who observed regular networks of defects in high angle G.B.

The crystal-like structure of G.B., generally a superlattice of the crystal lattice was demonstrated at special misorientations by electron [6] and X-ray diffraction [7]. Applying these concepts to semiconductors and using tetracoordinated atoms, Hornstra firstly proposed a detailed atomistic description of G.Bs [12] [13]. This pioneer work which has introduced the concept of structural unit, was also developed from computer simulations on cubic metals by Sutton and Vitek [14]. The lack of direct experimental tools to study the atomistic structure of G.Bs had prevented testing these models until recently. However with high resolution electron microscopy and the development of diffraction techniques it is now possible to make a direct comparison with the models [15].

This paper is divided into three parts. Firstly the geometrical G.B. models are presented. Then the dislocation model is discussed with particular emphasis on the non uniqueness of such a description. Finally the structural unit model in diamond cubic structure is compared with a few recent experimental results.

2. CRYSTALLOGRAPHIC OR GEOMETRICAL DESCRIPTION OF A G.B.

A G.B is an interface separating two identical crystals : I and II. The G.B state depends, in addition to standard thermodynamical variables, on geometrical variables. The variables are the following :

i) Three angular variables specify the rotation axis, $\vec{\Omega}$, and the rotation angle θ
ii) Two angular variables specify the orientation of the G.B. normal, $\vec{N}$
iii) Three translation variables define the rigid body translation, $\vec{\tau}$, between crystal I and II
iiii) An additional translation variable defines the G.B. location. In structures containing one atom type, this additional parameter is not independent of $\vec{\tau}$.

Therefore the total number of independent geometrical parameters is either 8 or 9.

The transformation crystal I → crystal II is however non unique. One introduces the space operator (an isometry) $(\Omega_\theta/\vec{\tau})$ defined by [16] :

$$\vec{r}_{II} = (\Omega_\theta/\vec{\tau})\ \vec{r}_I = \Omega_\theta\vec{r}_I + \vec{\tau} \tag{1}$$

where Ω_θ is a rotation operation with rotation axis passing at /0,0,0/ site of both crystals. However every point r'_I deduced from r_I by any $(g/\vec{t})$ space operation of the crystal space group G_0 is equivalent and defines the same boundary [17]. The set of equivalent operators associated with the same boundary is called the coset of the boundary [18] and denoted by :

$$\{(\Omega_\theta/\vec{\tau})G_0\} \tag{2}$$

In the particular case of the diamond structure (Si, Ge) the space group is holohedral Fd3m with two fcc sublattices translated from each other by 1/4, 1/4, 1/4. In the sphalerite structure (GaAs, CdTe ...) the space group is hemihedral F $\bar{4}$ 3m with two different partial sublattices. In the Fd3m structure there are generally twentyfour different rotations with an angle of rotation θ not exceeding π that can bring two arbitrary grains to equivalent orientations.

In the past, grain boundaries have been classified as "special" and "general". Friedel [19] first proposed to associate "special" coherent boundaries to the presence of an undisturbed fraction of the structure when crossing the interface. Generalizing this idea, for "special" relative orientations a three dimensional lattice of coincident points is present when one considers the

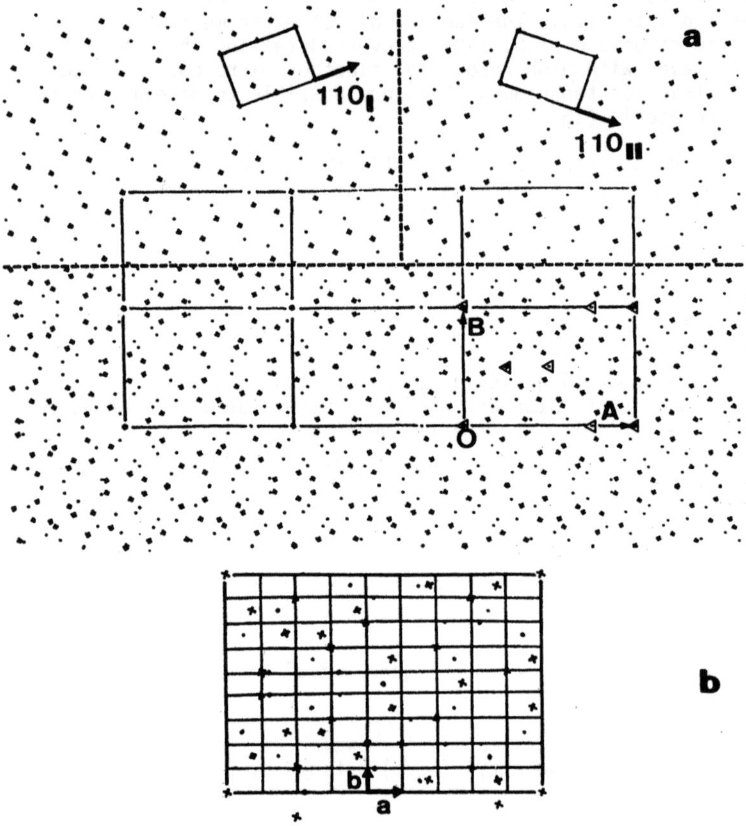

Fig. 1 a) Definition of the coincidence site lattice (CSL) by superposition of crystal I and II in a Σ=9 twin position (θ=38°94) as projected along the <011> direction in a diamond cubic lattice. The unit cell of the CSL is centered with A=<122>, B=1/2<411>, C=1/2<011>. The rigid body translation has been chosen to maximize the number of atomic sites in coincidence (4 per unit cell). A G.B has been formed along a mirror glide plane of the bicrystal giving a structure which contains this symmetry element. Crosses (+) are at level 0 and points (.) at level 1/2. Note that there are no exact coincident atoms close to the G.B interface. Coincident atoms are marked by Δ.
b) The enlargement of one unit cell of the CSL showing the DSC lattice in Σ=9 twin position. The DSC lattice is also centered with a=1/9<122>, b=1/18<411>, c=1/2<011>. Note that only one set of the two f.c.c. lattices composing the diamond lattice lies on the DSC lattice.

two interpenetrating infinite crystal I and II. This lattice is called the coincidence site lattice or CSL : it is usually constructed with a common lattice point at the origin of both crystals (fig. 1).

The ratio of density of crystal lattice sites to the density of coincidence sites defines Σ. Σ characterizes each CSL and any "special" or generalized twin position.

A coincidence site G.B plane is described by a plane intersecting the CSL lattice. A perfect unrelaxed bicrystal is produced when sites of crystal I are left on one side and sites of crystal II on the other side of the G.B plane. Such a description provides a general framework for defining the crystallographic aspects of a G.B. It is always possible for any misorientation to approximate the relative position of I and II, as precisely as required, by one coincidence position. For instance in cubic crystals the angle of coincidence misorientation and their Σ values are given by

$$\theta = 2\ tg^{-1}\ [\frac{m}{n}\ \sqrt{h^2 + k^2 + l^2}] \tag{3}$$

where m and n are integer, hkl define the rotation axis $\vec{\Omega}$, and

$$\Sigma = m^2 + n^2(h^2 + k^2 + l^2\ )\ \text{modulo}\ 2 \tag{4}$$

These values are tabulated in references [20] and [21]. For hexagonal crystals the CSLs have been examined by Delavignette [22].

The CSL can be generalized using coincident equivalent inner cell points instead of coincident lattice points. This is the so called "O-lattice" introduced by Bollmann [4]. For any given misorientation an O-lattice can be found as the inner points of the unit cell form a continuum. If $\underline{\underline{\Omega}}$ is the rotation matrix describing the transformation I → II, the O-lattice vectors $\vec{x}^{(o)}$ defined in crystal I are given by :

$$(\underline{\underline{I}} - \underline{\underline{\Omega}}^{-1})\ \vec{x}^{(o)} = \vec{b}_I \tag{5}$$

where $\underline{\underline{I}}$ is the matrix unity and $\vec{b}_I$ one of the crystal I translation. Although the matrix $\underline{\underline{\Omega}}$ depends on the particular choice among all the equivalent operations transforming I → II, it can be shown that the O-lattice is unique [18]. For the special coincident orientation however, the O-lattice should not be confused with the CSL : the CSL lattice is a sublattice of the O-lattice.

It is also necessary, to complete this crystallographic approach, to introduce the DSC lattice (fig. 1). It describes all the translations which leave invariant the bicrystal : any vector joining the lattice points of the two interpenetrating crystals is a DSC vector. Expressed in a different way the DSC lattice is the union of the translation group of crystal I by crystal II. It contains as subgroups the CSL, the O-lattice, the lattice I and the lattice II. It should be noted that during a transformation with a vector d_{DSC} of the DSC lattice :

$$\vec{\tau} = d_{DSC}$$

the coset defining the G.B is unchanged but the CSL is displaced owing to the displacement of coincident sites. For a tabulation of the CSL in cubic structures see [21]. For non special orientations the DSC lattice covers all space.

The above presentation has several limits :

- the CSL lattices and DSC are defined with reference to lattices points which should be clearly distinguished from the true atomic positions. Therefore the introduction of $\vec{t} \neq o$ modulo d_{DSC} will not change the CSL although it can bring new atomic positions into coincidence : the translation $\vec{t}$ = 1/4 [111] in diamond cubic crystals is one of these special translations.
- The O-lattice and DSC consider only translational symmetry which is crystallographically insufficient to describe any structures. In particular non symmorphic structures (including mirror glide planes and screw axes) should be characterized by their complete symmetry elements.

These limitations are clearly overcome by following the crystallographic methodology [18] [23]. This approach will be developed further during this course by R.C. Pond.

It is clear that a purely geometrical description, although giving a firm basis for definition of a bicrystal and a G.B, cannot predict what will be really observed. This approach is however fruitful as it can predict :

- the symmetry dictated extrema of the energy functional [18]
- the number of variants giving equivalent interfaces and their symmetry relation [23]
- new types of defects at intersections of domains having equivalent interfaces [23].

If one wants now to progress in this description it is important to introduce some energetical criterium : this was the original idea of the dislocation model.

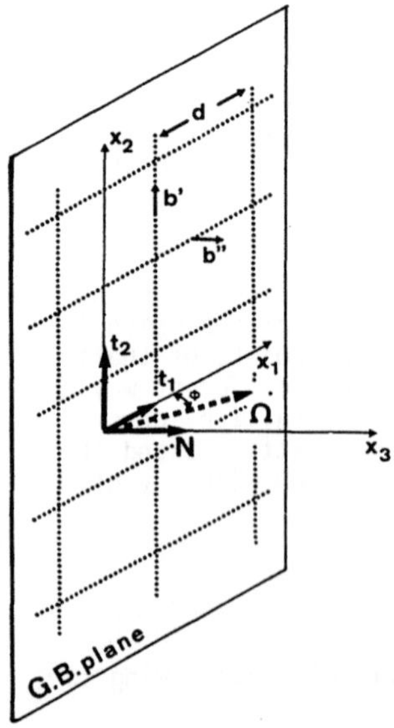

Fig. 2 Schematic representation of the G.B plane, with the rotation axis Ω and the two dislocation families which are the minimum necessary to accomodate a disorientation between crystal I and II.

3. DISLOCATION MODEL OF G.B

The dislocation theory and the extension of the elasticity theory to include plastic deformation enable calculation of stresses, strains and as a consequence elastic energy stored by the crystal after deformation. Can this theory be extended to G.Bs ?

Formally it has been shown by Bilby et al. [24] and Kröner [25] that the average surface dislocation density within a G.B may be expressed by

$$\bar{\underline{\underline{\alpha}}}^S = \vec{N} \times \underline{\underline{D}}_{II} - \vec{N} \times \underline{\underline{D}}_I \tag{6}$$

$$\text{or } \bar{\alpha}_{ij} = -\varepsilon_{ikl}\, n_k\, D^{II}_{jl} + \varepsilon_{ikl}\, n_k D^{I}_{jl} \tag{7}$$

where $\bar{\alpha}_{ij}$ gives for a Burgers vector component j and a dislocation along the i direction ($\alpha_{ij} = \frac{1}{d}\, t_i b_j$) the average number of dislocations per unit length in the G.B. plane (fig. 2). ε_{ijk} is the usual permutation symbol.

$\underline{\underline{D}}$ is a tensor describing the deformation from a reference lattice crystal I or crystal II, $\vec{N}$ is the unit vector normal to the interface. As an example this reduces, if I is taken as a reference lattice, and if the transformation I → II is taken as a rotation θ around an axis Ω to : [27]

$$\bar{\underline{\underline{\alpha}}}^S = \begin{pmatrix} -\sin\phi\sin\theta & 1-\cos\theta & \cos\phi\sin\theta \\ \sin^2\phi(1-\cos\theta) & -\sin\phi\sin\theta & -\sin\phi\cos\phi(1-\cos\theta) \\ 0 & 0 & 0 \end{pmatrix} \begin{matrix}\text{disloc.} \\ \text{direction}\end{matrix} \tag{8}$$

Burgers vector direction

However if one changes the reference lattice and takes the median lattice as a reference (same $\vec{\Omega}$ axis but at $\pm \frac{\theta}{2}$ from I and II), then the new average surface dislocation density becomes :

$$\bar{\underline{\underline{\alpha}}}^S = \begin{pmatrix} 2\sin\phi\sin\frac{\theta}{2} & 0 & -2\cos\phi\sin\frac{\theta}{2} \\ 0 & 2\sin\phi\sin\frac{\theta}{2} & 0 \\ 0 & 0 & 0 \end{pmatrix} \tag{9}$$

The tensor (9) reduces to the well known Frank formula :

$$\vec{b} = 2\sin\frac{\theta}{2}\, \vec{r}_{G.B} \wedge \vec{N} \tag{10}$$

where $\vec{r}_{GB}$ is one G.B plane vector and $\vec{b}$ the total Burgers vector crossing $\vec{r}_{GB}$. From (9) it is easy to deduce that for a pure tilt G.B (ϕ=0) one family of dislocations along x_1 with b_3 Burgers vector is accomodating the rotation (pure edge dislocation). Conversely for a pure screw G.B ($\phi = \frac{\pi}{2}$) there are at least two families of dislocations along x_1 and x_2 with b_1 and b_2 Burgers vectors respectively (pure screw dislocations).

It is clear when comparing (8) and (9) that the dislocation density is not uniquely defined : in this case it is due to a change in the reference lattice and it can be shown that these Burgers vectors can be deduced from one another easily by the transformation from one lattice reference to the other and it should not affect the energy calculation. However the usual linear elasticity theory has not yet been extended to large rotations and the energy cannot be calculated unless $\cos\theta \sim 1$, [25], which is definitely not the case for large angle G.Bs.

Moreover any transformation, $\underline{I}$, which leaves the lattice invariant could be applied to $\underline{\underline{D}}$ and would give a different expression for $\bar{\underline{\underline{\alpha}}}^{S}$. For example a mirror twin can be described as a pure shear parallel to the G.B plane as well as a rotation. In the former case the dislocation content is zero, as $\underline{\underline{D}}_{I}=I$ (identity) and

$$\underline{\underline{D}}_{II} = \left(\left(\begin{matrix} 1 & 0 & 0 \\ 0 & 1 & 0 \\ 0 & g & 1 \end{matrix}\right)\right) \tag{11}$$

In this example this result states that the twin position is chosen as the reference position (no internal stresses remain).

A second factor of non-uniqueness comes from the decomposition of the continuous surface dislocation density into quantized Burgers vectors. At this stage it is necessary to introduce the discrete nature of the material. For energetical reason it will be admitted that a continuous dislocation distribution is not stable : in fact discrete jumps of translation ($\vec{b}$), generally separated by "good crystal" will be favored. The $\vec{b}$ vectors will tend to be lattice vectors.

The Burgers vector is generally determined by the closure failure of a circuit constructed in a particular reference lattice and corresponding to a closed circuit constructed in the defected crystal. Such a Burgers circuit construction is obvious for isolated dislocations or low angle G.Bs: each individual dislocation is well separated from the following by a "good crystal" region. A natural reference crystal is the crystal lattice itself in a position to be defined (median, I or II). As for low angle G.Bs, the approximations involved in the elasticity theory are satisfied, the elastic field can be completely solved for any given dislocation array [27] [28]. It has been shown that there is a local minimum in the energy versus misorientation when the dislocations are exactly equally spaced, i.e. when the misorientation corresponds to a coincidence site misorientation.

In a high angle G.B, taking the crystal lattice as the reference crystal, it is possible to draw a Burgers circuit passing through the G.B plane at coincidence sites : it defines the $\vec{b}$ content enclosed between two coincidence sites [29]. Bollmann [4] gave a more general analysis using the O-lattice : the circuit cut the G.B plane at O-points which are necessarily coincidence sites. This approach gives a description of the G.B in terms of primary dislocations. This is illustrated on a {112} $\Sigma=9$ pure tilt symmetrical boundary in the diamond structure (fig. 3). A Burgers circuit drawn in crystal II defines in the G.B plane a total Burgers vector content of $\langle 110\rangle_{II}$ taken between two coincident points AB along 1/2<411>. However this vector can be decomposed into two $1/2\langle 110\rangle_{I}$ Lomer type dislocations between AA' or A'B if defined along a non planar set of coincidence sites.

It can be physically more suitable to treat the bicrystal as being close to a "special" twin position (see formula 11). Then the reference lattice with a low Σ value is the exact twin position which is supposed to contain no internal stresses. Any small deviation from this special orientation can then be treated as an additional small angle G.B to which the above analysis can be applied. The Burgers vectors in this case are given by any translation of the bicrystal lattice, i.e. they belong to the DSC lattice. These dislocations

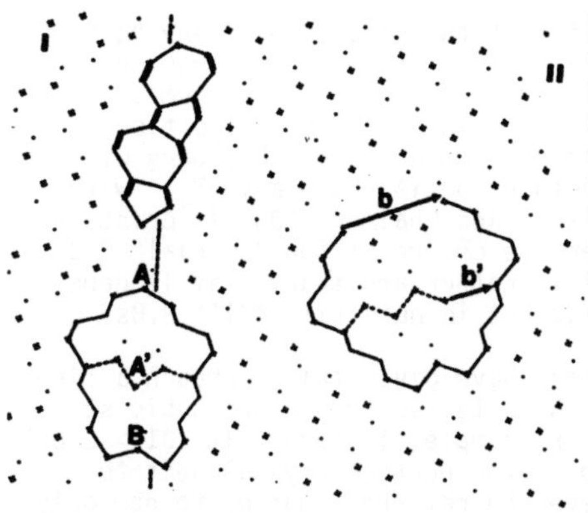

Fig. 3 Burgers circuit crossing a {112} Σ=9 twin G.B in a diamond structure. Points A and B are not exactly coincident in both lattices but are equivalent : they can be used to cross the interface. The circuit ABA gives a <110> Burgers vector which can be dissociated into two 1/2<110> in a smaller circuit A A' A. This latter circuit enclosed a Lomer type dislocation with its typical 5-7 atom ring.

are called secondary dislocations or relaxations. Once more the distribution of these secondary relaxations should be periodic for energy minimization. Using the special position as a reference, complete elastic calculation (at least isotropic) is possible to simulate the additional effect of these secondary relaxations. It is important to realize that these relaxations are due to a perturbation in the regular array of primary dislocations accommodating the deviation from the "special" position. Similarly to a sub G.B, the G.B is decomposed into a series of "good portions" (the "special" twin position) separated by perturbated regions where the distance between primary dislocations is changed by the introduction of a small translation $\vec{d}_{DSC}$.

The non uniqueness of the Burgers vector distribution to accommodate large disorientation is rather similar to the problem encountered in deformed crystals. Accommodation of a given plastic deformation is also non-unique : an infinite number of dislocation distributions is possible. However the success of the dislocation approach was due to the fact that the number of Burgers vector types is very limited in a given structure. It is the aim of the structural unit model to try to see if, in fact, any G.B can also be composed by mixing a limited number of well characterized structural units.

4. THE STRUCTURAL UNIT MODEL

This model is based on a geometrical description of the local atomic configuration at the G.B interface. Depending which aspect is emphasized there are three slightly different approaches :
i) the first approach insists on the particular bonding conditions of the material. In diamond structure (Si-Ge), for instance, covalent bonds with a sp^3 hybridization ensure tetracoordination with strongly oriented bonds. The perfect structure can then be described by a series of elementary polyhedra, the faces of which are in the form of hexagonal chair-shaped cycles. It is postulated that, although these bonds are flexible, the tetracoordination will tend to be preserved even at defect cores (dislocation and G.B).

If this coordination number is accepted as the main constraint the number of geometrical models of a given simple twin boundary is very limited. This was the basic idea of the first models proposed by Hornstra [12-13]. This approach is however incomplete as already recognized by Hornstra himself : some of the high angle G.Bs do contain dangling bonds (i.e. does not respect the tetracoordination). It is sometimes possible to reconstruct these bonds and restore the tetracoordination as in the partial $\frac{1}{2}$<011> dislocations [34], but then the number of configurations is much larger.

ii) A second approach is based on the fact that the structure of any "general" G.B should be described in terms of a mixture of distinct structural units characteristic of low Σ-value twin boundaries (energy criterium). A G.B with misorientation θ for instance will contain in different proportions structural α and β elements. The α unit comes from a special boundary with misorientation θ_i^o and the β unit from another special boundary θ_{i+1}^o with $\theta_i^o < \theta < \theta_{i+1}^o$. This idea introduced by Bishop and Chalmers [30] is cristallographically inevitable (fig. 4). However the choice of the "special" G.Bs which are important to build such a model is rather arbitrary : it is generally believed that only 2 or 3 are sufficient in pure tilt <011> G.Bs.

Atomistic calculations,in f.c.c. structures have given this approach a firm basis [31] and it is likely that it could be extended to diamond cubic structures as proposed by Vaudin et al [32]. For example, in pure tilt <011> G.Bs they proposed only two structural units plus the perfect crystal unit to build any symmetrical G.B. It should however be remarked that up to now only symmetrical G.Bs have been analyzed : new structural units should be involved in more general G.Bs.

iii) A third approach is given by generalizing the dislocation model of a G.B. Each dislocation is also characterized by one or several typical structural units : the edges of this unit define the smallest Burgers circuit around the core defect. For example a Lomer dislocation is characterized by typical 5 and 7 atom contiguous rings [33], a dissociated 60° dislocation contain two special units with reconstructed bonds [34] plus one unit with a boat shaped 6 atom ring characteristic of a stacking fault. In <011> dislocations more than nine structural units have been proposed [15]. A G.B with <011> tilt axis is therefore composed of a mixture of such distorted units. In this case a good knowledge of all the dislocation core structure encountered in low angle G.B will be enough to describe any G.B symmetrical or asymmetrical. This approach is however limited as there is no proof that new structural units are not stabilized in high angle G.Bs.

The structural unit model has the advantage of being closely related to the dislocation model. Each unit can be associated with a typical dislocation core structure if the reference structure is chosen to be the perfect crystal : however the boat-shaped 6 atom ring characteristic of the stacking fault as well as of the Σ=3 should be added. These units form the primary dislocations previously introduced.If the boundary is close to a low Σ_α twin G.B, it will be composed of a majority of α units including periodically a minority of a different β unit coming from a different twin (or a different dislocation core). These β units can then be regarded as centres of secondary dislocations (fig. 4).

A complete atlas of "stable" units existing in any G.B of a diamond structure is not yet available. Very special cases have been discussed such as pure tilt <011> or <001> G.B, for which some experimental results have been obtained (see below).

A further complication arises from the possible multiplicity of the structural units as pointed out by Vitek et al [35]. Computer calculations show that different units having the same "Burgers vector" have very similar energies : they generally differ by the number of atoms contained in the unit. Therefore emission or absorption of point defects change the unit. In diamond cubic structures this phenomenon was envisaged for a long time in partial dislocations : the 30° partial, as well as the 60° non dissociated dislocation cores do contain two possible core structures (so-called shuffle and glide set [34]). The exact atom content of any unit is difficult to determine experimentally, therefore alternative structures should always be considered.

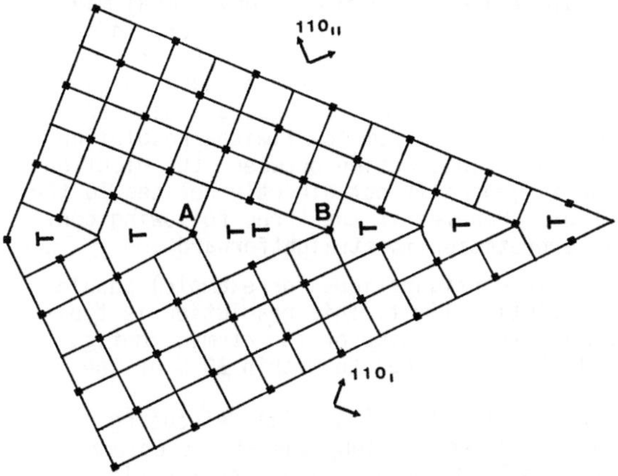

FIG. 4 Model of a secondary dislocation in a {120} Σ=5 twin θ=36°87°. F.C. Cubic structure projected along <001> axis. The misorientation is slightly larger than θ=36° 87, introducing periodically a new structural unit (β-unit in the text). They correspond to an additional DSC dislocation with Burgers vector a/5 <210>. Such configurations were clearly recognized in computer simulations by Vitek [31]. Locally the part AB corresponds to a {130} Σ=5 twin position.

5. EXPERIMENTAL METHODS USED TO STUDY G.B ATOMISTIC STRUCTURE

Both X-ray and electron diffraction have been employed in G.B studies [36-37] . The periodic strain field associated with the G.B gives rise to extra reflections : they are coming from few atomic layers on either side of the G.B interface and form in the reciprocal space rods parallel to the G.B normal. For $\vec{t}$ = 0, the diffraction pattern should be the reciprocal of the CSL plane parallel to the G.B plane. If $\vec{t} \neq 0$ several symmetry elements can be destroyed and the extinction rules change. In any case the structural information comes from structure factor measurement. The extinction rules give the projected space group and the unit cell of the relaxed bicrystal [37]. The relative position of the rods to the main reflections of both crystals enables investigation of the local relaxation of atomic plane spacing close to the interface [38]. X-ray diffraction experiments have the advantage of avoiding double diffraction : it is however necessary to have a very intense X-ray source as the scattering is weak (synchrotron sources are required). On the other hand an electron diffraction pattern must be carefully analyzed. For a particular geometry (symmetrical pure tilt G.B viewed along the interface normal) double diffraction is minimized and comes only from interlayer diffraction : these double diffraction spots can be clearly recongnized by a small tilt away from the normal or by varying the high voltage [37]. The diffraction methods have however several limitations. The X-ray diffraction pattern integrates over a large G.B plane area and some of the symmetry elements destroyed by rigid body displacement $\vec{t}$, may reappear if all the variants are present. Moreover phases of the diffracted beams cannot be determined and local spatial determination of the atom position is difficult. In a very simple case and using some a priori information Budai et al. were however able to localize the atoms in a Σ=5 pure twist G.B in gold [36]. The electron diffraction pattern is obtained from a much smaller area but may also suffer from the same limitations ; the structure factor determination apart from the symmetry determination is much more difficult.

Direct imaging method using high resolution electron microscopy (HREM) may sometimes overcome these limitations and at least give complementary information. This technique is nowadays widely spread for studying G.Bs in semiconductors since the pioneer work done by Krivanek et al [39]. An HREM image is a "deformed" two dimensional projection of the atomic structure projected along the viewing axis. The "deformation" comes from the limited resolution of the microscope, and from a very complex interaction between the electrons and the crystal. Therefore the image is generally not directly related to the projected potential, and not invertible. However if the three following conditions are satisfied, the image interpretation is straightforward :

i) the details to be interpreted (atomic configurations for example) should be larger than the point to point resolution limit. This resolution is typically 2.2 Å at 200 keV and 1.5 Å at 400 keV. If however the atomic configuration is known, atom positions could be determined to within 10 % of the resolution limit.
ii) imaging conditions should be well controlled : special defocusing distances, very thin specimen and very well oriented along the atomic columns.
iii) a G.B should be observed end-on ; only defects having a strain field independent of the z axis (observation axis) can be clearly resolved. The atomic columns should be displaced as a whole. This confines the observation to pure tilt G.Bs observed along their common axis.

In cubic diamond structures several situations are encountered depending on the point to point resolution [40] :

i) Resolution limited to 3 Å : <011> is the only available viewing axis. Each dot in the image characterizes a column pair which is not resolved. This enables Burgers vector determination, measurement of $\vec{t}$ and periodicity along one direction of the G.B and observation of microfacetting.
ii) Resolution limited to 2 Å : <011> and <001> projections are possible. Atomic columns are resolved only for <001> projection. Atomic position of the atom pairs in the <011> case can be determined. Information about structural unit becomes possible.
iii) Resolution limited to 1.4 Å : <011>, <001>, <111> and <112> directions are observable. Atomic columns are resolved in all these projections. Complete information about the structural units may be obtained (atom number and location).

Most of the experimental work already published concerns conditions i) and ii), in the near future HREM work will be employed extensively under conditions iii). Therefore the present results should be considered mostly as preliminary investigations although very interesting results have already been obtained.
It should be emphasized that simultaneous application of diffraction and direct atomic imaging method is certainly one of the most promising ways of solving the atomic structure of G.Bs.
Complementary information is also available from ordinary electron microscope imaging. In particular the α-fringe contrast of an inclined G.B gives directly the magnitude and sign of $\alpha=\vec{g}.\vec{t}$ where $\vec{g}$ is the common diffraction vector of the two crystals. A combination of several non-coplanar $\vec{g}$ vectors gives the direction of $\vec{t}$. This result can then be compared with that obtained by HREM.

6. SELECTED EXPERIMENTAL RESULTS

G.Bs in silicon and germanium are commonly observed. No differences have ever been found between either material. In order to illustrate the previous concepts some of the results obtained on pure tilt <011> G.Bs in both mate-

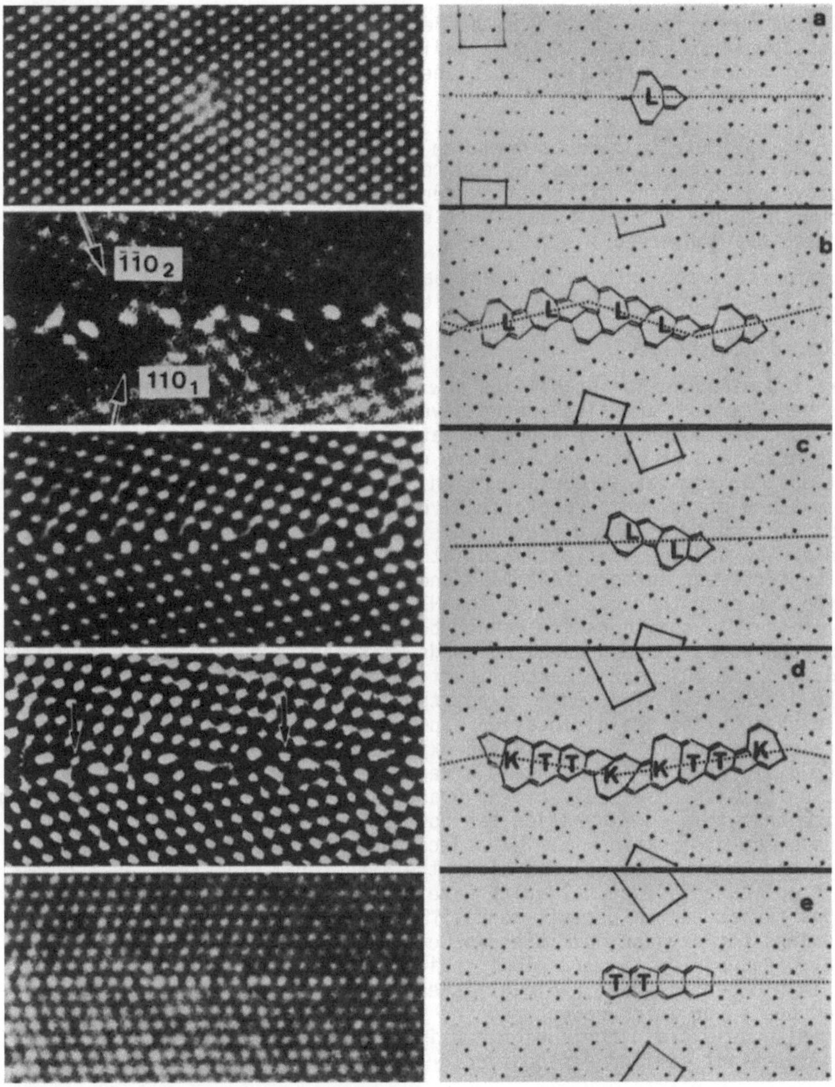

Fig. 5 Experimental HREM images of several symmetrical þure tilt G.Bs in Ge or Si as viewed along <011> axis. a) sub grain boundary, b) {255} Σ=27, c) {122} Σ=9, d) {233} Σ=11, e) {111} Σ=3. A scheme of all the structures as deduced from these observations is outlined. Note that these G.Bs are formed of few structural units : the L-unit (from Lomer dislocation), the T-unit from the mirror twin or fault. No relaxation has been introduced on these structures: the relaxation would tend to form more closely related units (K, L or T-units).

image b) = courtesy of Vaudin, Cunningham and Ast [32]
image d) = courtesy of Papon, Petit, Sylvestre, Bacmann [48]

rials will be summarized. These G.Bs differ by their rotation angle θ : by convention symmetric <011> G.Bs, whose median plane is {110} are characterized by $\theta<90°$ and those with a median plane {100} by $\theta > 90°$.

6.1 Sub G.Bs (symmetrical and asymmetrical) $\theta < 10°$ or $>170°$. A comprehensive study has been performed by Bourret et al [41] . Due to HREM it has been discovered that many different Burgers vectors exist in low angle GBs. Apart from the usual 60° and Lomer (pure edge) 1/2 <110> dislocation, three other $\vec{b}$ have been observed :

b' = <111> dissociated into three Frank partials 1/3 <111>
b" = 1/2 <211> dissociated into three partials forming a characteristic hook shape arrangement
b"' = <100> non-dissociated.

Therefore the dissociation stabilizes new configurations in low angle G.Bs, moreover very large $\vec{b}$ are also present even undissociated. Symmetric G.Bs with $\theta < 10°$, are composed of a regular array of Lomer dislocations (fig. 5) whereas symmetric G.Bs with $\theta > 170°$ are composed of a regular array of b"' dislocation. For asymmetric configurations the exact G.B plane-determines the $\vec{b}$s (or mixture of $\vec{b}$s) which are effectively present [41].
The structural units of the different dislocation cores are not yet completely determined. The 60° dislocation dissociated into two partials has been studied by different authors [42] [43] and was found to be mainly of the glide type. However the accuracy of the HREM observations is limited and up to 50 % of shuffle configuration cannot be excluded [40]. The structural units in both cases contain dangling bonds which are supposed to be reconstructed. In the 30° partial, reconstruction should double the periodicity along the dislocation line. This is not the case in the 30° partial where however a shear along the dislocation appears. It should be stressed that reconstruction has not yet been observed experimentally.
The Lomer dislocation should contain a very simple structural unit hereafter called L-unit, with completely closed 5 and 7 atom rings. It is disappointing to realize that this configuration has not yet been observed in low angle G.Bs: the Lomer dislocation is always decorated by some impurities (more likely oxygen) which changes completely the structural unit [44].
Various complicated structural units have been proposed for other dislocation types (Frank partial, stair-rods) [15]. However, these dislocations act as good sinks for oxygen impurities in Ge and Si and they are never observed undecorated except for the 1/3 <100> stair rod, the model of which is still unclear in the central part [46]. All these hypothetical models have very disturbed bonds for which reconstruction is either difficult or not possible : this can explain their high sensitivity to impurity decoration.
Clear and distinct similar dislocations are recognized up to $\theta \sim 10$-$12°$.

6.2 {255} $\Sigma = 27$, $\theta = 31°59$ (see fig.5b). For small θ values and up to $\Sigma=27$, the symmetrical G.B is always composed of an array of L-type structural units mixed with u structural unit (perfect crystal). At {255} $\Sigma=27$, this is still true, but facetting starts to occur with L and L' structural units [32]. Here ' denotes a mirror glide operation. However it is surprising that the sequence is LL'u(LL')' and not a simple LuL' : as a consequence the projected periodicity is equal to that given by the CSL and not to half the periodicity of the projected CSL along <011>. The facetting tends to form {111} facets along one of the crystal.

6.3 {122} $\Sigma=9$, $\theta = 38°94$ (see fig.5c). It is one of the most studied G.Bs [39,47]. The structure still contains L units in a zig zag array LL'LL' ... Although also slightly sensitive to impurities specially on the 5 atom rings this sequence is clearly resolved by HREM. For angles larger than 38°94 a different structural unit has to be introduced : the T-unit from the following $\Sigma = 3$ twin.

6.4 {233} Σ=11, θ = 50°48 (see fig.5d). In this case, although composed of a regular array mixing L and T units, the observed periodicity is twice the CSL periodicity [48]. This surprizing result is confirmed by electron diffraction : a glide plane is present in the G.B plane giving a double periodicity. A sequence containing LL'TTL'L is compatible with this periodicity and with the rigid body translation obtained by the α-fringes method. Once more, facets with a symmetrical arrangement penetrating alternatively in crystal I and II seem to be energetically favored. The facets are along a {111} plane of one crystal as in the Σ = 27 case.

6.5 {111} Σ=3, θ = 70°53 (see fig. 5e). The mirror twin frequently encountered in polycristalline silicon is composed as previously proposed of an array of T-units [47]. It is a boat-shaped six membered ring similar to the one existing in stacking faults, as suggested by Hornstra [33].

6.6 {111} Σ = 3 + 1° (ref.45). This G.B close to Σ=3 twin has additional periodic secondary dislocations with a 1/3<111> Burgers vector perpendicular to the G.B plane. They correspond to the insertion of a new structural unit F characteristic of a Frank dislocation. The sequence is then TTTFTTT. This unit has dangling bonds similar to those existing in a 30° partial and therefore can be reconstructed. It is not clear if this F unit is used for much larger angle twin positions.

6.7 θ > 70°53. The {112} Σ=3 has been studied by the α-fringe method [49] : several models have been proposed with or without dangling bonds. However they are not compatible with recent diffraction measurements which show a double periodicity compared to the CSL and in particular in the <011> direction [50].

Preliminary results on a particularly "general" G.B show that microfacetting in three dimensions is present (Σ=11, θ=127°). These configurations are particularly difficult to study : they are non-periodic and the G.B plane is variable [51]. It is interesting to note that such G.Bs are strongly electrically active when compared to symmetrical twin configurations. Similarly no electron beam induced current (EBIC) signal is observed at symmetric Σ = 3 and Σ = 9 G.Bs, although they are electrically active as soon as they contain additional defects [52].

7. CONCLUSION

The experimental results obtained on G.B structures have largely supported the crystallographic approach. The periodicity of the two adjacent crystals tend to be imposed to the G.B plane : for a number of "special" or twin orientations the G.B is really periodic. At an arbitrary position the G.B interface is non-periodic but can be decomposed in periodic areas separated by linear defects. A good description of these defects is given by the concept of secondary relaxation : they are equivalent to a dislocation whose Burgers vector belongs to the DSC lattice. The HREM has given some confidence in the structural unit model. At present only a few of these units are known for the diamond structure. Dislocation core units play an important role but it is likely that new units can be stablilized in G.Bs. It is clear that energy minimization is the major criterium which decides what kind of configuration should really exist. It is also possible that metastable configurations do exist giving a further variety to G.B structures. The experimental examples which have been presented previously show that surprising configurations tend sometimes to develop : double periodicity (when compared to CSL lattice) is frequently encountered, within each period the succession of units is difficult to predict, and very often microfacetting occurs with a general tendency to follow one of the {111} dense planes of one crystal. The case of asymmetric G.Bs should in the future be considered more carefully in view of their great electrical activity : it would be very interesting to associate well characterized electrical properties with each structural unit.

REFERENCES

1 W. Rosenhaim, J.C.W. Humphrey : J. Iron Steel Inst. 87, 219 (1913)
2 M. Rühle : Electron Microscopy 1980, Ed. P. Brederoo and G. Boom p. 366 Leiden (1980)
3 F.L. Vogel, W.G. Pfann, H.E. Corey, E.E. Thomas, Phys. Rev. 90, 489 (1953)
4 W. Bollmann, Crystal defects and crystalline interfaces, Springer : Berlin (1970)
5 T. Schober, R.W. Balluffi : Phil Mag 24, 165 (1971)
6 C.B. Carter, A.M. Donald, S. Sass : Phil Mag. A39, 533 (1979)
7 S.S. Sass, T.Y. Tan, R.W. Balluffi : Phil Mag. 31, 559 (1975)
8 F. Hargreaves, R.J. Hills, J. Inst. Metals, 41, 217 (1929)
9 N.F. Mott, Proc. Phys. Soc. 60, 391 (1948)
10 J.M. Burgers : Proc. Kon. Ned. Akad V. Wet. Amsterdam 42, 293 (1939)
11 D.G. Brandon, B. Ralph, S. Ranganathan, M.S. Wald : Acta Metall. 12, 813, (1964)
12 J. Hornstra : Physica 25, 409 (1959)
13 J. Hornstra : Physica 26, 198 (1960)
14 A.P. Sutton, V. Vitek : Phil. Trans. R. Soc. Lond. A309, 1 (1983)
15 A. Bourret, C. d'Anterroches : J. de Physique 43, n° 10 C1-1 (1982)
16 F. Seitz : Ann Math. Stat. 37, 17 (1936)
17 D. Gratias, R. Portier, M. Fayard, M. Guymond : Acta Cryst. A35, 885 (1979)
18 D. Gratias, R. Portier : J. de Physique C6, 15 (1982)
19 G. Friedel (1926) : Leçons de Cristallographie, Reprinted Paris-Blanchard (1964)
20 H. Grimmer, W. Bollmann, D.H. Warrington, Acta Cryst. A30, 197 (1974)
21 A.H. King : Acta Metall. 30, 419 (1982)
22 P. Delavignette : J. de Physique 43, C6-1 (1982)
23 R.C. Pond : Phil Mag. A 47, L49 (1983)
24 B.A. Bilby, R. Bullough, E. Smith : Proc. Roy. Soc. London, A231, 263 (1955)
25 E. Kröner : Kontinuumstheorie der Versetzungen und Eigenspannungen Springer Verlag Berlin (1958)
26 R.C. Pond : Inst. Phys. Conf. Ser. n° 67, The Institute of Physics 1983, p. 59
27 G. Saada : Acta Metall. 27, 921 (1979)
28 W.T. Read, W. Shockley : Phys. Rev. 78, 275 (1950)
29 R. Bullough : Phil.Mag. 12, 1139 (1965)
30 G.H. Bishop, B. Chalmers : Phil. Mag. 29, 515 (1971)
31 V. Vitek : J. de Physique Aussois 1984 (in press)
32 M.D. Vaudin, B. Cunningham, D.G. Ast : Scripta Met. 17, 191 (1983)
33 J. Hornstra : J. Phys. Chem. Solids, 3, 129 (1958)
34 S. Marklund : Phys. Stat. Sol. (b) 100, 77 (1980)
35 V. Vitek, A.P. Sutton, G.J. Wang, D. Schwartz : Scripta Metall. 17, 183 (1983)

36 J. Budai, P.D. Bristowe, S.L. Sass : Acta Met. 31, 699 (1983)
37 A.M. Papon, M.P. Petit, J.J. Bacmann : Phil. Mag. 49, 573 (1984)
38 P. Lamarre, S.L. Sass, Scripta Met. 17, 1141 (1983)
39 O.L. Krivanek, S. Isoda, K. Kobayashi : Phil. Mag. 36, 931 (1977)
40 A. Bourret, J. Thibault-Desseaux, C. d'Anterroches, J.M. Penisson, A. de Crécy : J. of Microscopy 129, 337 (1983)
41 A. Bourret, J. Desseaux : Phil. Mag. A39, 405 (1979)
42 A. Bourret, J. Desseaux, C. d'Anterroches : Inst. Phys. Conf. Ser. 60, The Institute of Physics, p. 9 (1981)
43 G. Anstis, P.B. Hirsch, C. Humphreys, J. Hutchinson, A. Ourmazd, Inst. Phys. Conf. Ser. 60, The Institute of Physics, p. 15 (1981)
44 A. Bourret, J. Desseaux, A. Renault : Phil. Mag. A45, 1 (1982)
45 A. Bourret, C. d'Anterroches, J.M. Penisson : J. de Physique 43, C6, 83 (1983)
46 A. Bourret, J. Desseaux-Thibault, F. Lançon : J. de Physique 44, C4, 15 (1983)
47 C. d'Anterroches, A. Bourret : Phil. Mag. in press (1984)
48 A.M. Papon, M. Petit, G. Silvestre, J.J. Bacmann: J. Micros. Spectro. Elec. 8, 135 (1983)
49 D.S. Vlachavas, R.C. Pond : Inst. Phys. Conf. Ser. 60, The Institute of Physics, p. 159 (1981)
50 A.M. Papon, M.P. Petit, J. Bacmann (private communication)
51 A. Bourret : J. de Physique to be published MRS Meeting, Strasbourg 1984
52 C. Dianteill : Thèse de l'Université de Toulouse (1983).

Computer Calculations of Grain Boundary Energies in Germanium and Silicon

H.J. Möller and H.H. Singer

Arbeitsbereich Halbleitertechnologie, Technische Universität Hamburg-Harburg
Eißendorfer Straße 38, D-2000 Hamburg 90, Fed. Rep. of Germany

Abstract - This paper reports results of computer calculations of the energy of coincidence and near-coincidence <011> tilt boundaries. Different valence potential functions are used and conclusions as to the reconstruction of dangling bonds are discussed. Geometrical models of dislocations in Σ3 and Σ9 (near-coincidence) tilt boundaries are developed and first results on their energy are presented.

1 Introduction

It is not necessary to emphasize the importance of polycrystalline semiconductors like solar cells, integrated circuits or of sensor applications. There is no doubt that the electronic properties of these materials are mainly determined by the energy levels in the band gap which are introduced by the grain boundaries. However, there is still a controversy about the origin of the trapping states. Several possibilities are discussed: the atomic structure of a grain boundary in a covalently bonded crystal is supposed to contain broken bonds which may give rise to energy states in the band gap. Alternatively one can assume, that the grain boundary structure is sufficiently reconstructed and segregated impurity atoms determine the electronic properties. The situation might even be more complicated if the atomic structure of a grain boundary is dependent on the degree of impurity contamination and the distribution of the energy levels is a balance of structural and impurity defects in the grain boundary. At present the experimental investigations of the electronic properties of bicrystals containing special grain boundaries [1-4] and of polycrystalline semiconductors [5-7] indicate that impurity segregation plays a role but do not allow to determine to what extent. In view of the technological importance of that question it is very desirable to learn what the electronic structure of a pure arbitrary grain boundary is in order to distinguish it from the modification introduced by impurity atoms.

The computer calculation of the energy levels arising from the topological disorder introduced in the perfect lattice by a grain boundary or a dislocation requires that the atomic structure of the defect is already given. Therefore the procedure involves in general the development of the atomic arrangement by minimizing the total energy of the defect. Because of the covalent bonding of the atoms in a semiconductor the energy calculations have to begin with some initial atomic configurations which are usually derived from geometrical considerations. Since it seems hopeless at present to give a general description of the atomic structure of an arbitrary grain boundary the interest has focussed for the beginning on some special cases: the low angle and coincidence grain boundaries. The most thoroughly investigated structures are those of <011> tilt boundaries and a number of geometrical models have been proposed [8-10]. Apart from the first and higher order twins all structures contained dangling bonds. However, alternative structures with reconstructed dangling bonds are possible. Since it is of vital importance in view of the electronic properties of the grain boundaries whether the reconstructed structures are the stable ones, comparative energy calculations are necessary. Unfortunately two difficulties arise: the dangling bond energy is not known with sufficient accuracy and at present there are a number of different atomic potentials available [12-15], used for the computer calculations, which give substantially different strain energy values.

Therefore it is the aim of this paper to discuss how the different valence potentials influence the considerations concerning the structure of some <011> and <001> tilt boundaries. In the second part the investigations shall be extended to near-coincidence grain boundaries and some structural aspects of grain boundary dislocations will be presented.

2 Energy Calculations - Valence Potentials and Methods

The methods used for the energy calculations of grain boundaries have been applied already successfully to the determination of the core structure of dislocations in germanium, silicon and diamond [15-17]. A common feature is that the energy of the defect (grain boundary, dislocation etc.) is calculated using semi-empirical valence potentials which allow the computation of the strain energy of displaced lattice atoms. Since in covalently bonded crystals the energy depends not only on the distance of neighboring atoms but also on the angles between bonds formed by nearest-neighbor atoms, the evaluation requires potentials which take into account changes in bond lengths, interbond angles, dihedral angles etc. The valence potentials currently in use describe the crystal energy for small (harmonic) distortions by force constants which are obtained from properties such as elastic constants and phonon frequencies. In general, however, larger deviations and distortions of bond lengths and angles are to be expected in the core of dislocations or at grain boundaries, therefore functions with reasonable anharmonic terms have to be added. A number of different potentials are discussed in the literature, therefore it shall be only described briefly which potential functions have been used in our calculations.

a) Keating potential. This potential has been used in the form given in [8,12]

$$E = \frac{3\alpha}{16r_o^2}\sum_i (r_i\, r_i - r_o^2)^2 + \frac{3\beta}{8r_o^2}\sum_{i,j>i} (r_i r_j + \tfrac{1}{3}r_o^2)^2$$

where all atomic forces are resolved into nearest-neighbor bond stretching and bond bending (with some stretching component) forces. r_i are the bond vectors about each atom, r_o the equilibrium bond length. The potential contains two parameter α,β which were fitted to the long wavelength elastic constants c_{11}, c_{12}, c_{44} with an accuracy of better than 1%. A shortcoming of this potential is that it is harmonic and overestimates the strain energy when bonds are distorted above the limit where harmonicity is a good approximation. For instance the k=o transverse optic mode frequency deviates already by about 5% [18].

b) BARAFF et.al. [14] have argued that a calculation of larger bond distortions requires a potential that reproduces phonon frequencies especially for larger wave vectors. To account for that they have proposed another set of the bond stretching and bending parameters α,β. One can notice that this modification essentially corresponds to a reduction of the bond bending forces by a factor of 3 and consequently should favor bond bending compared with bond breaking.

c) Modified Keating potential. To account for anharmonic bond distortions KOIZUMI and NINOMIYA [13] suggested a formula which they used in their treatment of amorphous germanium. When expanding their equation for small atomic displacements the Keating potential is reproduced. Three more free parameters ($\gamma, \delta, \varepsilon$) are introduced now which were derived from the fitting to third order elastic constants.

e) Lifson-Warshel potential [19]. Recently a new type of potential field has been applied to defect structures in silicon and diamond, which has proved successful in dealing with highly deformed chemical structures such as straine hydrocarbons. ALTMANN et.al. [15] have derived a set of parameters from the phonon spectrum and the stacking fault energies of silicon and applied it to calculate the core structure of dislocations. In the present investigation it has been used in the following form with

e1) a Morse potential to account for bond stretching

$$E_b = \sum_i D_b \{ \exp(-2\alpha(\Delta r_i) - 2\exp(-\alpha(\Delta r_i))\}$$

with $\Delta r_i = r_i - r_o$ the bond length change, D_b bond formation energy and $\alpha = (K_b/2D_b)^{1/2}$ In fitting the force field to the phonon dispersion curve only a harmonic approximation is needed, but in application to defect structures the complete Morse potential is used;

e2) a bond bending term

$$E_\Theta = \sum_i \{\frac{1}{2} K_\Theta \ (\Delta\Theta_i)^2 + \frac{1}{2} K_q \ (\Delta q_i)^2 + K'_\Theta \Delta\Theta_i\}$$

which contains the usual harmonic approximation in the bonding bending angles $\Delta\Theta_i = \Theta_i - \Theta_o$ and a Urey-Bradley-type interaction between two atoms bound to a common atom and separated by a distance q_i. If the coordination number of an atom is not four, K'_Θ is set equal to zero;

e3) a torsional potential

$$E_\Phi = \sum_i \{\frac{1}{18} K_\Phi \ (1 + \cos 3\Phi_i) + K_{\Theta\Theta} \ \Delta\Theta_{i2}\Delta\Theta_{i3}\cos\Theta_i\}$$

where, for a chain of atoms 1-4, Φ_i is the dihedral angle between the planes through the atoms 1,2,3 and 2,3,4, and $\Delta\Theta_{i2}$, $\Delta\Theta_{i3}$ are the bond angle changes at the atoms 2,3;

e4) a van der Waals interaction

$$E_w = \sum_i \{8.28 \times 10^5 \varepsilon \exp(-r/0.073\ r^*) - 2.25\varepsilon(r^*/r)^6\}$$

where the summation is over all nonbonded pairs. A full discussion of the forc field and its application to dislocation cores is given in [15,20]. It is wor mentioning here that the full phonon spectrum is reproduced by this valence potential which is not the case for all the other potentials used so far.

The energy of a grain boundary now is obtained by applying the different valence potentials to a cluster of 200-500 atoms containing the grain boundar under consideration. At the beginning of the procedure all atoms were fixed at the positions calculated from the orientation difference of the two crysta halves whereas the atoms in the boundary plane were connected according to th geometrical model of the boundary. The energy was minimized then by shifting all atoms in the direction of the largest energy gradient. The surface atoms parallel to the boundary plane (usually 5-10 atomic layers away) were kept fixed and periodic boundary conditions were adopted along the rotation axis and perpendicular to it along the boundary plane. In general it was necessary to introduce relative translations of the two crystal lattices perpendicular and parallel to the boundary plane in order to arrive at the lowest energy configuration. Dangling bond atoms were saturated by an atom which was allowed to adjust freely so that this free bond did not yield any additional strain energy. The total energy then is the sum of the strain energy plus the energy of the broken bonds E_{db}. Since there is still a great uncertainty as to the absolute value of E_{db} [19] the determination of the total energy is difficult.

3 Results

The misorientation of the adjacent grains for which coincidence grain boundaries occur can be calculated from Bollmann's O-lattice theory, but the arrangement of atoms in the grain boundary plane has to be developed simply by trying to obtain a good fitting. There is certainly some arbitrariness in this procedure and in some cases transmission electron microscopy observations actually have shown that the proposed structures [8-11, 21] have to be modified. However, the essential features of the boundary structures always remained: the <011> tilt coincidence boundaries are built up from a periodic array of five, six and seven membered rings. CSL-boundaries with misorientations $\Theta<70.5°$ (twin orientation) were all well reconstructed, whereas boundaries with larger angles all contained unsaturated bonds whose reconstruction would involve large bond distortions. In order to decide whether a reconstruction is favorable, energy calculations have already been perfomed. In the case of a $\Sigma 11$ grain boundary [9,22] and an incoherent twin in silicon [21] computer simulations have shown that reconstructions are possible, however, in view of the possible differences in the strain energies when using various valence potentials, one must admit that the problem is far from settled.

In the present study the strain energies and relative translations of the structures of the Σ 33, 19, 9, 17, and 11 <011> tilt boundaries proposed in [8] have been calculated both for germanium and silicon using the valence potentials discussed in the first section. The results are listed in table 1 and show that energy differences of up to 50% are obtained.

Table 1 - calculated strain energies (in J/m^2) for silicon and germanium (in brackets) using different valence potentials and translation vectors $t_{\parallel}$ (parallel) and $t_{\perp}$ (perpendicular to the boundary plane, in Å). $\Sigma 17$, and $\Sigma 11$ contain additionally the energy E_{db} of 2 and 4 dangling bonds, resp. (E_{db} = 0.5 - 1eV)

Σ	33	19	9	17*	11*
Θ	20.1	26.5	38.9	93.4	129.5
γ (a)	1.46	1.26	0.98	1.03	1.57
	(1.43)	(1.24)	(0.95)	(1.00)	(1.23)
γ (b)	1.09	0.74	0.65	0.69	1.21
	(0.94)	(0.67)	(0.61)	(0.65)	(0.81)
γ (c)	1.21	0.75	0.68	0.64	1.12
	(1.02)	(0.69)	(0.64)	(0.59)	(0.74)
γ (d)	1.02	0.67	0.64	0.65	1.05
$t_{\parallel}$	0	0	0.11	0.23	0.29
$t_{\perp}$	0.42	0.92	0.74	0.73	0.81

The harmonic Keating (*a*) and the Lifson-Warshel potential (*d*) generally yield the highest and lowest values, resp. Obviously when the anharmonicity of the potentials is taken into account the strain energies come out lower. These potentials are still not correct for large deviations from the equilibrium values, however, this should not change the total energies significantly since in general only a few bonds are distorted considerably.

In [9, 22] energy values were presented for Σ33, 19, 9, and 11 which were calculated using a modified Keating potential (*c*) and are in agreement with the present study. It has been proposed that for the Σ11 grain boundary a reconstruction of the 4 dangling bonds along the <011> direction is possible. The calculation of the total energies seemed to indicate that a partial reconstruction where every second bond-pair is reconstructed is most favorable (table 2). However, the lower strain energy values (*d*) obtained in this study reduce the differences in the total energies of partially and fully reconstructed structures, especially if the dangling bond energy is quite large (E_{db} = 1 eV).

Table 2 - total energies (γ in J/m^2) for a Σ11 boundary depending on the degree of reconstruction and dangling bond energies E_{db} [in eV].

E_{db}	0.5 (c)	1.0	0.5 (d)	1.0
no reconstruction	1.9	2.6	1.7	2.4
part. reconstruction	1.3	1.7	1.2	1.5
tot. reconstruction	1.8	1.8	1.5	1.5

One must admit that at present the valence potential calculations are still not accurate enough to decide whether a total reconstruction of dangling bonds is preferred.

The calculations are only performed yet for the very particular case of coincidence boundaries where a good atomic matching is achieved which might favor bond reconstruction. Therefore the attempt is made to extend the calculations to more general grain boundaries, namely to near-coincidence boundaries.

4 Dislocations in Coincidence Grain Boundaries

It is likely to assume, and TEM investigations seem to confirm it, that to account for small deviations from a coincidence orientation, dislocations are incorporated in the grain boundary comparable to a small angle boundary. In Bollmann's O-lattice theory a geometrical description of grain boundary dislocation is given which shows that the coincidence boundary structure can only be preserved if the Burger's vectors of the secondary dislocations are chosen from the corresponding DSC-lattice. Experimental observations, however, not only revealed dislocations with DSC-Burger's vectors but also dissociated grain boundary dislocations [23], therefore the assumption of "bounding structure preserving" Burger's vectors $\underline{b}$ cannot generally be valid.

In isotropic approximation the energy of a dislocation is given by

$$E_d = kb^2 \ln(r/rc) + E_c$$

where k depends on the dislocation character, E_c is the core energy, r_c the core radius and r is approximately the radius of a cylinder over which the elastic strain field extends. For small deviations $\Delta\Theta < 10^\circ$ the energy (per area) γ of an assembly of parallel dislocations can be approximated by

$$\gamma = kb\Theta\,(1 - \ln 2\pi\Theta + E_c/kb^2)$$

γ is both dependent on b and E_c therefore it is likely that a secondary dislocation type is chosen for which γ (not E_d) is lowest. Since for larger deviations $\Delta\Theta$ the core energy E_c should become more important, we attempted to investigate this point in more detail for the simple cases of dislocations in a Σ3 and Σ9 <011> tilt boundary.

Since the energy calculations require a geometrical model of the atomic structure of the defect a short outline of the modelling procedure shall be given first. It can best be described in the DSC-lattice which contains both crystal lattices as a sublattice. (Relative shifts of the two crystal lattices resulting from the real grain boundary structure are not considered at the moment). For the <011> rotation axis the DSC lattice has a body centered orthorhombic unit cell (k_1, k_2 integer numbers, $\Sigma = 2k_1^2 + k_2^2$) with

$\underline{d}_1 = a\,[k_2\ k_1\ \bar{k}_1]/\Sigma$ $\quad\quad$ $\underline{d}_2 = a\,[2\bar{k}_1\ k_2\ \bar{k}_2]/\,2\Sigma$

$\underline{d}_3 = a\,[011]/2$

Formally a dislocation is introduced (see Fig. 1) by making a cut along the grain boundary plane, shifting lattice 1 with respect to lattice 2 along the cut by the Burger's vector (equal to a DSC-vector) and fitting the crystal halves together so that the grain boundary structure is preserved. A complication arises since in general the DSC lattice is shifted by a crystal lattice vector $\underline{a}$ upon a relative translation of the two crystal lattices. Then the grain boundary structure cannot be preserved along the original trace of the grain boundary plane, but is displaced by a vector $\underline{a}$ and a step in the boundary plane of height $\underline{h} = \underline{a} - \underline{b}$ arises.

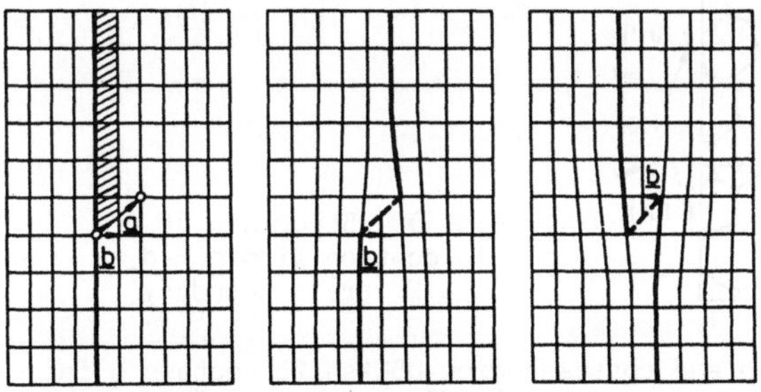

Fig. 1 Dislocation with a DSC Burger's vector $\underline{b}$ and a DSC shift vector $\underline{a}$ in a grain boundary

An example is given for the simple case of a Σ3 twin boundary ($k_2 = k_1 = 1$, $\Theta = 70.5^\circ$). The two shortest DSC Burger's vectors and the corresponding DSC shift vectors $\underline{a}$ are given in table 3, the developed structures of the corresponding grain boundary dislocation are presented in Fig. 2a,b. The shortest b vector of Fig. 2b produces a larger step in the boundary as compared to $\underline{b} = \underline{d}_1 + \underline{d}_2$ and contains a bond (dashed line) which is strongly distorted if a reconstruction is energetically favorable at all. The core of the dislocation with the longer b-vector obviously fits better into the boundary structure. The latter is a Lomer type dislocation ($b = \frac{a}{2}[01\bar{1}]$) and has already been observed experimentally in small angle [4] and in Σ9 and Σ11 <011> tilt boundaries.

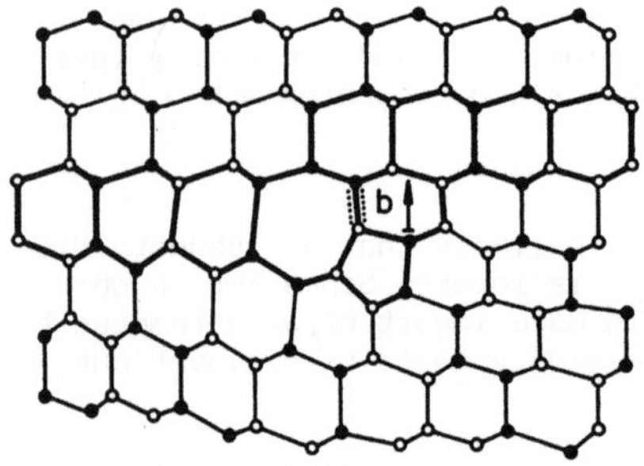

Fig. 2a - Core structure of a dislocation in a $\Sigma 3$ twin boundary with $\underline{b} = \underline{d}_1$ (dashed line indicates possible bond reconstruction)

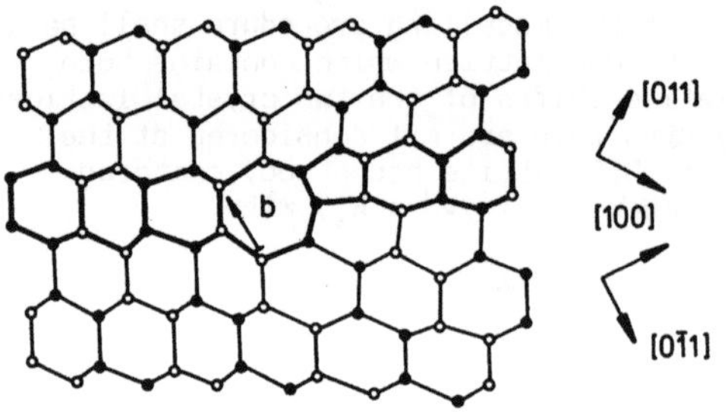

Fig. 2b - Core structure of a dislocation in a $\Sigma 3$ twin boundary with $\underline{b} = \underline{d}_1 + \underline{d}_2$

Another example is given for the case of a $\Sigma 9$ boundary ($k_1 = 2$, $k_2 = 1$, $\Theta = 38.9°$). The shortest DSC - Burger's vector again yields a large step in the boundary plane (table 3, Fig. 3) and the core of the dislocation contains two dangling bonds. With increasing b-length the step $\underline{a}$ decreases and consequently the dislocation core extends over a smaller area.

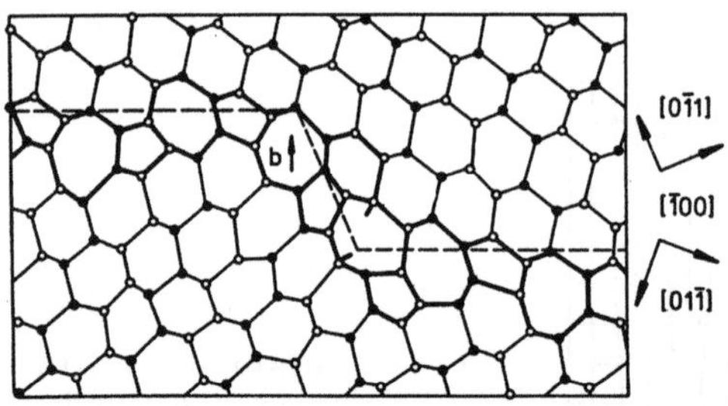

Fig. 3 - Structure of a dislocation with $\underline{b} = \underline{d}_1$ in a $\Sigma 9$ grain boundary

Table 3 - DSC-Burger's vectors $\underline{b}$ and corresponding DSC-shift vectors $\underline{a}$ for a $\Sigma 3$ and $\Sigma 9$ coincidence grain boundary. $\underline{a}_1 = a[\bar{1}00]$ and $\underline{a}_2 = a[0\bar{1}1]/2$ (a lattice constant)

	$\underline{b}$	b/d_2	$\underline{a}$	$\lvert\underline{a}\rvert/a_2$
$\Sigma 3$	$\underline{d}_1$	$\sqrt{2}$	a_2	1
	$\underline{d}_2 + \underline{d}_1$	$\sqrt{3}$	$-\underline{a}_2$	1
$\Sigma 9$	$\underline{d}_1$	$\sqrt{2}$	$2\underline{a}_2$	2
	$\underline{d}_2 - \underline{d}_1$	$\sqrt{3}$	$2\underline{a}_2$	2
	$\underline{d}_2 + \underline{d}_1$	$\sqrt{3}$	$\underline{a}_1 + \underline{a}_2$	$\sqrt{3}$
	$2\underline{d}_2 - \underline{d}_1$	$\sqrt{6}$	$\underline{a}_1 + a_2$	$\sqrt{3}$
	$2\underline{d}_2 + \underline{d}_1$	$\sqrt{6}$	$\underline{a}_2$	1

In Fig. 4a a dislocation with $\underline{b} = \underline{d}_1 + 2\underline{d}_2$ is shown which again contains a reconstructed bond (dashed line) or alternatively a broken bond. This dislocation has been identified recently by a TEM investigation in a near-coincidence Σ9 grain boundary ($\Delta\Theta \approx -3^{\circ}$) grown in a germanium bicrystal [24]. One can assume that this type has the lowest energy which is why it has attracted our interest. A full discussion of the results will be given in a following paper, therefore only a brief outline of the procedure of the energy calculation will be given here. A large unit cell containing two dislocations of opposite Burger's vector was constructed. This ensured that periodic boundary conditions could be adopted. Far away from the dislocation cores the displacements of the atoms from their positions in an unstrained bicrystal (containing a Σ9 boundary) were calculated using isotropic elasticity theory. The atoms outside at the surface were frozen to their initial positions whereas all the other atoms were allowed to relax until the strain energy was minimal. If the two dislocations were sufficiently widely separated (> 6a) the total energy (calculated with a modified Keating potential) became about 1.1 J/m² (E_{db} = 0.5 eV) or 1.39 J/m² (E_{db} = 1 eV) compared to 0.65 J/m² of the perfect Σ9 grain boundary in silicon. If we attribute the difference approximately to the core energy of the dislocation pair we obtain 0.32 eV/Å (0.53 eV/Å resp.) or about 0.15 eV/Å (0.26 eV/Å) for each dislocation. This is less than the calculated core energies of glide dislocations in a perfect silicon lattice [16] E_c > 0.5 eV/Å and may be due to the fact that the core of a dislocation in an already disturbed structure (of the grain boundary) does not introduce so much additional strain energy as compared to the perfect lattice. Since the excess energies are rather small it is certainly too early at this point clearly to distinguish between different core structures, especially in view of a possible bond reconstruction.

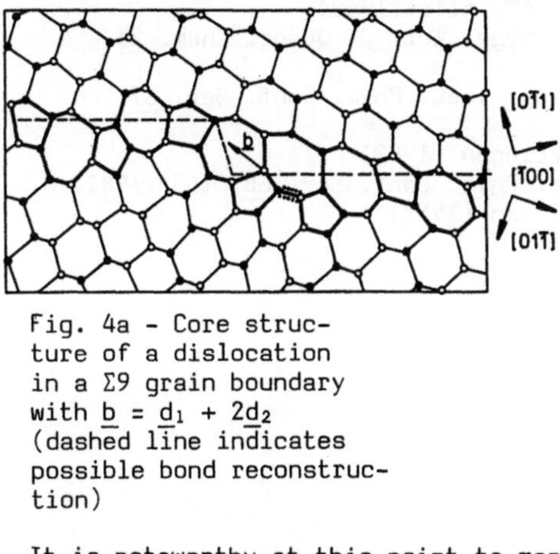

Fig. 4a - Core structure of a dislocation in a Σ9 grain boundary with $\underline{b} = \underline{d}_1 + 2\underline{d}_2$ (dashed line indicates possible bond reconstruction)

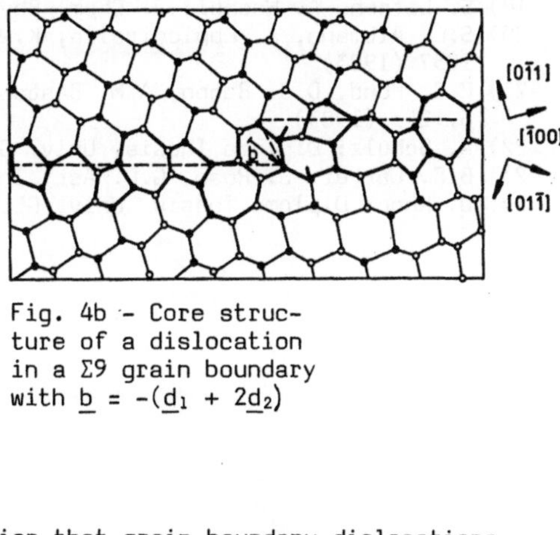

Fig. 4b - Core structure of a dislocation in a Σ9 grain boundary with $\underline{b} = -(\underline{d}_1 + 2\underline{d}_2)$

It is noteworthy at this point to mention that grain boundary dislocations with opposite signs of the Burger's vector in general have a different core structure since the step vector $\underline{a}$ also changes its sign. An example is given in Fig. 4b where $\underline{b} = -(\underline{d}_1 + 2\underline{d}_2)$. Whereas the first type decreases the rotation angle of the two crystal lattices the dislocation with the opposite sign of the Burger's vectors increases it (or vice versa). Because of the different core structures of the dislocation pair in Fig. 4a,b it is certainly not correct to attribute half of the total strain energy to each dislocation, however, for the discussion given above the approximate values are sufficient. In any case the results seem to indicate that secondary dislocations are not necessarily the ones with the shortest DSC-Burger's vector, but the core energies or a dissociation of the dislocations have to be considered, too.

Acknowledgements - The financial support of the Deutsche Forschungsgemeinschaft for one of the authors (H.H. Singer) is gratefully acknowledged.

5 References

1) X.J. Wu, V. Szielko, P. Haasen: J. Phys. 43, Cl - 135 (1982)
2) J. Werner, H. Strunk: J. Phys. 43, Cl - 89 (1982)
3) A. Broniatowski: J. Phys. 43, Cl - 63 (1982)
4) A. Bourret, C. d'Antesroches: J. Phys. 43, Cl - 1 (1982)
5) R. Podbielski, H.J. Möller: Proceed. MRS Europe Conf. 1984, J. Phys., in press
6) C.H. Seager, D.J. Sharp, J.K.G. Panitz, J.I. Hanoka: J. Phys. 43, Cl-103 (1982)
7) L.L. Kazmerski: J. Vac. Sci. Technol. 20, 423 (1982)
8) H.J. Möller: Phil. Mag. A43, 1045 (1981)
9) H.J. Möller: J. Phys. 43, Cl - 33 (1982)
10) R.C. Pond, D. Vlachavas: Proceed. Roy. Soc. A386, 95 (1983)
11) A.M. Papon, M. Petit, J.J. Bacmann: Phil. Mag. 49, 573 (1984)
12) P.N. Keating: Phys. Rev. 145, 637 (1966)
13) H. Korizumi, T. Ninomiya: J. Phys. Soc. Jap. 44, 898 (1978)
14) G.A. Baraff, E.O. Kane, M. Schlüter: Phys. Rev. B21, 5662 (1980)
15) S.L. Altmann, K.W. Lodge, A. Lapiccirella: Phil. Mag. A47, 827 (1983)
16) S. Marklund: Phys. Stat. Sol. B100, 77 (1980)
17) R. Jones: J. Phys. 40, C6 - 33 (1979)
18) R.M. Martin: Phys. Rev. B1, 4005 (1970)
19) S. Lifson, A. Washel: J. Chem. Phys. 49, 5116 (1968)
20) S.L. Altmann, A. Lapiccirella, K.W. Lodge: Int. J. Quant. Chem. 23, 1057 (1983)
21) R.C. Pond, D.J. Bacon, A.M. Bastaweesy: Inst. Phys. Conf. Ser. 67, p. 253 (1983)
22) A. Schulz: Diploma Thesis, Univ. Göttingen (1982)
23) B.C. Carter, J. Rose, G.D. Ast: Inst. Phys. Conf. Ser. 60, 153 (1981)
24) U. Knop: Diploma Thesis, Univ. Göttingen (1982)

The Geometrical Character of Extended Interfacial Defects in Semiconducting Materials

R.C. Pond

Department of Metallurgy and Materials Science, The University of Liverpool
P.O. Box 147, Liverpool L69 3BX, United Kingdom

ABSTRACT The geometrical character of extended defects in important semiconductor materials is described. A crystallographic approach is taken which emphasizes the inter-relation of bulk, surface and interfacial defects, and which also shows that the character of defects existing in a given material depends on its symmetry. In particular, it is shown that the existence of screw-rotation or mirror glideplanes in a crystal's space-group, as is the case for diamond structure materials for example, leads to important consequences. In such materials a special type of surface and interfacial step can arise. Moreover, in materials with structures which are derived from high symmetry structures having screw axes and glide planes, such as sphalerite which is derived from diamond, these special steps can also arise but are associated with antisite domain boundaries. The dependence of the nature of interfacial defects on the symmetry of a material is illustrated by comparing corresponding interfaces in Al, diamond and sphalerite structure materials. Interphase interfaces are also considered; it is shown that interfacial defects arise in interfaces where the adjacent crystals have different symmetry. This is illustrated by the discussion of defects in GaAs:Ge and $NiSi_2$:Si epitaxial interfaces.

1. INTRODUCTION

The transport properties of polycrystalline semiconductor materials can be radically different to their counterparts in single crystals. These consequences of the presence of interfaces may be advantageous or disadvantageous. For example, in multiple quantum well devices the special environment at heterojunctions is responsible for the confinement of a two dimensional electron gas, the special properties of which can be exploited. On the other hand, grain boundaries in solar cells are known to act as recombination centres and thereby reduce the efficiency of such devices. Interfaces have such remarkable effects on transport properties because additional mechanisms or significant changes to the parameters of diffusive or scattering mechanisms, for example, can arise. Some of these mechanisms are known to be very sensitive to thermodynamical variables such as impurity segregation to interfaces, and other authors in this volume will be describing the physical properties of interfaces from a macroscopic point of view. However, our current understanding of interfacial processes at the atomic level is incomplete, although there is accumulating evidence that interfacial defects have central roles. Thus, the object of the present article is to develop the theory of the geometrical character of extended defects in semiconductor interfaces. It is intended that this discussion should complement the theoretical and experimental studies of relaxed interfacial structures treated by Möller and Bourret in this volume, and provide a link between the atomic structure considerations of these latter authors and the discussion of macroscopic interfacial properties by others.

Before embarking on the theory of interfacial defects we outline two instances where experimental observations indicate that the magnitudes of important transport properties are governed by mechanisms occuring at interfacial defects. The first property is electrical conductivity across interfaces; considerable experimental evidence (see the sections by Broniatowski and Werner in this volume) has been obtained which supports the electrostatic potential barrier model of an electrically charged boundary. According to this model, interfacial electronic states, having energies within the forbidden band of the adjacent crystals, are filled with electrons (in 'n' type material), resulting in the bending of bands in the vicinity of the interfaces where the majority carriers are depleted. The origin of these special electronic states may be the distorted or dangling bonds in the interfacial region. Microscopical observations , particularly using the EBIC technique of electron microscopy (see the chapter by Donolato), support the view that these electronic states may be associated with interfacial dislocations, or possibly special points analogous to kinks on such defects.

The second observation, relevant to the recrystallisation of polycrystalline semiconductors, is the report by Smith and Tan [1] of enhanced grain growth in polysilicon due to the effects of doping and annealing atmospheres. These authors annealed intrinsic and phosphorous doped small grain size polysilicon in N_2 and O_2 ambients. They found that the final grain size was greater for (a) doped material, and (b) for material annealed in O_2. The explanation suggested by these authors concerns the effects of the experimental variables on the mechanism of grain boundary migration. This subject has been reviewed by Smith [2], and there is considerable evidence that the diffusive transfer of material from one grain to its neighbour occurs at either interfacial dislocations or interfacial steps, or both. As will be discussed later, interfacial dislocations usually have steps associated with their cores, and, in general, their movement along an interface is by combined glide and climb. If the mechanism of boundary migration is the movement of steps or dislocations, enhanced grain growth can be expected to result from increased dislocation mobility. The authors suggest that the enhancing effect of O_2 ambients is due to the formation of a SiO_2 surface film which injects interstitials into the polysilicon during the oxidation process thereby accelerating the rate of climb of interfacial dislocations. They also suggest that the predominant enhancing effect of the phosphorous in solution in the bulk of the crystals is to lower the Fermi level relative to electronic energy levels associated with kink-like configurations on interfacial dislocations.

2. EXTENDED DEFECTS IN CRYSTALLINE MATERIALS

The crystalline state is an ordered arrangement of matter which exhibits symmetry to a greater or lesser extent. Defects in crystalline materials are discontinuities of this order, and hence the nature of defects which can exist in a particular material depends on the symmetry of the unperturbed material. Defects can arise in the bulk or on the surface of a single crystal, or in interfaces between crystals, and the geometrical characters of these different types of defect are closely inter-related. The object of the present article is to illustrate this inter-relationship for materials used in semiconductor devices.

A unified theory of bulk, surface and interfacial defects has been developed by Pond and coworkers [3,4,5,6]. This approach is based on the principle of symmetry compensation which states that if symmetry is reduced at one structural level, it arises and is preserved at another. In other words, if any modification is made to a symmetrical structure, thereby reducing its symmetry, crystallographically equivalent variants of the modified structure exist, and these are inter-related by the symmetry operations supressed by the modification. The equivalent variants may coexist in a single crystal, on the surface of a crystal, or in an interface, and will be separated by extended defects. Thus, in order to predict the geometrical nature of the defects which can exist in a given structure it is necessary to identify the sequence of events in which the structure of interest is imagined to have been created from the most symmetric precursor conceivable. Each stage in the sequence of modifications which causes the symmetry to be lowered will lead to the existence of equivalent variants, and hence the possibility of extended defects separating these. We shall illustrate this procedure for single crystals before proceeding to bicrystals.

3. BULK AND SURFACE DEFECTS IN SINGLE CRYSTALS

The sequence of stages for identifying bulk and surface defects is indicated schematically in Fig.1(a). The most symmetric precursor for a single crystal is its lattice; this is an infinite array of points having identical surroundings. The spacegroup of the lattice is designated Φ^*, the asterisk signifying that this group has the maximum possible symmetry.

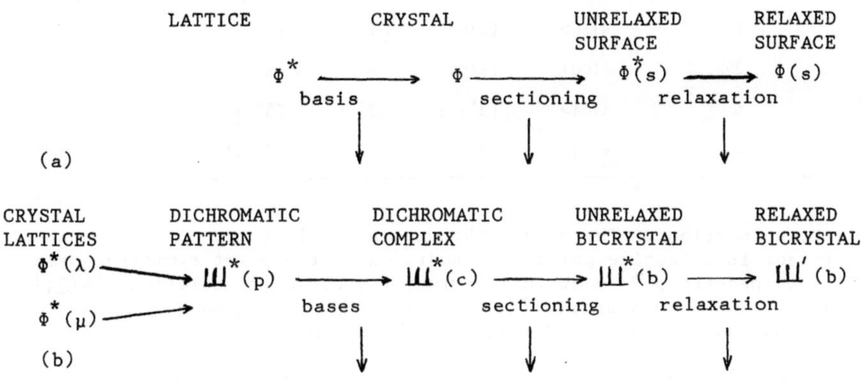

Fig. 1 a) Schematic illustration of the descent in symmetry associated with the imaginary sequence of events leading to the creation of a relaxed single crystal surface structure.
b) The analogous procedure for creating a relaxed bicrystal. Horizontal arrows represent the progression from one stage to the next, and vertical arrows correspond to the creation of crystallographic variants arising from any reduction of symmetry

We note that lattice spacegroups are always symmorphic, i.e. do not contain screw-rotation axes or mirror-glide planes in their generating set, and are holosymmetric, i.e. exhibit maximum symmetry. In the case of fcc lattices Φ^* = Fm$\bar{3}$m, and the 48 point symmetry operations in this group are listed in Table 1. These operations are conveniently expressed in matrix formulation as [D_i/0], where D_i represents the ith orthogonal operation such as a rotation, reflection or inversion, and the zero signifies that no additional translation is necessary in order to leave

Table 1 Corresponding symmetry operations in $Fm\bar{3}m$ and $Fd\bar{3}m$. Positive and negative superscripts to symmetry operations represent clockwise and anticlockwise rotations respectively

Fm$\bar{3}$m				Fd$\bar{3}$m	
Symmetry Operation	No. of Operations	Orientation	Matrix	Symmetry Operation	Matrix
1	1	-	[1/0]	1	[1/0]
$\bar{4}^+$	3	<100>	$[\bar{4}^+/0]$	$\bar{4}^+$	$[\bar{4}^+/0]$
$\bar{4}^-$	3	<100>	$[\bar{4}^-/0]$	$\bar{4}^-$	$[\bar{4}^-/0]$
2	3	<100>	[2/0]	2	[2/0]
m	6	{110}	[m/0]	m	[m/0]
3^+	4	<111>	$[3^+/0]$	3^+	$[3^+/0]$
3^-	4	<111>	$[3^-/0]$	3^-	$[3^-/0]$
$\bar{1}$	1	-	$[\bar{1}/0]$	$\bar{1}$	$[\bar{1}/\underline{\alpha}]$
4^+	3	<100>	$[4^+/0]$	4_1^+	$[4^+/\underline{\alpha}]$
4^-	3	<100>	$[4^-/0]$	4_1^-	$[4^-/\underline{\alpha}]$
2	6	<110>	[2/0]	2	$[2/\underline{\alpha}]$
m	3	{100}	[m/0]	d	$[m/\underline{\alpha}]$
$\bar{3}^+$	4	<111>	$[\bar{3}^+/0]$	$\bar{3}^+$	$[\bar{3}^+/\underline{\alpha}]$
$\bar{3}^-$	4	<111>	$[\bar{3}^-/0]$	$\bar{3}^-$	$[\bar{3}^-/\underline{\alpha}]$

the lattice invariant. The origin through which all these operations are taken to act is a lattice point. In addition to the point symmetry operations, the spacegroup Φ^* contains the infinite set of translations $[E/\underline{\tau}_j]$ where E is the identity matrix (i.e. representing the symmetry operation 1) and $\underline{\tau}_j$ is the jth translation vector of the lattice.

A perfect infinite crystal can be created from its lattice by placing a basis, i.e. an atom or group of atoms, at each lattice point. (Let the kth atom in the basis set be located by the vector $\underline{r}_k$ with respect to a lattice point.) This represents the second stage of the sequence, and the spacegroup of the crystal is designated Φ and may exhibit lower symmetry than the lattice. According to the principle of symmetry compensation, all symmetry operations suppressed in going from one stage to the next become exchange operations relating crystallographically equivalent variants. In the present case these are designated basis-variants, and these may co-exist in a single imperfect crystal. For example, crystals having the spacegroup Φ = F23 are based on fcc lattices but the nature of the basis suppresses, for example, mirror symmetry on {110}. The consequence is that merohedral twins can exist in such crystals, where the mirror operation on {110}, which becomes a basis-exchange operation, leaves the lattice invariant but relates bases in twin orientation.

In the case of simple structures such as Al, where the basis is a single atom ($\underline{r}_1$=0) no basis-variants exist. In the case of crystals having the diamond structure the basis is two atoms with $\underline{r}_1=0$ and $\underline{r}_2=\frac{1}{4}[111]$; the spacegroup, Φ, is $Fd\bar{3}m$ which is isomorphous with $Fm\bar{3}m$ (i.e. there is a one to one correspondence between the operations in the two groups), and it follows therefore that no basis-variants exist. However the diamond structure is non-symmorphic and the corresponding operations in the two isomorphous spacegroups are shown in Table 1. It can be seen that half of the operations in the two groups are identical, whilst the remainder have a supplementary translation $\underline{\alpha}$, where $\underline{\alpha} = \frac{1}{4}[111]$ (modulo $\underline{\tau}_j$) for the diamond structure. (It is most important to note that the orthogonal part of the operation $[D_i/\underline{\alpha}_i]$ is taken to act through the same origin, i.e. a lattice point, as for the isomorphous symmorphic structure; it can then be seen that these matrix formulations of the symmetry operations leave the nonsymmorphic crystal invariant.) It is instructive to examine the operations in the group $Fd\bar{3}m$ further. Fig. 2 is a (001) projection of a diamond crystal and four unit cells have been indicated in the centre of the drawing. It is convenient to regard this structure as the interpenetration of two sets of atoms, one depicted by circular symbols and the other by square symbols. The first set is an fcc array of atoms located at the lattice points; we designate these the fcc set. The second is an identical set displaced from the first by $\underline{\alpha}$, and we designate these the tetrahedral set. We note that the first 24 operations in Table 1, i.e. those common to $Fm\bar{3}m$ and $Fd\bar{3}m$ and therefore having zero supplementary displacement, relate atoms within the fcc or tetrahedral sets, but do not relate atoms of one set to those of the other. The latter 24 operations, which have the supplementary displacement $\underline{\alpha}$ in the group $Fd\bar{3}m$ are either nonsymmorphic operations, i.e. $4_1^+, 4_1^-$, d, or symmorphic operations which do not act through the chosen origin, i.e. $\bar{1}, 2, \bar{3}^+, \bar{3}^-$. These operations interrelate atoms in the two sets, but do not relate atoms within the same set.

Crystals having the sphalerite structure can be described formally as having an fcc lattice and a basis consisting of one atom at $\underline{r}_1=0$ and a second atom, different to the first, at $\underline{r}_2=\frac{1}{4}[111]$. Thus $\Phi^* = Fm\bar{3}m$ and Φ

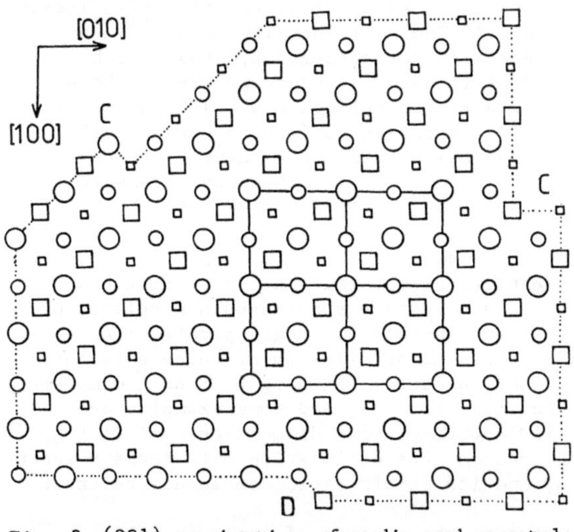

Fig. 2 (001) projection of a diamond crystal; circular symbols represent atomic sites at levels 0 (large) and $\frac{1}{2}$ (small), and square symbols represent sites at $\frac{1}{4}$ (small) and $\frac{3}{4}$ (large). Complete and demi-steps on unrelaxed crystal surfaces are indicated by C and D respectively

= $F\bar{4}3m$, and it follows that two basis-variants exist. (The number of variants and exchange operations inter-relating them can be found by the use of group decomposition methods [4]). However, for the present purposes it is more convenient to regard the sphalerite structure as a derivative of the diamond structure, i.e. to take the most symmetric precursor, Φ^*, to be $Fd\bar{3}m$, and Φ as $F\bar{4}3m$. We note that the symmorphic point group $\bar{4}3m$ comprises the first 24 operations in Table 1, i.e. those which relate atoms within the fcc and tetrahedral sets. Thus, the other 24 operations in Table 1 become basis-exchange operations inter-relating the two basis-variants. It can be seen that since these operations interchange atoms at fcc and tetrahedral sites the basis-variants are anti-site domains, and the extended defect which would separate coexisting variants would be an anti-site domain boundary.

We now turn our attention briefly to line-defects in single crystals. These fit into the present scheme in a crystallographically trivial sense in so far as they are extended defects relating identical variants which coexist in an imperfect crystal. Perfect dislocations are associated with any translation operation $[E/\underline{\tau}_i]$, although only dislocations with primitive Burgers vectors are generally stable. For completeness we note that disclinations [7] and dispirations [8] are associated with rotation and screw operations respectively, but these are not thought to be physically important except in the cases of crystals with low elastic modulus such as liquid crystals.

In this section we consider the nature of unrelaxed surfaces on single crystals. We can create an arbitrary unrelaxed crystal surface from an infinite perfect crystal by sectioning the crystal on a plane having the required orientation and location, and subsequently discarding all atomic sites to one side of this plane. This processs corresponds to the third stage of the scheme depicted in fig.1(a). (In the case of crystals having more than one atom in the basis we note that sectioning will produce different surface structures depending on the location of the sectioning plane except for surfaces where there exists a set of vectors $(\underline{r}_i+\underline{\tau}_i)$ each of which is parallel to the surface.) The semi-infinite crystal with the selected unrelaxed surface can be assigned a spacegroup, $\Phi^*(s)$, which can only have two, or fewer, independent translation symmetry axes. Those operations belonging to Φ which do not also belong to $\Phi^*(s)$ become surface-exchange operations and relate the chosen surface to energetically degenerate surface-variants. We can now consider the nature of steps which can exist on crystal surfaces, and will find that crystals with non-symmorphic spacegroups can exhibit a special type of step. First, we consider a type of step, which can exist on any rational surface, for all types of crystal. We designate such features complete steps, and note that they join surfaces which are identical except that the sectioning has been carried out at different levels. Such surface structures are related by any vector $\underline{\tau}_i$ belonging to the set of surface-exchange operations, i.e. any translation vector having a component parallel to $\underline{n}$, where $\underline{n}$ is the outward pointing unit normal to the surface. The height of such complete surface steps, designated h^s_c. is equal to $\underline{n}.\underline{\tau}_i$, and some examples are illustrated on the (100) and (110) surfaces of a diamond crystal in fig.2. We note that such steps can be created if a dislocation with Burgers vector $\underline{\tau}_i$ reaches the free surface, $\underline{n}$, and its strain field completely relaxes.

A second type of step can exist on the surface of non-symmorphic crystals. For example, in the case of diamond structure materials the set of surface-exchange operations may include non-symmorphic operations, and these may relate equivalent surfaces more closely spaced than h^s_c. An example is shown on the (100) surface in fig.2; here the degenerate surface structures on either side of the step are related by the operations 4_1^+, 4_1^- and d which are perpendicular to the surface and the step height is only half of the minimum complete step height for this surface. Note that these nonsymmorphic operations relate a surface consisting of fcc atoms on one side of the step to a surface consisting of tetrahedral sites on the other. Such steps are designated surface demi-steps, and their height, h_d^s, is given by $n.(\alpha + \tau_1)$. It has been shown [9] that demi-steps can exist only on the {hko} surfaces of diamond crystals where h and k (taken as having no common factors) are mixed (i.e. odd and even) integers. (We note that for { hko } surfaces where h and k are both odd the unrelaxed surface structure is comprised of both fcc and tetrahedral atoms, i.e. there is always a vector $(\underline{r} + \underline{\tau}_1)$ parallel to the surface. When h and k are mixed the surface is comprised of either fcc or tetrahedral atoms, but not both simultaneously).

We now consider the nature of surface steps on sphalerite-structure materials. It will be recalled that Φ for such crystals is the symmorphic spacegroup $F\bar{4}3m$, and it therefore follows that only complete steps are expected to exist on such crystal surfaces. However, it will also be recalled that the sphalerite structure is a derivative of the diamond structure, and that the last 24 operations in Table 1 are basis-exchange operations relating anti-site domains. It follows (see [9] for details) that demi-steps can exist on the same surfaces as for diamond except that these now delineate the emergence from the bulk of anti-site domain boundaries. In other words, anti-site domain boundaries emerging onto {hko} surfaces of sphalerite structure materials separate energetically degenerate surface structures. In the cases where h and k are mixed, the emergence of the domain boundary is associated with a demi-step. Such a demi-step is illustrated schematically in Fig. 4.

Finally in this section we consider briefly the last stage of the scheme depicted in Fig.1(a). This corresponds to allowing the unrelaxed surface to relax. Relaxation may be accompanied by a reduction of symmetry from $\Phi^*(s)$ to $\Phi(s)$, the latter being the spacegroup of the fully relaxed structure. As before, equivalent relaxation-variants can arise inter-related by relaxation-exchange operations, and the variants may coexist on the same surface and be separated by extended defects. An important distinction between this and previous stages of dissymmetrisation is that any reduction of symmetry cannot be predicted on the basis of crystallographic arguments; a physical theory is necessary in order to predict the nature of surface relaxation. Currently, no comprehensive theory exists and our knowledge of surface relaxation is based primarily on experimental observation [10]. Once the nature of surface relaxation on a given surface has been established, crystallographic methods can be used to identify the number of relaxation-variants, the relaxation-exchange operations and the nature of the extended defects separating equivalent variants. Analogous relaxation defects can also arise in the bulk of single crystals, for example stacking faults may exist in some materials. However, as for surface relaxation effects, a physical theory is necessary in order to predict the nature of stable or metastable defects of this type.

Before considering the nature of defects in bicrystals we consider briefly the morphology of single crystals. Most treatments of morphology are based on the Wulff theory [11] which shows that the morphology of perfect crystals must be consistent with the point symmetry of Φ. However, the principle of symmetry compensation implies that the morphology of an imperfect crystal can be consistent with Φ^* rather than Φ, because de-

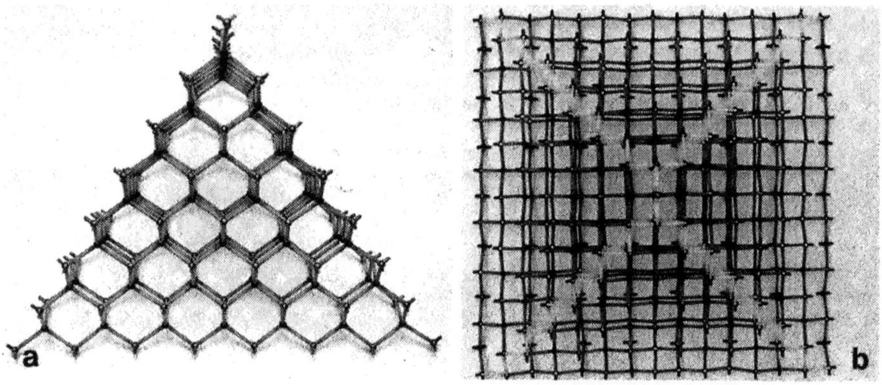

Fig. 3 a) Model of a perfect crystal of sphalerite having {111} surfaces, showing tetrahedral morphology, and viewed along <110>.
b) View along <100> of an imperfect crystal exhibiting anti-site domain boundaries (light coloured bonds) parallel to {001}; the morphology is octahedral

fects, which separate basis-variants, can always be introduced in such a way as to compensate for any reduction of symmetry. This is illustrated in Fig.3(a) which shows the view along <110> of a model of a perfect sphalerite crystal having {111} faces; the crystal morphology is tetrahedral consistent with the point symmetry of $\Phi = F\bar{4}3m$. However, the introduction of anti-site domain boundaries on {100} planes separating basis-variants enables octahedral morphology consistent with $\Phi^* = Fm\bar{3}m$ as shown in Fig. 3(b).

4. INTERFACIAL DEFECTS

In the foregoing discussion a treatment of defects in single crystals was developed which emphasized the inter-relation of bulk and surface defects. For example, the nature of complete and demi-steps, and their relationship to dislocations, dispirations, and anti-site domain boundaries was explained in terms of the crystal's spacegroup. In the present section we consider the nature of interfacial defects, and the approach adopted is analogous to that used previously. This will be expressed in terms of bicrystal symmetry, and will make use of the principle of symmetry compensation, and will emphasize the inter-relation of bulk, surface and interfacial defects. Both interphase interfaces and grain boundaries can be treated, and the methodology is shown schematically in Fig. 1(b). The starting point is again the identification of the most symmetric precursor conceivable. Since two crystals are now involved, this initial configuration is obtained by bringing the two crystal lattices together. We designate one lattice white, having a spacegroup $\Phi^*(\lambda)$, and the other black with spacegroup $\Phi^*(\mu)$, and the configuration obtained by allowing the lattices to interpenetrate is called the dichromatic pattern and its spacegroup is designated Ш*(p). In the case of interphase boundaries, such as those where the adjacent crystals have an epitaxial alignment and small misfit between their lattice parameters, the initial choice of lattice parameters is chosen so that the dichromatic pattern has exact translation symmetry. In the course of the treatment, the character of possible interfacial defects will be elucidated, and any actual misfit between the crystals can be accommodated in real interfaces by arrays of appropriate defects. Similarly, in the case of grain boundaries, the relative orientation of the two lattices is taken to be such as to lead to a coincidence site lattice (csl), and any deviation from this ideal in an actual boundary can be accommodated by interfacial defects.

The extent of the analogy between the treatment of single and bicrystals is apparent from Fig. 1. It can be seen that the creation of a fully relaxed interfacial structure is imagined to be a four stage process starting with the dichromatic pattern. The next stage is to place bases at the black and white lattice points to obtain the dichromatic complex, or pattern of the inter-penetrating structures, and the spacegroup of this configuration is designated Ш*(c). An ideal unrelaxed (maximum symmetry) bicrystal can be created from this complex by sectioning; the orientation and location of the interfacial plane are chosen and black bases discarded on one side of this plane and white on the other. The spacegroup of the unrelaxed bicrystal is designated Ш*(b), and the final stage is to allow this structure to relax, with potential reduction of symmetry, to the relaxed bicrystal spacegroup Ш(b). As was the case for single crystals, basis-variants, morphological-variants, and relaxation-variants arise as a consequence of dissymmetrisation at each of the corresponding stages, and the number of variants and the exchange operations inter-relating them can be found by group decomposition. However, two noteworthy differences between the single and bicrystal cases are apparent. The first is that an additional set of variants, called orientation-variants, arises in the bicrystal case if any symmetry is suppressed when the dichromatic pattern is formed. For example, there are six orientation-variants of $\Sigma=5$ related cubic lattices corresponding to clockwise rotations of 36.9° about the six equivalent <001> directions in the white lattice. The second difference is that, in the case of grain boundaries, symmetry operations not present in the component crystal lattices can arise in the dichromatic pattern (and hence subsequent stages) if the relative orientation has special forms such as a mirror reflection or diad rotation. Such symmetry operations are designated antisymmetry operations and distinguished from ordinary operations (i.e. those which are common to the black and white components) by primed symbols. We note that the principle of symmetry compensation applies to both ordinary and antisymmetry operations, and that analytical methods for the determination and decomposition of dichromatic spacegroups have been presented [4,6].

In the remainder of this section we shall discuss interfacial defects in particular interphase and grain boundaries. However, we shall not do this by presenting a detailed analysis following the scheme of Fig.1(b) in each case, since these have appeared in the literature recently. Instead, we shall illustrate the main principles by reference to particular cases, and emphasize the analogy, which is implicit in Fig. 1, between bulk and surface defects in single crystals and those found in bicrystals. The discussion is limited to the geometrical character of interfacial defects since a physical theory is necessary in order to predict relaxation effects. The nature of interfacial relaxation has been studied experimentally and theoretically, and these topics are discussed in this volume by Bourret and Möller respectively.

5. GaAs:Ge AND $NiSi_2$:Si EPITAXIAL INTERFACES

GaAs:Ge epitaxial layers have been studied as model systems in the development of multiple quantum well devices. GaAs, which has the sphalerite structure, has been grown epitaxially on Ge substrates of various orientations. The two crystals grow with parallel orientation and have very similar lattice parameters. The present crystallographic method has been applied to this bicrystal system [9,12], and it has been shown that when the Ge substrate orientation is {hko} anti-site disorder can arise in the GaAs overlayer. The anti-site boundaries are nucleated at the interface and separate energetically degenerate interfacial domains, and this conclusion is entirely consistent with experimental observations.

In the terminology of the present work, the dichromatic complex has symmetry $Ш^*(c) = F\bar{4}3m$. The Ge structure has higher symmetry, $\Phi(\lambda) = Fd\bar{3}m$, but the GaAs structure when interpenetrated with Ge reduces the overall symmetry. It follows that two basis-variants of the interprenetrating structure exist, and these are inter-related by the last 24 operations in Table 1. It will be recalled that these basis-exchange operations leave Ge invariant but relate anti-site domains of GaAs. On sectioning these two complexes it is found that equivalent structures can coexist on the same interface if the basis-exchange operations also leave the orientation of $\underline{n}$ invariant. It follows that degenerate interfacial domains can exist on {hko} interfaces since individual operations in the basis-exchange set leave the orientation of these normals invariant. This result can be readily rationalised in terms of the different surfaces (i.e. (hkl) and $(\bar{h}\bar{k}\bar{l})$) on GaAs and Ge. {hko} surfaces of Ge, where h and k are both odd, comprise atoms in both fcc and tetrahedral sites, and energetically degenerate interfacial structures arise whether the GaAs nucleates from either one type of atomic site or the other. In the cases where h and k are mixed, the Ge surface is comprised of either fcc or tetrahedral atoms but not both. These surfaces can be demi-stepped, and the only way that the GaAs overlayer can grow contiguously over such steps with degenerate interfacial energy on either side is by introducing an antisite domain boundary emanating from the demi-step. This is shown schematically in Fig. 4. Complete steps on the surface of the substrate can always be accommodated by complementary steps on GaAs, but demi-steps, such as that shown in Fig. 4 on a (100) substrate, can only be accommodated by introduction of anti-site disorder. The existence of such disorder is, of course, detrimental to the electronic mobility parallel to the interface since antiphase boundaries act as charged scattering centres.

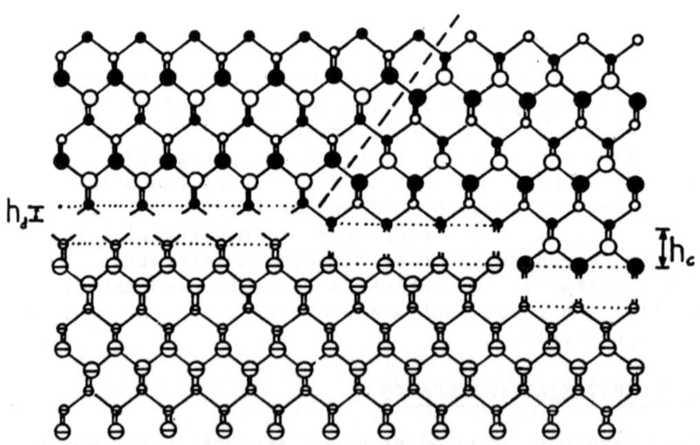

Fig. 4 (110) projection of epitaxial GaAs and Ge films having complementary (001) surfaces. The demi-step on the GaAs surface is associated with the emergence from the crystal bulk of an anti-site domain boundary, shown dashed

An observation closely related to that described above concerns $NiSi_2$:Si epitaxial interfaces [13,14]. $NiSi_2$ is a promising contact material for Si devices, and it is important that the interface should be free of defects in order to maximise conductivity across the interface. $NiSi_2$ overlayers, formed by evaporation of Ni followed by annealing, grow with parallel orientation to the substrate and similar lattice parameters. The structure of $NiSi_2$ is that of fluorite and has the spacegroup $\Phi(\mu) = Fm\bar{3}m$; the structure can be regarded as Ni atoms having displaced Si atoms from the fcc sites, and these displaced Si atoms occupy the empty tetrahedral sites in the original Si structure. The dichromatic groups $Ш^*(p)$, $Ш^*(c)$ and $Ш^*(b)$ are identical to those for the GaAs:Ge case, and hence the various sets of exchange operations are also identical. Moreover, it follows that energetically degenerate interfacial structures can coexist on {hko} interfaces. As previously, $NiSi_2$ can nucleate from either fcc or tetrahedral Si atoms on interfaces where h and k are odd. Interfaces where h and k are mixed may be demi-stepped and degenerate interfaces can form on either side of the step. However, unlike GaAs, $NiSi_2$ is holosymmetric and hence cannot exhibit anti-site domains. Instead, the extended defect separating equivalent interfacial domains is an interfacial dislocation with Burgers vector equal to α. This is depicted in Fig. 5 which shows complementary (100) surfaces for $NiSi_2$ and Si. The silicon substrate is demi-stepped, and in the r.h. part of the figure the accommodation of this demi-step as the 'extra half plane' of an interfacial dislocation with Burgers vector α is shown. It is readily seen that the interfacial structures on either side of the defect, in this and the previous case, are related by basis-exchange operations, e.g. 4_1^+, 4_1^-, d, and that these leave the orientation of $\underline{n}$ invariant. We also note that such dislocations can contribute to the accommodation of misfit on {hko} interfaces.

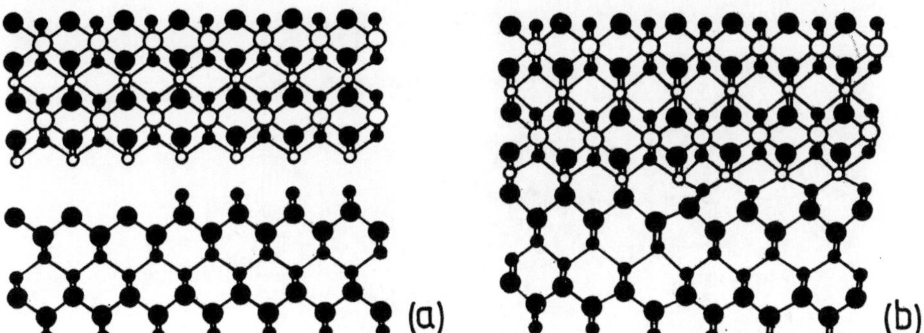

Fig. 5 a) Schematic projection along [110] of (001) surfaces in $NiSi_2$ (unstepped), and Si (demi-stepped).
b) shows the accommodation of the demi-step by ½<111> interfacial dislocation. Filled symbols represent Si, and large and small symbols correspond to ...ABABA... stacking along [110]

It is implicit in Fig. 1 that the morphology of a white precipitate inside a black matrix (or vice-versa) has point symmetry consistent with $Ш^*(c)$ in the case of perfect crystals, but can be consistent with $Ш^*(p)$ if appropriate defects are present to compensate for any symmetry reduction. Fig. 6 shows a (110) projection of a GaAs precipitate in a Ge matrix. The model has been constructed with an antisite domain boundary parallel to (001) in the GaAs precipitate in order to exhibit morphology

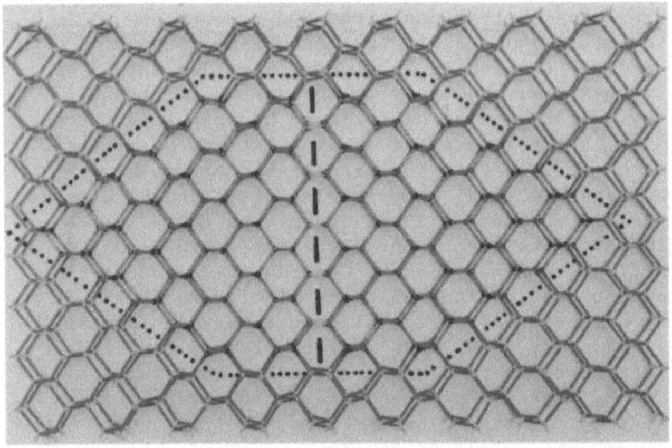

Fig. 6 (110) projection of an imperfect GaAs precipitate (anti-site domain boundary shown dashed) in a Ge matrix

consistent with $Ш^*(p)$. The analogous model for a precipitate of $NiSi_2$ is shown in Fig. 7, again exhibiting a mrophology consistent with $Ш^*(p)$. We note that electron microscope observations of $NiSi_2$ precipitates in Si reveal octahedral morphology, i.e. consistent with $\hat{Ш}^*(p)$ [15].

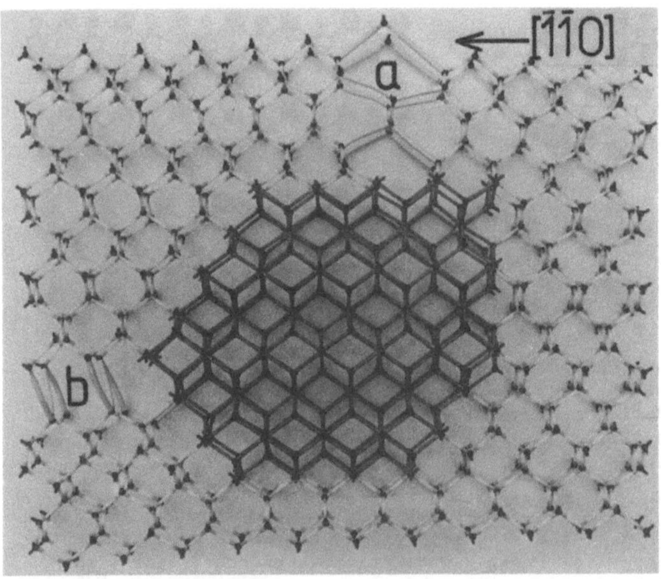

Fig. 7 ($1\bar{1}0$) projection of a $NiSi_2$ precipitate in Si. Dislocation a ($\underline{b}$ = $\frac{1}{4}[\bar{1}\bar{1}1]$)separates (basis-variant) domains on the same interfacial plane, ($00\bar{1}$), and b ($\underline{b}$ = $\frac{1}{4}[1\bar{1}1]$)separates basis variants on symmetry related interfaces, ($\bar{1}\bar{1}1$) and ($\bar{1}\bar{1}\bar{1}$)

6. GRAIN BOUNDARY DEFECTS

In this section we shall consider grain boundaries between $\Sigma=5$ related crystals, and compare the nature of defects anticipated when the adjacent crystals have simple fcc, diamond, and sphalerite structures. In all cases the component lattices are fcc and the relative orientation can be taken as a rotation of 53.1° about a common <001> axis. Fig. 8(a) shows the dichromatic pattern which has the spacegroup $Ш^{*}(p) = I\frac{4}{m}m'm'$. Now, as has been explained already in this volume by Bourret, the DSC lattice is given by the union of the black and white lattices. Vectors of the DSC lattice, designated $\underline{b}^{i}$ here, are the possible Burgers vectors of perfect interfacial dislocations, irrespective of crystal structure. These dislocations are analogous to perfect dislocations in single crystals, and arise because there is an additional degree of freedom in bicrystals, i.e. that for a given relative orientation, the adjacent crystals can be displaced with respect to each other. It has already been mentioned that the core of such defects may be associated with interfacial steps. We now show that such DSC dislocations can be rationalised in terms of the accommodation of complete surface steps on the complementary surfaces which are brought together to form an interface; this approach was first developed by King and Smith [16]. Consider a DSC dislocation with $\underline{b}^{i} = \frac{1}{10}[\bar{3}10]$ (using the coordinate system of the black crystal) in a (130) interface, i.e. so that $\underline{b}^{i}$ is parallel to the interface and hence the defect is glissile. First, we consider the two crystals to be separated at the interface and introduce complete surface steps with height $h_{c}^{s}(\lambda) = \underline{n}.\underline{\tau}(\lambda)_{i}$ on the white crystal ($\underline{n}$ is taken as pointing into the black crystal) and $h_{c}^{s}(\mu) = \underline{n}.\underline{\tau}(\mu)_{j}$ on the black crystal, where $\underline{\tau}(\lambda)_{i}$ and $\underline{\tau}(\mu)_{j}$ have been chosen such that $\underline{b}^{i} = \underline{\tau}(\lambda)_{i} - R\underline{\tau}(\mu)_{j} = \frac{1}{10}[\bar{3}10]$ (R is the coordinate transformation matrix relating the black and white crystals). The situation is depicted in Fig. 9 for the case where the component crystals have the diamond structure. Now imagine that the two crystals are brought together in such a way that on the left of the steps coincident sites are formed for example. It can be seen that the same interfacial structure can only be created to the right of the steps if the crystals are relatively sheared by $\underline{b}^{i}$, i.e. if an edge dislocation is created. It is easy to show that the interfacial step associated with the dislocation has a height, $h_{\perp}^{i}$, equal to $\frac{1}{2}(h_{c}^{s}(\lambda) + h_{c}^{s}(\mu))$. The climb component fo the interfacial dislocation will be given by $h_{c}^{s}(\lambda) - h_{c}^{s}(\mu)$, which is zero in the present case. Now consider a dislocation with the same Burgers vector in a $(\bar{1}20)$ interface between $\Sigma=5$ related crystals, i.e. such that the dislocation can only move along the interface by a combina-

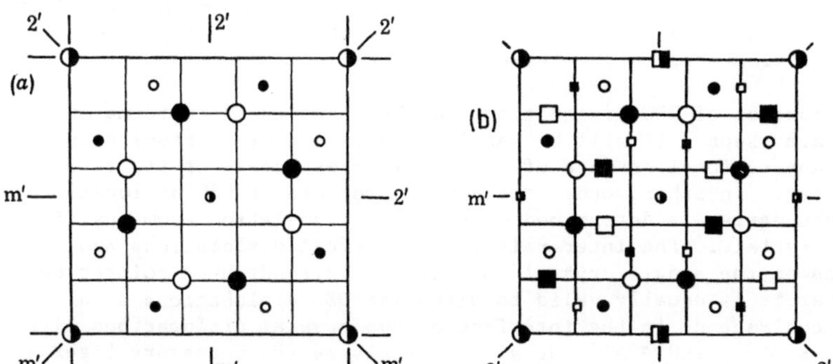

Fig. 8 a) Dichromatic pattern, and b) dichromatic complex for $\Sigma=5$ related diamond crystals; the spacegroups are $I\frac{4}{m}m'm'$ and $I\frac{4_1}{a}m'd'$ respectively

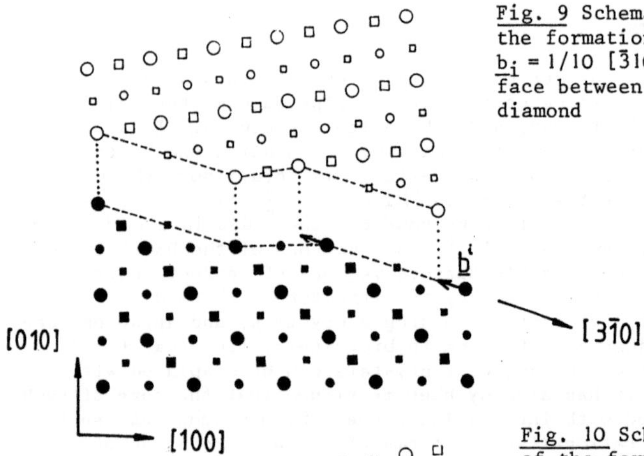

Fig. 9 Schematic (001) projection of the formation of a DSC dislocation, $\underline{b}_i = 1/10\ [\bar{3}10]$, in a (130) interface between Σ=5 related crystals of diamond

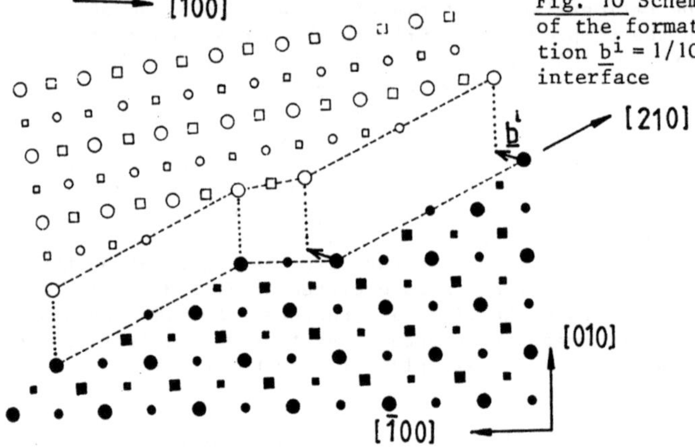

Fig. 10 Schematic (001) projection of the formation of a DSC dislocation $\underline{b}^i = 1/10\ [\bar{3}10]$, in a $(\bar{1}20)$ interface

tion of glide and climb. Fig. 10 shows the two crystals with surface steps $h_c^s(\lambda)$ and $h_c^s(\mu)$ separated at the interface. If the same matching procedure as for the (130) boundary is carried out, it can be seen that the same shear, $\underline{b}^i$, is necessary to form the interface on the right hand side of the steps, but this shear now has components parallel and perpendicular to the interface. The perpendicular component, i.e. the climb component, is given by $h_c^s(\lambda) - h_c^s(\mu)$ which is now non zero, and the interfacial step associated with the dislocation core, $h_\perp^i$, is given by $\frac{1}{2}(h_c^s(\lambda) + h_c^s(\mu))$.

The above discussion of DSC dislocations has been expressed in terms of complete surface steps. It will be recalled that complete surface steps can exist on any rational surface of a crystal irrespective of its structure or symmetry. In other words, the Burgers vectors of DSC dislocations in periodic boundaries is determined only by the translation symmetry of the adjacent crystals. The inter-relation of perfect dislocations and complete steps on the single crystal surfaces has already been pointed out. It follows that it is equally valid to visualise DSC dislocations as arising from the coalescence in the interface of two crystal dislocations with Burgers vectors $\underline{\tau}(\lambda)_i$ and $\underline{\tau}(\mu)_j$, or alternatively as the necessary interfacial remnant when a dislocation from one crystal penetrates through the interface into the other. Finally, we note that in the GaAs:Ge and $NiSi_2$:Si cases discussed previously, the DSC lattice is identical to the crystal lattices.

We now consider the second stage in the scheme; the dichromatic complex. For structurally simple materials like Al the spacegroup of the complex is the same as that of the pattern, i.e. $\text{Ш}^*(p) = \text{Ш}^*(c) = I4/mm'm'$ and hence no basis-variants exist. Unrelaxed interfacial structures can be obtained by sectioning, and, analogously to the case for single crystals, complete interfacial steps of height $h_c^I = \underline{n}.\underline{\tau}(csl)_k$ can exist, where $\underline{\tau}(csl)_k$ is the kth translation vector of the CSL. Since $\text{Ш}^*(p)$ must be symmorphic, only complete interfacial steps can exist in periodic Al interfaces (unless the step is associated with a DSC dislocation or a change of interfacial structure). Relaxation modes at grain boundaries in Al have been studied theoretically and experimentally [17]. The decomposition of DSC dislocations into components which separate degenerate interfacial relaxation-variants has been observed using transmission electron microscopy.

The dichromatic complex for $\Sigma=5$ related crystals having the diamond structure is shown in Fig. 8(b). The complex is isomorphous to the pattern shown in Fig. 8(a), and hence no basis-variants exist, but it is nonsymmorphic having the spacegroup $I4_1/am'd'$. The ordinary 4_1 axis is the common [001] screw-axis of the two crystals, and the ordinary mirror-glide plane, a, is parallel to (001). This latter symmetry element arises from the coincident diamond mirror-glide planes, $[m_{(001)}|\underline{\alpha}(\lambda)]$ and $[m_{(001)}|\underline{\alpha}(\mu)]$. It can be seen that although $\alpha(\lambda)$ and $\alpha(\mu)$ are no longer common after the rotation R, common vectors $\underline{\alpha}(\lambda) + \underline{\tau}(\lambda)_1$ and $\underline{\alpha}(\mu) + \underline{\tau}(\mu)_m$ exist (i.e. the vector $\frac{1}{4}$<311> joining coincident fcc sites to coincident tetrahedral sites in Fig. 8(b)). The vector $\underline{\alpha}(\lambda) + \underline{\tau}(\lambda)_1$ is designated $\underline{\beta}$ and corresponds to the supplementary displacement associated with nonsymmorphic operations in the dichromatic complex, such as the a mirror-glide plane on (001), i.e. the operation $[m_{(001)}/\underline{\beta}]$. Table 2 shows the 16 point symmetry operations present in $I4_1/am'd'$; the operations in the table have been nested into groups of four, the first group being ordinary operations (i.e. relating sites of the same colour) which relate fcc sites to other fcc sites and tetrahedral sites to other tetrahedral sites. The second group consists of antisymmetry operations (i.e. relating sites of opposite colour) also relating fcc and tetrahedral sites in the same way as the first group. The third and fourth groups are respectively ordinary and antisymmetry operations relating fcc sites to tetrahedral sites. (We note that, as in the case of Table 1, the orthogonal part of all symmetry operations is taken to act through coincident fcc sites.)

The significance of a nonsymmorphic complex spacegroup, $\text{Ш}^*(c)$, is analogous to the consequences of nonsymmorphic symmetry in a crystal spacegroup, Φ, i.e. that interfacial demi-steps can exist on certain interfaces. Demi-steps can exist on interfaces where the morphological-exchange operations leave the orientation of $\underline{n}$ invariant. Consider for example a {310} boundary (i.e. (010) or (100) in CSL coordinates); the spacegroup $\text{Ш}^*(b)$ is $p2_1'am'$ and it can be seen by decomposing $\text{Ш}^*(c)$ with respect to $\text{Ш}^*(b)$ that none of morphological-exchange operations leave $\underline{n}$ invariant, and hence demi-steps cannot exist on such interfaces. A complete step, with height $h_c^I = \underline{n}.\underline{\tau}(csl)$, on such an interface is illustrated in Fig. 11(a), and we note that the interface is comprised of both fcc and tetrahedral sites and that these are inter-related, for example, by the a mirror-glide plane. On the other hand, for a particular choice of location, the group $\text{Ш}^*(b)$ for a {210} interface (i.e. (110) in CSL coordinates) is $c12'1$, and the a mirror-glide plane (i.e. the operation $[m_{(001)}|\underline{\beta}]$ is a morphological-exchange operation which leaves the orientation of $\underline{n}$ invariant but shifts its origin by $\underline{\beta}$. Such a demi-step on a {210} interface is shown in Fig. 11(b), and we note that the height of the step, h_d^I, is given by $\underline{n}.\underline{\beta}$. We also note that the interfacial structures comprise coincident fcc sites

Table 2 Symmetry operations in the group $I\frac{4}{a}\frac{m'}{1}d'$

Symmetry Operation	Orientation (CSL indexing)	Matrix
1	-	[1/0]
2	[001]	[2/0]
$\bar{4}^+$	[001]	[$\bar{4}^+$/0]
$\bar{4}^-$	[001]	[$\bar{4}^-$/0]
2'	[110]	[2'/0]
2'	[1$\bar{1}$0]	[2'/0]
m'	(100)	[m'/0]
m'	(010)	[m'/0]
4_1^+	[001]	[4^+/β]
4_1^-	[001]	[4^-/β]
$\bar{1}$	-	[$\bar{1}$/β]
a	(001)	[m/β]
$2_1'$	[100]	[2'/β]
$2_1'$	[010]	[2'/β]
d'	(110)	[m'/β]
d'	(1$\bar{1}$0)	[m'/β]

on one side of the demi-step and coincident tetrahedral sites on the other. Relaxation at grain boundaries in Si and Ge has been studied experimentally and theoretically (see the articles by Bourret and Möller in this volume for example). Detailed investigation of Σ=5 {310} interfaces in Ge has been carried out by Bacmann et al. [18] for example, and interfacial dislocations separating energetically degenerate relaxation-variants have been observed.

We next consider interfacial features in Σ=5 related crystals having the sphalerite structure. As explained previously, sphalerite can be regarded as a derivative of the holosymmetric structure in which the atoms occupying the fcc and tetrahedral sites are no longer equivalent. It is therefore convenient to take the most symmetric precursor to be the dichromatic complex for diamond structure crystals. The second stage corresponds to obtaining the dichromatic complex for sphalerite structures, and this can be visualised using Fig. 8(b) if the square and circular symbols are no longer regarded as equivalent. The symmetry of the resulting complex, Ш^*(c), is I$\bar{4}$m'2', and the point operations in this group are the first eight operations in Table 2. The last eight operations, which previously related fcc to tetrahedral sites, become basis-exchange operations. These exchange operations inter-relate two equivalent dichromatic complexes, the second being obtained from the first by reversing the atomic occupation of the fcc and tetrahedral sites in both crystals, i.e. by creating the

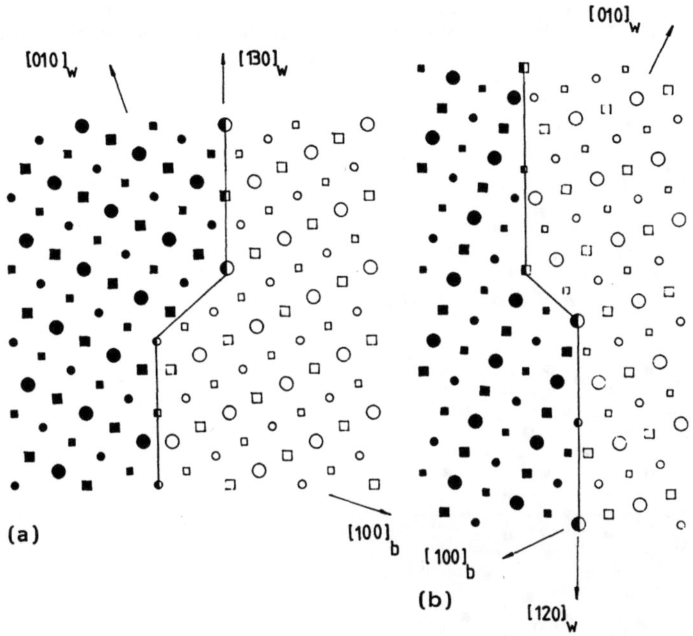

Fig. 11 Schematic (001) projection of interfacial steps, in a diamond structure bicrystal, (a) complete step on {310} interface, b) demi-step on {210} interface

second complex with anti-site related crystals. Since the complex now has a symmorphic spacegroup only complete interfacial steps, with height $h^i_c = \underline{n}.\underline{\tau}(csl)_k$ can exist. However, analogously to the case of single crystals, it is possible that basis-variant structures can coexist on the same interface if basis-exchange operations leave $\underline{n}$ invariant. For example, in the case of ($\bar{1}$30) boundaries, for which Ш*(b) = p11m' (c.f. Ш*(b) for the corresponding boundary in diamond is p2$_1$'am'), it is seen that the a mirror-glide plane $[m_{(001)}|\underline{\beta}]$ is an operation which leaves the orientation of $\underline{n}$ invariant and shifts its origin parallel to the sectioning plane. It follows that anti-site domain boundaries can penetrate ($\bar{1}$30) interfaces, and separate energetically degenerate interfacial domains as is depicted in Fig. 12(a). In the case of a ($\bar{1}$20) interface *(b) = p1 (cf. c12'1 for diamond); the same basis-exchange operation as above now leaves the orientation of $\underline{n}$ invariant but shifts its origin by $\underline{\beta}$, thereby introducing an interfacial demi-step at the intersection of the anti-site domain boundary and the grain boundary, as is depicted schematically in Fig. 12(b). No theoretical or experimental studies of interfacial relaxation in sphalerite structure materials have been reported so far.

Finally we note that the morphology of a grain embedded in a matrix of the same material but having a different orientation depends on whether or not these crystals are perfect. In the case of perfect crystals the precipitate morphology must be consistent with the point symmetry of Ш*(c), but if defects are present to compensate for any loss of symmetry the morphology can be consistent with the point symmetry of Ш*(p) [5]. Kalongi and Cahn [19] have introduced the notion of the Wulff group, i.e. the monochromatic point group which expresses minimum morphological symmetry of

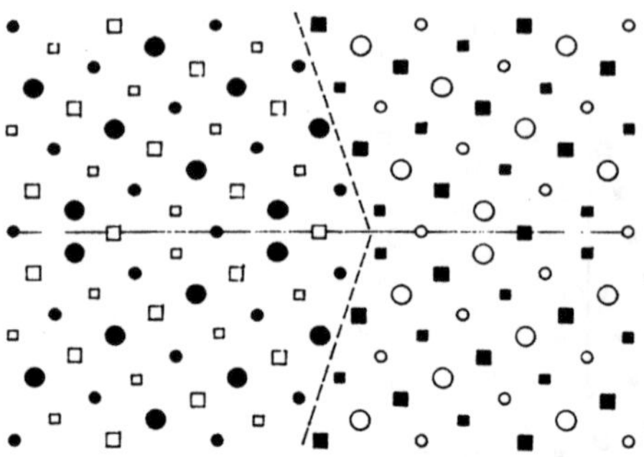

(a)

Fig. 12 Schematic (001) projection of anti-site domain boundaries penetrating grain boundaries in sphalerite structure materials, (a) no interfacial step arises at {310} boundaries, (b) an interfacial demi-step is necessary at {210} interfaces

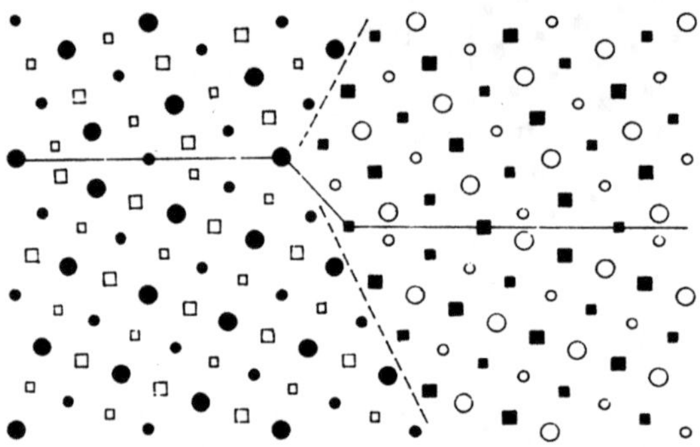

(b)

a precipitate. They point out that such a monochromatic group can be obtained by multiplying all the antisymmetry operations present in Ш*(c) by $\bar{1}$ and ignoring the colour reversing character of the resulting operations (this takes into account the fact that an included precipitate can only have outward pointing interfacial normals). However, their treatment is restricted to the case of perfect crystals and is therefore distinct from the conclusions reached here. We note that since Ш*(p) is always a centrosymmetric group the multiplication of antisymmetry operations by $\bar{1}$ is not necessary; the Wulff group of imperfect precipitates is simply the point symmetry of Ш*(p) with the colour reversing nature of any antisymmetry operations present being ignored.

7. CONCLUSIONS

We have described a unified approach to the geometrical character of extended defects that exist in the bulk and on the surface of single crystals and in homo- and heterointerfaces. It has been shown that the natures of defects existing in a particular material depend on the symmetry of that material, and that there is a close inter-relation between the character of bulk, surface and interfacial defects. Using crystallographic arguments it is possible to predict only the geometrical character of defects. In order to establish the actual atomic structure of defects a physical theory, which takes into account the nature of the bonding in the particular material, or sophisticated experimental observations, is required. A complete understanding of interfacial properties, and the mechanistic role played by defects in these properties, will require this atomic level appreciation, and much theoretical and experimental research is anticipated in order to obtain this knowledge.

ACKNOWLEDGEMENT

The author wishes to thank Mr. A.M. Bastaweesy for many helpful conversations.

REFERENCES

1. D.A. Smith and T.Y.Tan: Proc. Mat. Res. Soc., Boston, 1981, North-Holland, 5, 65-70
2. D.A. Smith, C.M. Rae and C.R.M. Grovenor: "Grain Boundary Migration", in Grain Boundary Structure and Kinetics, A.S.M. Ohio, 1980, 337-371
3. R.C. Pond and W. Bollmann: Phil. Trans. R. Soc. Lon., 292,449 (1979)
4. R.C. Pond and D.S. Vlachavas: Proc. R. Soc. Lond., A386, 95, (1983)
5. R.C. Pond: Phil. Mag. A47, L49, (1983)
6. R.C. Pond: Inst. Phys. Conf. Ser. No. 67, 59, (1983)
7. M. Kleman: "The General Theory of Disclinations", in Dislocations in Solids, North-Holland, Amsterdam, 243-297, (1980).
8. W.F. Harris: Phil. Mag. 22, 949 (1970)
9. R.C. Pond, J.P. Gowers and B.A. Joyce: Surf. Sci., (Proc. ECOSS-6, York, 1984), in press
10. R.H. Williams: Surf. Sci. 132 122, (1983)
11. M.J. Buerger: Elementary Crystallography, M.I.T. Press (1978)
12. R.C. Pond, J.P. Gowers, D.B. Holt, B.A. Joyce, J.H. Neave and P.K. Larsen: Mat. Res. Soc. Symp. Proc. 25 273,423 (1984).
13. D. Cherns and R.C. Pond: Mat. Res. Soc. Symp. Proc. 25, (1984).
14. R.C. Pond and D. Cherns: Surf. Sci. (Proc. ECOSS-6, York, 1984) in press
15. P.D. Augustus: Inst. Phys. Conf. Ser. No. 67, 229, (1983).
16. A.H. King and D.A. Smith: Acta Cryst. A36, 335-343
17. R.C. Pond and V. Vitek: Proc. R. Soc. Lond. A357, 453 (1977)
18. J.J. Bacmann: J. de Phys. 43, C6-93 (1982)
19. G. Kalongi and J.W. Cahn: J. de Phys. 43, C6-25 (1982)

Grain Boundary Segregation. Grain Boundary Diffusion

Marc Aucouturier

Laboratoire de Physique des Solides, CNRS-1, Pl. A. Briand
F-92195 Meudon Cedex, France

INTRODUCTION

In polycrystalline materials, grain boundaries differ from the monocrystalline bulk by : i) their structure, ii) their electronic properties, iii) their chemical composition, iv) the atomic mobility of all the species.

This paper will focus on the third and fourth point . As both properties involve atomic interaction and transport, they are not independent. Segregation and precipitation phenomena will be described first, for the knowledge of local chemistry is necessary to get a consistent expression of the diffusion fluxes.

The approach of the phenomena will be intentionally macroscopic and thermodynamic, in order to list the tools available for the exploitation of the experimental data, mostly macroscopic themselves. When atomic models have been proposed, they will be mentioned "in fine", as a complementary approach.

Most of the concepts of the macroscopic approach may be extrapolated to the properties of any kind of interfaces and sometimes dislocations; the extension will not be done here.

I - GRAIN BOUNDARY SEGREGATION

I.1. Definitions

If the general significance of segragation is "chemical heterogeneities in the solid state", impurity segregation can be the consequence of [1] :

i) a partition of impurities between the solid and the liquid during solidification.

ii) a precipitation of impurities during cooling or annealing of a supersaturated material.

iii) a pollution of the material during or after elaboration by adsorption, absorption and diffusion of impurities on and from the surface into the bulk.

iv) an impurity enrichment on the defects (surface, grain boundaries, interfaces, dislocations) resulting from an interaction between these defects and the solute.

The grain boundary segregation involved in the present article belongs to the fourth category; it is in fact linked to grain boundary precipitation, which will be briefly described in a first section.

As a preliminary point, segregation during solidification is one of the most common ways to produce a pollution of a material by non-expected impurities; it is the result of a difference of the equilibrium composition of the liquid and of the solid C_L and C_S at the solid/liquid interface. The

Table I Segregation ratio of some impurities in silicon

	Ia	IIa	IIIa	IVa	Va	VIa	VIIa		VIII		Ib	IIb	IIIb	IVb	Vb	VIb	VIIb	O
1	1 H ?																	2 He
2	3 Li 1.10^{-2} *	4 Be											5 B 8.10^{-1}	6 C 7.10^{-2}	7 N 7.10^{-4}	8 O 1.25	9 F	10 Ne
3	11 Na 2.10^{-3} *	12 Mg 2.10^{-3} *											13 Al 2.10^{-3}	14 Si	15 P $3.5.10^{-1}$	16 S 1.10^{-5}	17 Cl	18 A
4	19 K *	20 Ca *	21 Sc	22 Ti 5.10^{-6}	23 V 5.10^{-6}	24 Cr 5.10^{-6}	25 Mn 1.10^{-5} *	26 Fe 1.10^{-5}	27 Co 1.10^{-5}	28 Ni 1.10^{-4}	29 Cu 4.10^{-4}	30 Zn 1.10^{-5} *	31 Ga 8.10^{-3} *	32 Ge 3.10^{-1}	33 As 3.10^{-1}	34 Se *	35 Br	36 Kr
5	37 Rb *	38 Sr *	39 Y	40 Zr 2.10^{-8}	41 Nb 5.10^{-7}	42 Mo 5.10^{-8}	43 Tc	44 Ru	45 Rh	46 Pd 5.10^{-5}	47 Ag 1.10^{-4}	48 Cd 1.10^{-6} *	49 In 4.10^{-4} *	50 Sn 2.10^{-2}	51 Sb 2.10^{-2} *	52 Te *	53 I	54 Xe
6	55 Cs *	56 Ba *	57-71 T.R.	72 Hf	73 Ta 2.10^{-8}	74 W 2.10^{-8}	75 Re	76 Os	77 Ir	78 Pt	79 Au 3.10^{-5}	80 Hg *	81 Tl 2.10^{-4}	82 Pb	83 Bi 7.10^{-4} *	84 Po	85 At	86 Rn
7	87 Fr	88 Ra	89 Ac	90 Th	91 Pa	92 U												

* Elements presenting an equilibrium vapor pressure greater than one torr at 1410°C

segregation ratio C_S/C_L is more or less correlated to the solid solubility of the given impurity : the purifying effect of a progressive solidification is more efficient (small C_S/C_L) for impurities with a low solubility. Table I shows for instance that progressive solidification of silicon is very efficient for the elimination of metallic impurities but that a preexisting pollution with O, C, B, P, As, Sb is not modified [2].

The purification efficiency of solidification is also controlled by the cooling rate and the cooling geometry; for instance, in the elaboration of polycrystalline silicon ingots, a convex shape of the solidification front is preferable as impurities are rejected to the upper peripheric part of the ingot [2] [3].

I.2. <u>Parameters controlling heterogeneous (grain boundary) precipitation</u>
Precipitation is the thermodynamical consequence of the shape of the excess free enthalpy (GIBBS free energy) versus concentration curve (fig. 1); when the atomic fraction X is larger that X_α, the minimum free enthalpy is obtained by splitting the solution into two phases with composition X_α and X_β [4]. The driving force for precipitation is ΔG_V (chemical potential difference of B element between the α and β phases). The driving force compensates the surface energy $\sigma(\alpha\beta)$ of the precipitate; the activation energy for the creation of a spherical incoherent nucleus is [5]:

$$\Delta G^* = \frac{16\pi\sigma(\alpha\beta)}{3\Delta G_V^2}$$

If the nucleus is coherent, ΔG_V is replaced by $\Delta G_V + \xi$, where ξ is the strain energy associated to the coherent nucleation. In the case of a grain boundary nucleus (fig. 2), the activation energy becomes :

$$\Delta G^* = \frac{32\pi\sigma(\alpha\beta)^3}{3\ G_V^2}\ f(\theta) \quad ; \quad f(\theta) = (2 - 3\cos\theta + \cos^3\theta)/4$$

Qualitatively, the activation energy is decreased if θ is decreased; this is obtained either in high energy boundaries ($\sigma_{\alpha\alpha}$ large) or for low energy

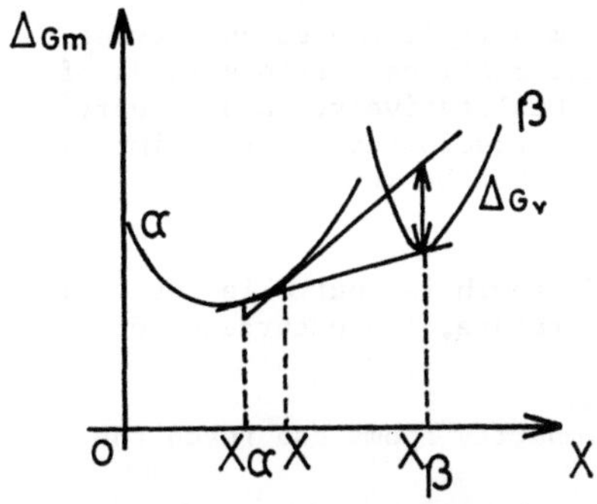

Fig. 1 : Excess free enthalpy versus concentration for a binary solid solution exhibiting precipitation

Fig. 2 : Schematic shape of a grain boundary precipitate

interphase boundaries ($\sigma_{\alpha\beta}$ small), that is for orientated precipitation. A more complete discussion can be found in ref. [6]; a qualitative enumeration of only some parameters will be given here.

I.2.1 Influence of temperature, solubility, diffusion

The solid solubility of a given impurity is an Arrhenius function of temperature [4]:

$$X_\ell = X_o \exp(-\Delta H_f/kT)$$

Fig. 3 [7] is an example of the variation with temperature of the solid solubilities of various impurities in silicon : oxygen and carbon have a very low solubility.

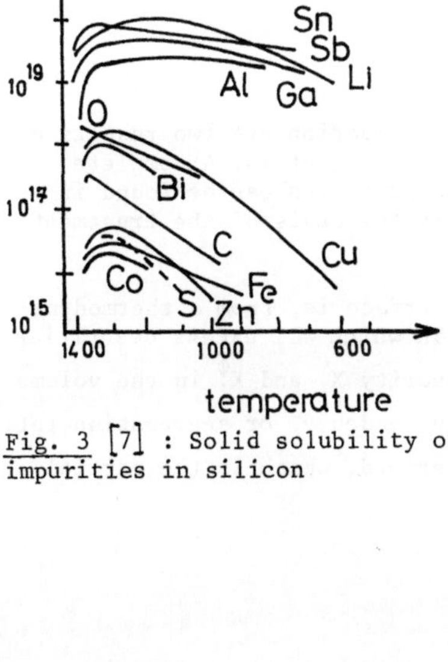

Fig. 3 [7] : Solid solubility of impurities in silicon

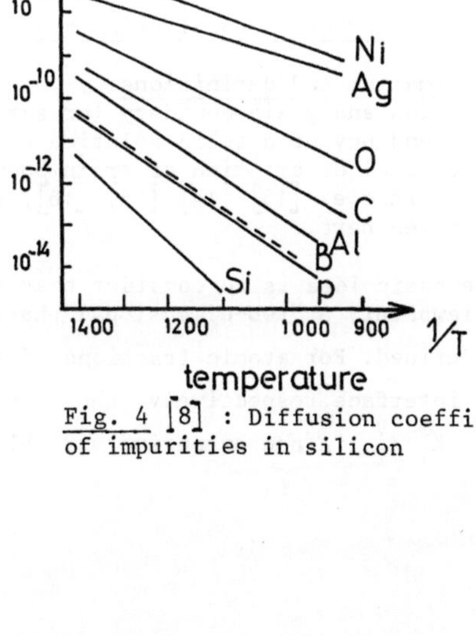

Fig. 4 [8] : Diffusion coefficients of impurities in silicon

In multicomponent solutions (more than one impurity), the solubility of each component is modified by solid solution interactions. Various kinds of "gettering effects" have no other explanation. Qualitatively, in a ternary solution for instance, the driving force for the precipitation of an impurity is increased by addition of a value porportional to the interaction energy between the two impurities.

The kinetics of precipitates nucleation and growth is controlled by solid state diffusion (fig. 4) [8]. For a given temperature, the occurrence of grain boundary precipitation needs :

i) a large enough driving force

ii) a diffusion coefficient such that the impurity atoms can reach the boundary within the annealing time.

Thus in a given material, precipitation can occur on bulk defects (dislocations) or randomly at low temperature (large driving force, short diffusion range) and on grain boundaries at higher temperatures (long diffusion range, smaller driving force). This effect is emphasized in multicomponent systems with strong solid solution interactions.

I.2.2. Influence of structure

The influence of crystallographic parameters on the precipitate nucleation has been extensively studied and discussed, especially regarding metals [6] [9]. However, no general rule can be applied, except for very simple cases such as the symmetrical Σ_3 boundaries of f.c.c. materials where precipitation is scarcely observed. Two factors seem to overcame the role of the grain boundary crystallography [9].

One is the role of grain boundary non periodic linear defects, often called extrinsic dislocations (E.G.B.D's), as they do not participate to the periodic structure until they are incorporated in this structure by accommodation or dispersion [9]. These act clearly as nucleation sites for precipitation, in metals [9] [10] or in silicon [11] [12], in all kinds of grain boundaries.

The second factor is the crystallography of the grains/grain boundary / nucleus system [6] [9]. The growth of precipitates, and sometimes their nucleation, are favored if epitaxial relations exist between the precipitate, one of the grains and the grain boundary plane.

I.3 Thermodynamics of equilibrium segregation

1.3.1 Formalism and definitions

Precipitation and grain boundary (or surface) segregation are two results of the same tendency of a solid solution to expel solute atoms. A complete thermodynamical description of grain boundary segregation can be found in extensive articles [13] [14] [15] [16], and only the basis of the treatment will be given here.

The basic idea is to consider that the interface is, from a thermodynamical viewpoint, a "two-dimensional phase" ϕ, in which all excess quantities can be defined. For atomic fractions of the impurity X_i^V and X_i^ϕ in the volume and the interface, respectively, the segregation tendency, or segregation ratio is : $\beta = X_i^\phi / X_i^V$. A new extensive variable is defined, which is the partial surface occupancy $\omega_i = \left(\frac{\partial\Omega}{\partial\Omega_i^{\phi}}\right)_{p,T,\sigma,X_i}$, where Ω is the area of the interface, σ the surface tension, and other variables have their usual meaning. The equilibrium is obtained by minimizing the Helmholtz free energy F_m^{ϕ} of the ϕ phase, in which a "compression" term σω appears (instead of minimizing the Gibbs free energy in the case of 3 D precipitation) (for simplification ω_i is supposed to be the same for all species).

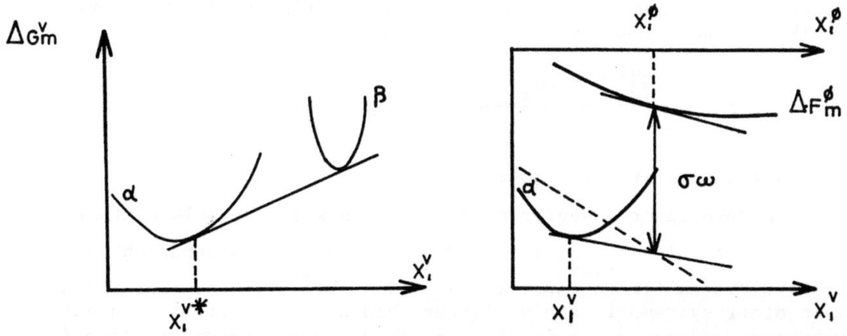

Fig. 5 : Diagrams of the excess free energies as a function of concentration (see text)

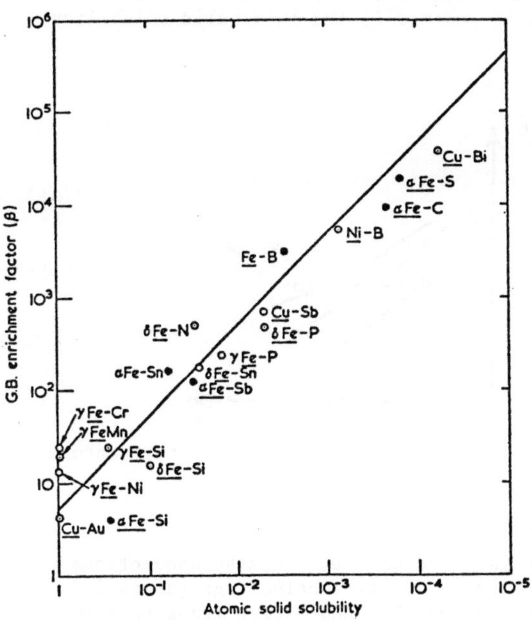

Fig. 6 : Correlation between the grain boundary segregation tendency and the bulk solid solubility [16]

Figure 5 shows why segregation, i.e. values of X_i^{ϕ} different from X_i^{γ}, occur for an extended range of bulk concentration, as far as this is smaller than the bulk solubility X_i^{V*}. It also indicates that a strong precipitation tendency (minimum of the G_m curve) is associated to a strong segregation tendency. The result is well known, at least for metals (fig. 6) [16]: the segregation tendency (grain boundary enrichment in fig. 6) is correlated to the solid solubility in all materials.

I.3.2 Segregation isotherms. Binary and ternary solutions

The basic driving force for segregation is the gain in free enthalpy ΔG_i^{oseg}. As for surface segregation, it may be the result either of elastic interactions [17] or of any kind of interactions, including electrical forces. The isotherm, i.e. the relation between the average ratio $Y_i^\phi = X_i^\phi / X_{i_{max}}^\phi$ and the bulk atomic fraction X_i^V has, in the case of an ideal binary solution, the same form as the classical adsorption isotherm. In non-ideal solutions, the regular solution approximation is used and an interaction factor α_{1s} between the solute (1) and the solvent (s) is introduced :

$$\frac{Y_1^\phi}{1 - Y_1^\phi} = X_1^V \exp\left(- \frac{\Delta G^{oseg} - 2\alpha_{1s} Y_1^\phi}{kT} \right)$$

The main consequences of this law are :

i) at a given temperature, segregation increases with the bulk content but reaches a saturation value $X_{i_{max}}^\phi$. This saturation coverage depends on the system; in the Ni-S system for instance it is $\sim$ 50 at % [20], i.e. half a monolayer, but in other systems it can be larger than a monolayer; it is usually not temperature-dependent except in the case of 2 D phase transformation (see below). The saturation value may not be reached if the bulk solid solubility X_i^* is small (fig. 7).

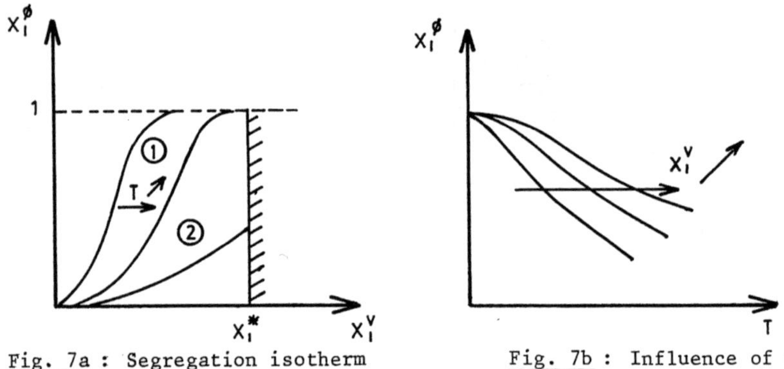

Fig. 7a : Segregation isotherm
1 : influence of temperature
2 : case of low bulk solubility

Fig. 7b : Influence of temperature on segregation

ii) the segregation driving force is given with the chemist conventions : negative values correspond to attractions. Thus attractive segregation is smaller for higher temperatures : increasing the temperature leads to impurity desegregation.

From several comparisons, it has been shown that the driving force is, in absolute value, larger for surface segregation than for grain boundary segregation, but has often the same sign. The study of surface segregation can thus indicate the trend of a system concerning grain boundary segregation but lead to overestimated effects.

In a ternary system (two impurities 1 and 2 in a solvent S), it is often realistic to consider that one impurity (1) has a small solubility as compared to the other. A full analysis of the ternary systems has been made [14] |15| with various assumptions. The results of two cases are given here:

i) in a regular solution model, with a competition of the segregating species on the grain-boundary sites (1 and 2 occupy the same atomic sites), the isotherm is :

$$\frac{Y_i^\phi}{I - Y_1^\phi - Y_2^\phi} = X_i \exp(-\Delta G_i^{seg}/kT)$$

Where $\Delta G_1 = \Delta G_i^o + \alpha_{1s}^\phi (1 - 2 Y_1^\phi) + \alpha'_{12} Y_2^\phi$

$$\Delta G_2 = \Delta G_2 + \alpha_{2s}^\phi (1 - 2 Y_2^\phi) + \alpha'_{12} Y_1^\phi$$

ΔG_i^o and α_{is} have the same meaning as above.

The new term $\alpha'_{12} Y_j^\phi$ expresses the influence of the segregated amount of one species upon the driving force of the other

$$\alpha'_{12} = \alpha_{12}^\phi - \alpha_{1s}^\phi - \alpha_{2s}^\phi$$

where α_{12}^ϕ describes the interaction between the two impurities. In absence of other data, α_{12}^ϕ can be related to the molar formation enthalpy of the (1,2) compound with $X_1^* X_2^*$ composition :

$$\alpha_{12}^\phi = \Delta H_{12}^f / X_1^* X_2^*$$

A negative value of α_{12} (i.e. a strong 1-2 interaction) results in an increased segregation of either impurity by the presence of the other

ii) in a regular solution model without competition of the segregation species (segregation on different sites), the isotherm is :

$$\frac{Y_i^\phi}{1 - Y_i^\phi} = X_i \exp(-\Delta G_i^{seg}/kT)$$

with $\Delta G_1^{seg} = \Delta G_1^o + \beta' \frac{Y_2^\phi}{b}$; $\Delta G_2^{seg} = \Delta G_2^o + \beta' \frac{Y_1^\phi}{a}$

a and b are fractions of area occupied by the segregation sites, and β' has a significance analogous to α', applied here for regular interstitial solutions. The effect of a negative β' is the same as above for $\alpha' < o$.

Figure 8 summarizes the cosegregation effects : attractive interactions stabilize the segregation; but the ternary component has also often other actions: it can lower the solubility limit of the studied impurity, either by stabilizing the precipitated phase, or by inferring the precipitation of a more stable phase (e.g. new carbides or oxides instead of SiC or SiO_2 in silicon). The effect can be paradoxically to decrease the segregated amount (fig. 8).

The competition on the segregation sites can lead also to a reduction of the maximum segregation.

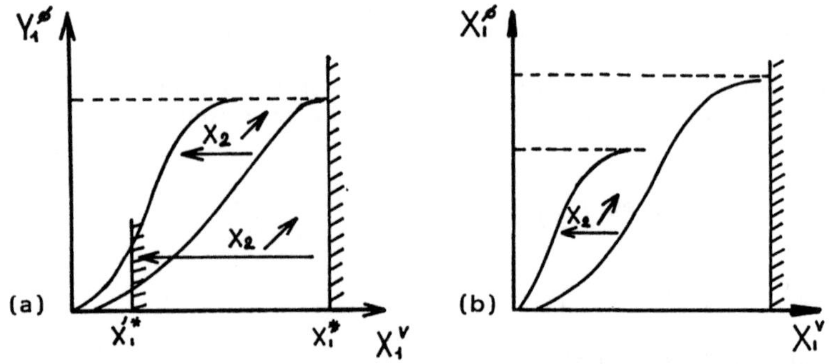

Fig. 8 : Effect of the addition of a second element on the segregation isotherm of an impurity; attractive interactions. a) without competition b) with competition on the segregation sites

In summary the segregation phenomena in a multicomponent system (which is always the case in practice) cannot be described and predicted unless a complete set of experiments is conducted, for various temperature and various bulk concentrations of the impurity under investigation.

I.3.3. Segregation kinetics.

The segregation kinetics is generally controlled by atomic diffusion. In the simple case of an ideal binary solution (no interaction), the variation of the grain-boundary concentration for a given time X_t^ϕ is [17] :

$$x = \frac{X_t^\phi - X_o^\phi}{X_\infty^\phi - X_o^\phi} = 1 - \exp s^2 \operatorname{erfcs}$$

s is a reduced time variable $\quad s = \frac{2\ (Dt))^{1/2}}{\beta d}$

d is the "width" of the segregation region.

$\beta = \frac{X^\phi}{X_t^V}$, where X_t^V is calculated in the "immediate vicinity" of the interface (first atomic layer ?).

A usual assumption is that of the local time-independent equilibirum :

$$\beta = \text{cte} = \frac{X_\infty^\phi}{X_i^V}$$

For short times and under the same assumptions, the formula simplifies to :

$$X_t^\phi = 4\,\frac{X_i^V}{d}\,\left(\frac{Dt}{\pi}\right)^{1/2}$$

or $\Gamma_t^\phi = 4\,C_i^V\,\left(\frac{Dt}{\pi}\right)^{1/2}$, if Γ_t^ϕ is the surface coverage of the grain boundary (in $At.cm^{-2}$) and C_i^V the bulk concentration (in $At.cm^{-3}$).

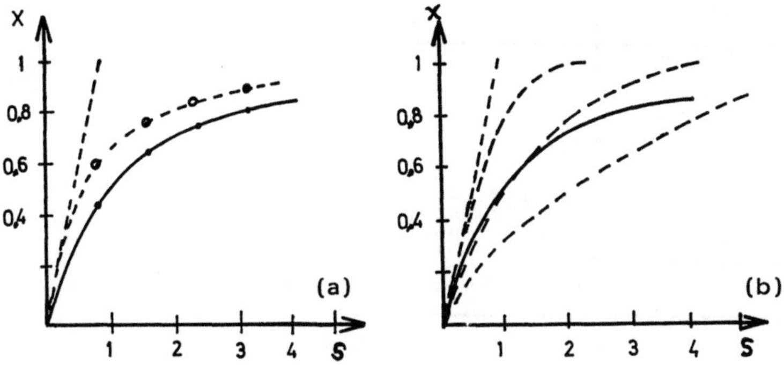

Fig. 9 : Segregation kinetics [17, 18, 19]
a) without interaction (ideal solution). Straight line : "rate limited"; full circles : β constant; open circles $\beta_{t=o}$ = 93, β_{∞} = 86

other data : $X_i^V = 10^{-3}$ $\Delta G/^{seg}/RT = 4.53$

b) with interactions (ternary system, regular solution) ... straight line : "rate limited"; full curve : ideal solution; dotted curves : for various values of interaction parameter

The unknown value d does not appear in the second formulation. Figure 9 a shows the shape of the x = f(s) curve for the short time approximation (straight line), for the analytical solution with β constant (full circles) and for a reasonable variation of β with the time (open circles) |20, 21|

More general models have been proposed [18,19], taking into account the interactions in a binary or ternary system and differing values of the diffusion coefficients of the various species. The segregation may be either accelerated or delayed (fig. 9b). Those kinetic processes may be very complex, leading to reverse fluxes to or from the grain boundary, and Kirkendall effects around the interfaces (pores or dislocation multiplication).

I.3.4 Early stages of precipitation

The thermodynamic formalism used here allows a new type of phase transformation. An extended phase rule can be applied to such systems [14]:

$w = C + 3 - \varphi - \psi$ (instead of $v = C + 2 - \varphi$)

is the variance of the system containing C components, φ three dimensional (3D) phases, and ψ two-dimensional (2D) phases.

A 2D-phase is considered either as the grain boundary core itself or as any precipitation confined to the grain boundary core, provided that the bulk solid solubility is not reached. An example in metals is the formation of a 2D sulfide NiS in the boundaries of a nickel-sulfur alloy [18, 20, 21]. Another example might be the "precipitation" on an atomic scale of silicon or germanium oxide observed by high resolution microscopy on grain boundary dislocations of Si or Ge [22].

The generalized phase rule and the free energy diagram (fig. 10) infer that, for a given temperature (and pressure) the 2D phase transformation

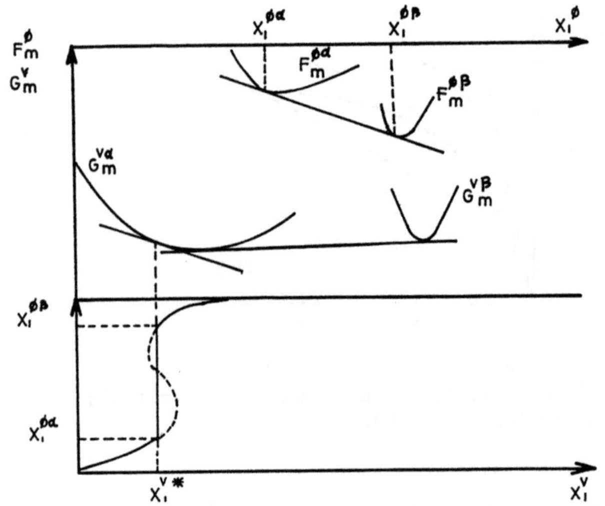

Fig. 10 : Free-energy versus concentrations and interface concentration versus bulk concentration diagrams for a 2D phase transformation

occurs for a given bulk concentration X_1^{V*}; X_1^{V*} is temperature-dependent exactly as a normal solid solubility. Another impurity (ternary system) may modify it, often toward lower values (attractive interaction).

At this stage, the limitation of the macroscopic thermodynamic treatment appears : the definition of a "2D-phase" depends on the extension of the core volume of the defect; as soon as core interactions are involved, macroscopic thermodynamics can no longer be applied. A structural approach must be taken.

I.4 Grain boundary segregation and structure

I.4.1 Influence of the grain boundary structure on segregation

Very few experimental data are available on this subject, although it is clear that segregation is structure dependent. A recent study on the Ni-S system [23] has shown, through the observation of a local electrochemical etching of thin foils for transmission electron microscopy, that the segregation of sulfur depends not only on the disorientation between the grains but more on the orientation of the grain boundary plane (fig. 11).

Similarly, the oxide micro precipitation observed in the core of silicon or germanium dislocations or grain boundaries [22] is influenced by the structure : the number of atoms involved in this segregation depends on the Burger's vector of the dislocation (fig. 12).

It has been qualitatively demonstrated by a radioisotope and high resolution autoradiography technique that the segregation of carbon and of hydrogen in polycrystalline silicon occurs only on a limited number of grain boundaries [24].

The predominant role of the grain boundary linear defects on impurity segregation has been clearly evidenced for metals [9] [10] . The mechanism involved in this process implies the influence of the long range strain field of these defects and on accelerated diffusion of the impurity along the boundary and the dislocations themselves ("in situ segregation").

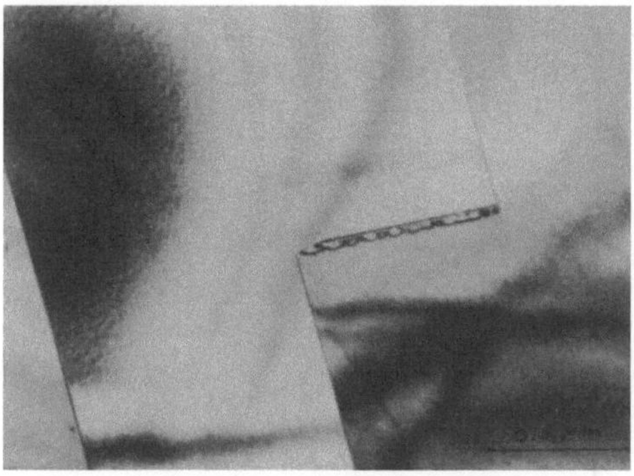

Fig. 11 : Local etching of a thin foil of Ni-S alloy indicating preferential segregation. Σ3 boundary [23]

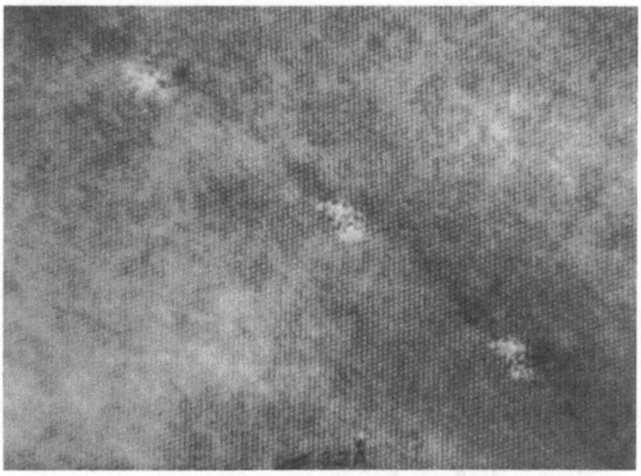

Fig. 12 : Oxide micro precipitation in dislocation cores of silicon [22]

On the other hand, it has been demonstrated that the segregation of impurities can either modify the equilibrium grain boundary structure [25] or stabilize (i.e. prevent the accomodation or the dispersion) of extrinsic grain boundary dislocations [9]. A comprehensive description of the grain boundary segregation should, instead of expressing it as an interaction between a given entity (the grain boundary) and solute impurities, consider that both the structural and chemical parameters are modified to minimize the total energy.

I.4.2 The "width" of the segregation

The macroscopic thermodynamics of segregation does not make any structural conclusion on the physical nature and space extension of the 2D-phases describing the interface, except their "incompressibility" in a direction normal to the interface [13].

In the case of metals, all the models and experimental observations agree with the fact that the volume around the boundaries in which a composition change occurs is, at equilibrium, not larger than 1 or 2 interatomic distances (fig. 13) [26] [27].

The case of ionic or covalent materials is more complicated. A space charge can extend to relatively large distances from the boundaries; segregation may be influenced by the space charge and, by return, modify it. Such enlarged segregation has not been observed in oxides [27] [28] up to now (fig. 14) and is still under discussion.

I.4.3 Modelling of the grain boundary segregation

Several atomistic approaches, more appropriate than crystallographic or geometrical approaches, are used.

The hard sphere model [29] estimates the excess volume of the free sites in the grain boundary core and gives a qualitative description of the influence of the coincidence sites density on the segregated amount. It can be considered as a starting point for more quantitative descriptions. The structural unit model [30] or energetic models [31] [32] [33] are essentially static computations aiming at the minimization of the grain boundary energy

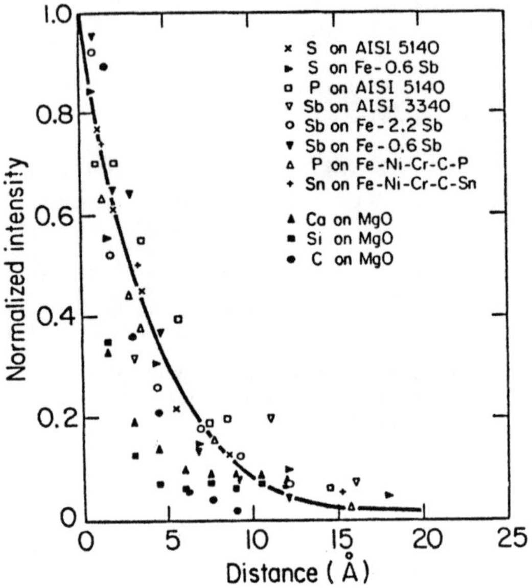

Fig. 13 : Compilation of results of Auger analysis of segregation in different materials [26] [27]

in presence of impurity atoms. The validity of these models depends on the approximation used for pair interaction potentials. They are for the moment applied only to special grain boundaries (small Σ , symetrical). In the case of non-metallic two-component matrixes (oxides), the computations may lead to the conclusion that given grain boundaries, unstable if the material is pure, are stabilized by the impurity [31, 32]. The influence of the space charge has to be accounted for when the impurity is aliovalent [33]. The

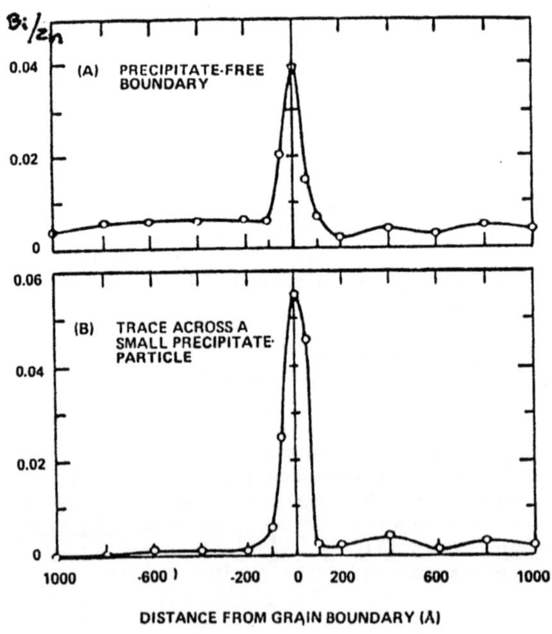

Fig. 14: STEM analysis of Bi segregation in ZnO grain boundary [28]

molecular dynamics approaches [34, 35] are also dependent on the approximations of the interatomic potentials, but produce more precise descriptions of the driving force, local strains. It is restricted to given special boundaries.

The calculation of local electronic densities [35] [36] through a molecular orbitals approach has given interesting results on metals, applicable to general boundaries, described as an array of polyhedra belonging to a group of finite number.

The two last approaches have not been applied, to our knowledge, to non-metallic materials. Multicomponent systems are, for the moment, not considered in the models.

I.5 "Non-equilibrium" segregation.
When the material itself is not in thermodynamic equilibrium, the segregation phenomena are strongly affected by irreversible processes such as bulk and grain boundary diffusion, defect creation, annihilation and movement. In semiconductors, such a situation is probable in non stabilized materials, such as : recrystallized thin films, rapidly solidified materials, implanted junctions, ...

It has, for instance been demonstrated in metals that the moving boundaries (recrystallization) may have a dragging effect on the bulk impurities [37, 38], leading to very large segregation amount and, reversely, to a modification of the grain boundary structure.

Another case, well described in Al alloys for instance, is the existence of fluxes of point defects (vacancies and interstitials) toward or away from the grain boundaries. Apparent enlarged segregation width effects can often be attributed to such phenomena.

Grain boundary segregation may also be a preliminary step to precipitation, the system goes for increasing time (at given temperature) through the sequence : segregation, 2D precipitation, 3D precipitation.

II MEASUREMENTS OF GRAIN BOUNDARY SEGREGATION, EXAMPLES OF RESULTS.

As for all phenomena incompletely described by theory, a correct analysis of segregation should be conducted through several experimental approaches.

II.1 Direct analysis.
Table II gives a list of some of the experimental means available to measure the grain boundary composition. First an analysis of the bulk has to be done by :
i) chemical analysis; sensitivity 10 to 100 ppm
ii) spectroscopic methodes; sensitivity 1 to 10 ppm
iii) nuclear analysis; sensitivity sometimes better than 1 ppb
iv) specific methods as nuclear reaction spectrometry, neutron scattering, I.R. absorption
v) electrical measurements (dopant concentrations).

II.1.1 Surface analysis (Table II a)
These techniques have a very high depth resolving power (a fraction of nm if combined to ion sputtering) and a good sensitivity (0.1 atomic layer or less). They have been extensively used for the study of surface segregation [39]. Some of them give information on the environment and bonding of the impurities (e.g. XPS, AES, ...).

Table II Direct analysis techniques

a) surface analysis

	Electron micro analysis	Auger Electron spectrometry (AES)	Ion back scattering	Photoelectron spectrometries XPS, UPS	SIMS	EELS (reflexion)	Glow discharge optical spectrometry
Elements analyzed	Z > 4	Z > 3	specific	Z > 2	all	all	all
Thickness analyzed	1 µm	0.5 to 30 nm (layer by layer)*	0.5 to 30 nm layer by layer*	0.5 to 30 nm 1 atomic layer*	1 nm 1 atomic layer*	variable	a few 0.1 nm
Sensitivity	$10^{-1} \sim 10^{-3}$	$10^{-2} \sim 10^{-3}$	$10^{-2} \sim 10^{-3}$	$10^{-2} \sim 10^{-3}$	$10^{-3} \sim 10^{-5}$	$10^{-2} \sim 10^{-3}$	$10^{-3} \sim 10^{-4}$
Surface analyzed	1 µm^2 $\sim$ cm^2	50 x 50 µm^2 or 5.10^{-2} µm^2	1 to 50 mm^2	1 to 50 mm^2	3 to 10 mm^2 or 20 x 20 µm^2	50 x 50 µm^2	1 cm^2
Lateral resolution (image)	$\sim$ 1 µm	30 nm	-	1 mm	0.5 µm	30 nm	-
Fracture analysis	quantitative (precipitates)	quantitative	quantitative	qualitative	qualitative	qualitative	?
Information on bonding	no	yes (fine structure of lines)	yes	yes	qualitative	no	no
Advantages		resolution	specificity	bonding	sensitivity light elements	all elements	sensitivity all elements
Drawbacks		sensitivity	sensitivity neavy equipt.	sensitivity	calibration	calibration	resolution

* by ion sputtering

Table II Direct analysis techniques

b) Local microanalysis, radioisotopes, local electrical activity

	T.E.M. (thin foils)	T.E.M. High resolution	Field ion microscope (atomic probe)	STEM (X rays)	STEM (EELS)	Radioisotopes	Local electrical activity LBIC	Local electrical activity EBIC	Local electrical activity STEBIC
Elements analyzed	precipitate structure	structure	all	Z > 4	∿ all	limited nimber	-	-	-
Thickness	100 nm ∿ 1mm (H.V.)	∿ 10nm	∿10 nm (tip)	1 ∿ 10 nm	∿ 20 nm	-	1 ∿ 100 µm	0.1 ∿ 1 µm	100 nm
Sensitivity	-	-	atoms	< atom layer	?	<atom layer	?	?	?
Surface	0.1µm^2 ∿1 µm^2	-	wire tip	0.1mm^2 ∿1 µm^2	0.1mm^2 0.1 µm^2	G.B. G.B.	cm^2	1 µm^2 cm^2	0.1mm^2 0.1 µm^2
Lateral resolution	1 to 3 nm	0.3 nm	atome	1 nm	1 nm	100nm 10 µm	1 10 µm	1 µm	?
Quantitative analysis	Yes (precipitates)	Yes (structure)	no	Yes	?	Yes	Yes	Yes	?
Type of investigation for G.B.	precipitation structure	core structure	atomic location	segregation precipitation	segregation	segregation diffusion	recombination	recombination	recombination
Advantages	resolution + structure	resolution + structure	resolution specificity	resolution + structure	resolution + structure	sensitivity Analysis of C, H	quantitative	resolution	resolution + structure
Drawbacks	Rhin foil	Thin foil orientation limited	preparation	Thin foil + quantification	Thin foil preparation + quantification	elements limited	resolution	resolution	quantification

Their application to grain boundary segregation is hindered by their limited lateral resolving power and volume concentration sensitivity. They become very powerful if it is possible to expose to the incident beam the surface of the grain boundary itself. The specimen must be fractured in situ under high vacuum. This method is easy in embrittled materials [13, 14, 18, 20, 27] ; in semiconductors a tedious statistical study of the fractures is necessary. An extensive investigation of the grain boundary segregation of Fe, Al, C, O, F in silicon has been conducted, using AES, SIMS, XPS, EELS techniques, combined with electrical (LBIC, EBIC) measurement and depth profiling (fig. 15) [40].

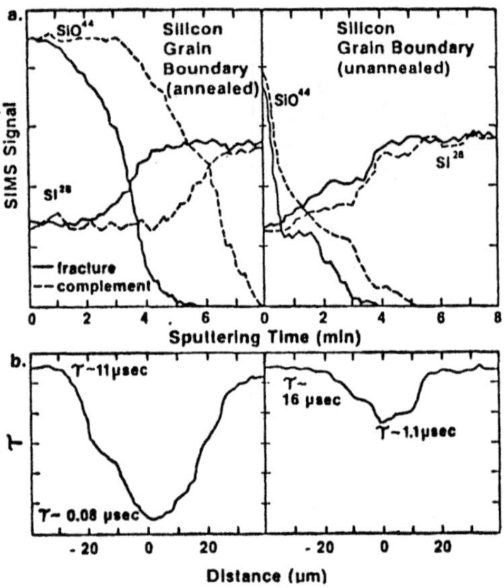

Fig. 15 : Si grain boundary (from ref.40)

a) SIMS depth profiling (oxygen segregation
b) minority carrier lifetime

II.1.2 Local microanalysis (Table II b)
These techniques, nearly all based on electron microscopy, have a high resolving power, and allow a description of the influence of the structure on segregation. Their main limitation up to now has been the quantitative expression of the results |23, 41, 42|.

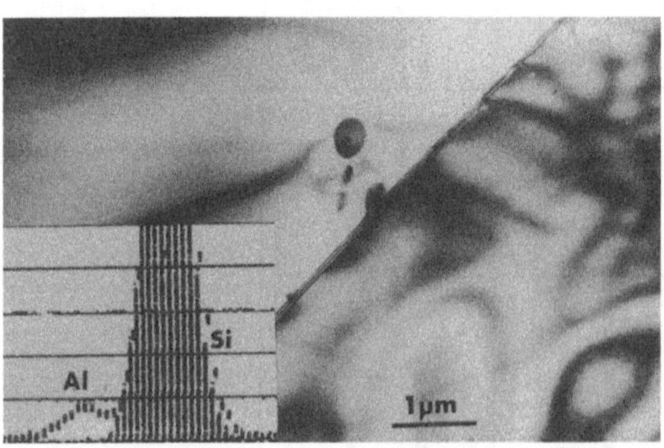

Fig. 16: STEM analysis of Al segregation in polycrystalline silicon (from ref.45)

In semiconductors, some examples are :
i) observation and measurement of oxygen segregation on germanium and silicon dislocation by high resolution electron microscopy combined with EELS by STEM [22,43].
ii) STEM measurements of P segregation in silicon [44]. An interaction energy of 0.3 ev, for segregation amounts of 0.1 to 0.3 atomic layers has been obtained.
iii) evidence of aluminum and aluminum oxide grain boundary segregation [45] (fig.16), or of silicon carbide precipitation [11] [12] in silicon, in relation with microstructure and electrical properties (EBIC) of the same boundaries.

II.1.3 Radioisotopes
Their advantage is a high sensitivity, even for species as C (radioisotope ^{14}C) or H (radioisotope ^{3}H). The autoradiography technique has been improved to reach a good resolving power and used to evidence the segregation of carbon and passivating hydrogen in polycrystalline silicon, in relation with electrical properties [24].

Quantitative measurements by particle counting, in a direction either perpendicular [46] or parallel [47] to the interface plane, give precise values of the segregated amount.

The limitation of the method is the availability of a 'good' radioisotope (β emitter, long lifetime) of the interesting impurity.

II.2 Indirect measurements

II.2.1 Macroscopic electrical properties
Carrier concentration measurements in semiconductors [48,49] may give an estimate of the dopant segregation provided the segregated dopant is supposed to be electrically inactive. On the contrary, capacity measurements may be used to give evidence of segregation related grain boundary states [50].

II.2.2 Local electrical properties
Local resistivity can be related to impurity segregation [51].

LBIC and EBIC techniques are used to investigate the recombination properties of the grain boundaries in relation to impurity segregation : for instance oxygen and various impurities in silicon and GaAs [40], carbon in silicon [11] [24], hydrogen in silicon [52, 53, 54] . This problem of the interaction between hydrogen and grain boundaries is critical for the interpretation of electrical passivation mechanisms. Recent exodiffusion results [55] indicate that the interaction energy is high.

II.2.3 Mechanical properties, grain boundary migration
It is well known that, in metals, the grain boundary cohesion (embrittlement) [14, 18] and other mechanical properties as microhardness are related to the local impurity content. Such effect has not been up to now evidenced in semiconductors.

The grain boundary mobility is also used to study segregation.

II.2.4 Electrochemical properties
Localized corrosion occurs on impurity decorated defects. Preferential etching can be a powerful tool to measure the segregation [23, 56, 57] . Quantitative measurements and TEM observations of electrochemical etching have been used to study the influence of the structure on segregation (fig.11).

II.2.5 Diffusion
Grain boundary diffusion is largely affected by segregation and diffusion measurements are used to quantify the segregation effects (see following sections).

III GRAIN BOUNDARY DIFFUSION

Preferential diffusion along grain boundaries is a well known phenomenon, described in a large number of articles (see ref. 58 for review). It is a necessary process for several metallurgical phenomena (recrystallization, sintering..)

III.1 Qualitative aspect

Intergranular diffusion is often detected when anomalously high values of diffusion coefficients are measured at low temperatures (fig. 17).

A more direct evidence is obtained by using radioisotope autoradiography [59]. Thanks to this technique, one can separate, in an experiment of diffusion form the free surface of the specimen, the successive depth ranges of volume and grain boundary diffusion. Autoradiography observation show clearly the influence of the grain boundary structure on preferential diffusion. The study of phosphorusdiffusion in silicon [60] [61] is an example of this technique.

In the particular case of the diffusion of a dopant in a semi-conductor (phosphorus in silicon), preferential diffusion can be evidenced by the revelation of the junction profile, either by chemical etching [60], or by EBIC technique [62].

From the qualitative investigations, the following general features can be deduced [58]:

i) grain boundary diffusion is a low temperature phenomenon (small activation energy)
ii) grain boundary diffusion is a short time phenomenon : for long time and high temperature annealing, the grain boundary penetration becomes of the same order as the bulk diffusion penetration.
iii) the grain boundary structure has a strong influence on grain boundary preferential diffusion : it is not accelerated in twins or low angle boundaries.

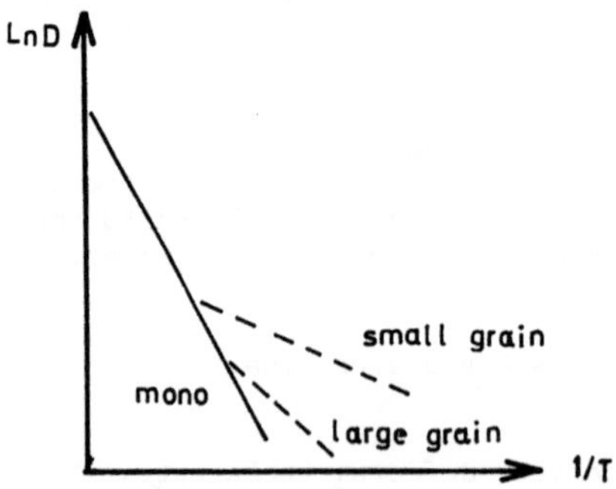

Fig. 17 : Effect of grain boundary diffusion on the Arrhenius plot LnD = f(1/T)

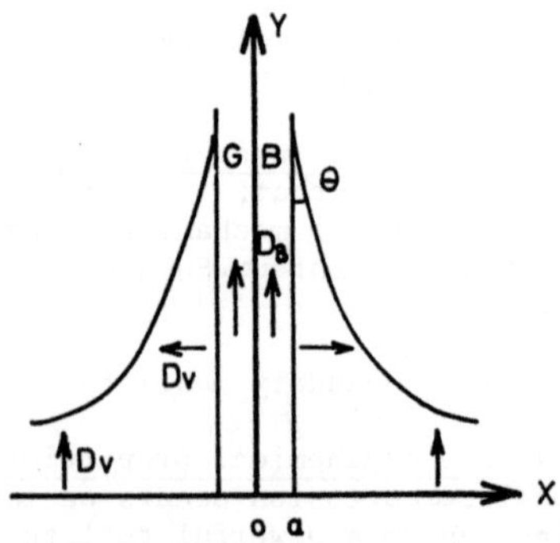

Fig. 18 : Fisher model for grain boundary diffusion

III.2 Phenomenology

III.2.1 The Fisher model

For quantitative description of the diffusion phenomena, a simple model has been proposed by Fisher [63] (fig. 18), used in nearly all investigations. The grain boundary is considered a slab of finite thickness $\delta = 2a$, perpendicular to the surface, in which the diffusivity D_B is different from the diffusivity D_V in the bulk.

Fick's low is written

$$\frac{\partial c}{\partial t} = D_V \, \Delta^2 C \quad \text{for} \quad |x| > a$$

$$\frac{\partial C_B}{\partial t} = D_B \frac{\partial^2 C_B}{\partial Y^2} + \frac{D_V}{a}\frac{\partial C}{\partial x} \quad \text{for} \quad |x| = a$$

The concentration C_B is supposed to be constant in the x direction inside the slab.

For self-diffusion, the concentration profile is continuous, thus $C_B = C_V$ for $|x| = a$.

For heterodiffusion, further developments of the model have supposed that

$$\frac{C_B}{C_V} = k \quad \text{for } |x| = a, \quad \text{with a local equilibrium hypothesis : } k = \left(\frac{C_B}{C_V}\right)_{\text{equil.}}$$

k is then the equilibrium segregation ratio.

Solutions of this model are usually expressed with reduced variables :

$$\xi = \frac{x - a}{D_V t} \qquad \eta = \frac{y}{D_V t} \qquad \beta = \frac{D_B}{D_V}\,\frac{ka}{D_V t}$$

ξ and η are reduced spatial variables for a diffusion time t.

The quantity β expresses the accelerated diffusion effect, and Fig. 19 shows its physical significance : "large β means a high grain boundary diffusivity with respect to the bulk diffusivity".

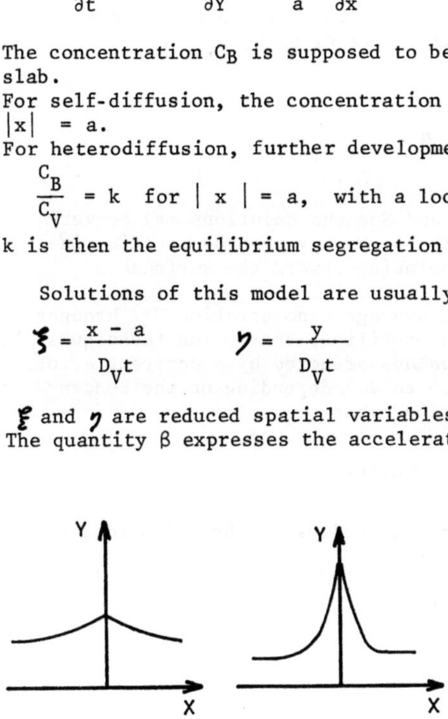

Fig. 19 : Significance of the reduced variable β; left : β small; right: β large

III.2.2 Solutions ot the Fisher model

A simple analytical solution has been given by Fisher [63] under the approximation of a very large D_B and with the boundary conditions :

$C = C_0$ for $y = 0$ and $t > 0$ (constant source)

$$C = C_0 \ \text{efrc} \ \frac{\xi}{2} \exp\left(- \Pi^{-1/4} \eta \, \beta^{-1/2}\right)$$

From this approximated solution, it appears that the "penetration depth" (y value for a given C/C_0 ratio) varies as $t^{1/4}$ (instead of $t^{1/2}$ for volume diffusion). This confirms that intergranular diffusion is a short term phenomenon.

Two exact solutions of the differential equations have been given for two different boundary conditions. Both are non-resolved integrals.

i) constant source ($C = C_o$ for $y = o$ and $t > o$) : Whipple solution [65].

ii) finite source (thin layer on the surface ; $\int_o^\infty C dy = Q$) Suzuoka solution [66].

They have been tabulated by a large number of authors (see ref. 58), depending on the experimental measurements used to obtain β.

Detailed comparisons of the three solutions have been made [58, 67] for different experimental cases. The main conclusions are :

i) The theoretical penetration profiles along the boundary C(o,y,t) derived from the three solutions are substantially different except for large values of $\eta\beta^{-1/2}$.

ii) The angle θ (fig. 18) between an isoconcentration curve (obtained for instance by chemical etching) and the boundary, is given by :

$$\mathrm{Cot}\theta = \left(\frac{\partial C}{\partial \xi} / \frac{\partial C}{\partial \eta} \right)_{\to o}$$

In Fisher solution it is independant of η

$$\mathrm{Cot}\theta = \Pi^{-1/4} \beta^{1/2}$$

The tabulated values of θ from Whipple and Suzuoka solutions may be very different for small η (< 5) and large β (> 10). θ may even be negative in Suzuoka solution (isoconcentration curve pointing toward the surface).

iii) If one measures as a function of y the average concentration $\bar{C}_{II}$ brought by diffusion to the depth y (average defth profiling, sectioning techniques..), and for $\beta > 5$, Fisher solution gives D_B values affected by a constant error factor : D_B (true) $=\alpha D_B$ (Fisher) ; $\alpha = 1.5$ to 2.7 depending on the concentration range. The analytical Fisher solution gives :

$$2\,ka\,D_B = k\,\delta\,D_B = \left(\frac{\partial L_n \bar{C}_{II}}{\partial\, y} \right)^{-2} \sqrt{\frac{4D_V}{t}} \times 0.564$$

For $\beta > 5$ and $\eta > 1$, Whipple and Suzuoka solutions can be adjusted respectively to :

$$k\,\delta\,D_B = \left(\frac{\partial L_n \bar{C}_{II}}{\partial y^{6/5}} \right)^{-5/3} \sqrt{\frac{4D_V}{t}} \times 0.661$$

and

$$k\,\delta\,D_B = \left(\frac{\partial L_n \bar{C}_{II}}{\partial y^{6/5}} \right)^{-5/3} \sqrt{\frac{4D_V}{t}} \times \beta^{0.013} \times 0.578$$

This means that both solutions give nearly the same result. This point is important as it is often difficult to know the actual boundary conditions : finite source conditions may not be respected for the grain boundary if the surface source is refilled by surface diffusion during the diffusion anneal.

It must be emphasized that all the solutions of the Fisher model allow the determination of the parameter β, that is, if D_V is known, of the quantity $k\delta D_B$. Assumptions or independent determinations of k and δ must be made to derive a value of the diffusion coefficient D_B.

III.2.3 Polycrystals

In polycrystals the preceding solutions are valid only if the grain boundaries are perpendicular to the surface and independent from each other. This means that the grain size is large as compared to $\sqrt{D_V t}$. If the grain size is still large but the boundaries not perpendicular to the surface (with an angle φ with the normal), one measures an apparent diffusion coefficient

$$D_B(\text{app.}) = D_B \cos^2\varphi$$

If, on the contrary, the grain size is very small as compared to $\sqrt{D_V t}$, the solutions are not valid; one measures a concentration profile typical of volume diffusion, with an apparent diffusion coefficient Deff. enhanced by the contribution of grain boundaries ;

$$D_{eff.} = D_V = f \cdot D_B$$

f is the fraction of diffusion sites belonging to the grain boundaries. f is evaluated by assuming a given boundary "width" δ.

III.2.4. Thermodynamic model

A model has been developed [68], using the same formalism as for segregation : the interface is a two-dimensional space (δ = o) which exchanges matter with the surrounding grains and in which a surface diffusion occurs. The concentration C_B is replaced by a surface occupancy ρ_B (in at cm^{-2}) and the segregation isotherm must be known. It can be shown [69] that the solutions are the same as for the Fisher model, but the quantity $k\,\delta\, D_B = \frac{C_B}{C_V}\,\delta D_B$ has to be replaced by $\frac{\rho_B}{C_V}\,D_B$ (same dimensions). ρ_B is, in fact, the result of, for instance, Auger electron spectroscopy measurements.

III.3 Mechanisms of grain boundary diffusion. Role of temperature and structure

III.3.1 Atomic model [70]

Starting from the formalism used for the atomic model for volume diffusion, and taking into account the specificity of the grain boundary structure for a given grain boundary, one can consider the atomic transport as the result of various jump frequencies :

Γ, for the volume jumps; Γ_i, for the jumps toward the grain boundary; Γ_o, for the jumps from the grain boundary; Γ'_1, Γ'_2, Γ'_3, ..., for the jumps from a grain boundary site to another grain boundary site. From the atomistic description of the grain boundary, an expression of the average intergranular jump frequency is obtained [70] :

$$\bar{\Gamma}' = f(\Gamma'_1, \Gamma'_2, \Gamma'_3, \ldots)$$

The solution of the differential equations leads to the determination of a quantity β', which is experimentally measured

$$\beta' = \frac{\Gamma_i}{\Gamma_o}\,\frac{\bar{\Gamma}'}{\Gamma}\,\frac{1}{2\sqrt{\Gamma t}}\,, \quad \text{to be compared to} \quad \beta = \frac{k\, D_B\, \delta}{2\, D_V \sqrt{D_V t}}$$

The segregation effect is expressed by the ratio $\frac{\Gamma_i}{\Gamma_o}$

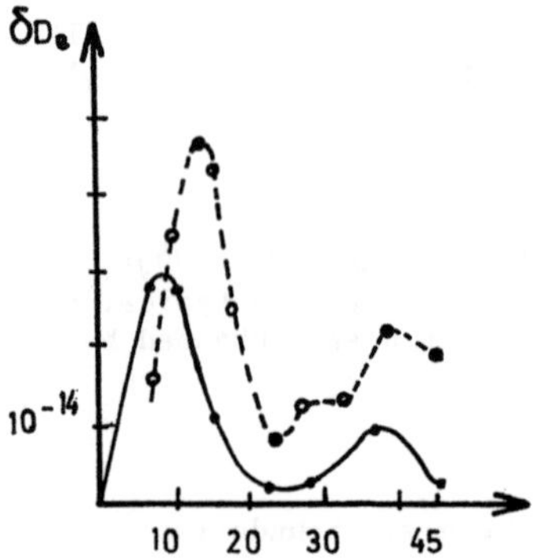

Fig. 20 (from ref. 71) Computed (open circles, arbitrary units) and measured (full circles) influence of the misorientation angle on the grain boundary diffusion in Al-grain boundary

III.3.2 Energetic model [71]
The energetic modeling of the atomic structure of the grain boundaries, applied to simple structures (symmetrical tilt or twist boundaries), allows a determination of the spatial energy distribution in a given boundary.

A "defect formation energy" on each point is defined and a diffusion path is obtained as the most probable path for an atom jumping from defect to defect. The diffusion coefficients estimated from this model are in good qualitative agreement with experience (fig. 20).

III.3.3 Molecular dynamics simulation [34]
Also applied to given simple boundaries, these simulations are able to estimate the formation and migration probabilities of point defects and consequently the diffusion paths and jump frequencies. Their validity depends on the approximation used for the pair interaction potentials and the size of the simulation cell. This method has given interesting results such as the prediction of a progressive transition of the grain boundary structure around 0.5 Tm and evidence for a vacancy mechanism for intergranular self-diffusion.

III.3.4 The problem of the grain boundary "width" [72]
The grain boundary "width" δ of the Fisher model is in fact [58] an "ad hoc" parameter without any physical significance; other models [68] [70] do not need this parameter but introduce at least one other space parameter : definition of the " jump distance" in atomic models, "grain boundary plane" in the thermodynamic model.

Experimental measurements on polycrystals give a parameter P_B (in units of cm^3s^{-1}), which represents the quantity of diffusing species rather than its actual mobility

$$P_B = \frac{C_B}{C_V} \delta D_B = \frac{\rho_B}{C_V} D_B \quad \text{for heterodiffusion}$$

$$P_B = \delta D_B \quad \text{for self-diffusion}$$

In self-diffusion measurements and for comparison purpose, the authors give to δ an arbitrary value of 0.5 nm, considered as a reasonable average value of the disordered region width around the boundaries, at least in metals. This assumption has been for instance useful to demonstrate that the grain boundary self-diffusion coefficient reaches values comparable with the liquid self-diffusion coefficient near the melting point [72] (this supposes that δ is not temperature dependent).

The problem is more complicated in semiconductors (ionic or covalent) in which the space charge region may have an influence.

The case of heterodiffusion remains ambiguous [47] [69] : it is usually assumed that the diffusion sites are identical to the segregation sites, although one may imagine that the diffusion path may avoid the stable segregated atoms (analogy with the hydrogen diffusion with trapping).

III.3.5 Influence of temperature

It is generally observed that the diffusion parameter is thermally activated :

$$P_B = P_{Bo} \exp(-Q'_B/kT),$$

or, for self diffusion $\delta D_B = (\delta D_B)_o \exp(-Q_B/kT)$

Q_B can be considered as a combination of the formation and migration energies of the grain boundary defects. One finds experimentally, for metals, values between 0.4 and 0.7 times the volume self diffusion activation energy, i.e. not far from the migration energy of bulk vacancies.

From the hypothesis of a vacancy mechanism, it is possible to propose a theoretical correlation between the grain boundary self-diffusion coefficient and the grain boundary energy γ [72, 73, 74]

$$\gamma = \frac{kT}{2L^2} L_n \frac{D_B}{D_V}$$

where L is a length related to the grain boundary structure ("average distance between boundary atoms").

III.3.6 Influence of structure

The grain boundary structure has a strong influence on intergranular diffusion, which has been evidenced in very early experiments [75] and, since, confirmed by numerous experimental and theoretical investigations. Figures 20 [71] and 21 [76] show that grain boundary penetration not only depends on the grain boundary misorientation and position but is also anisotropic (depending on the direction inside the grain boundary).

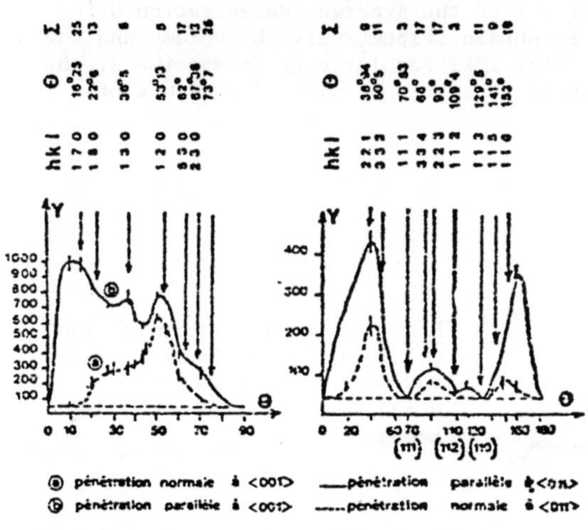

Fig. 21 : Penetration of Zn in Al bicrystals (symmetrical tilt boundaries) [76]

The atomic models might give an answer to this complicated question.

In the case of "random" polycrystals, one considers usually that the grain boundary diffusion coefficient obtained is an average value in which the contribution of the high energy boundaries is predominant, but many controversies in the literature might be explained by this effect of the structure.

III.4 Measurements of grain boundary diffusion - Some examples

III.4.1 Experimental conditions

The grain boundary diffusion has been measured in a great number of materials (see for instance ref. 58 and "Defects and Diffusion Data"), but the results are often scattered over a wide range. The origin of the discrepancies can be sometimes found in the experimental conditions : impurity content and stabilization annealing of the specimen, nature of the grain-boundaries, grain size, temperature and duration of the treatments.

The computation of the diffusion parameter P_B from experimental data needs the knowledge of the volume diffusion coefficient D_V. This is obtained either by extrapolation of literature data or by simultaneous measurement on the same specimen. Both methods are delicate as grain boundary diffusion is measurable in a temperature and time range (low temperature, short time) where volume diffusion is often negligible. The estimation of D_V becomes then a source of error.

Another condition is to choose the temperature and time to obtain $\sqrt{D_V t}$ values either very small or very large as compared to the grain size (§III.2.3)

Grain boundary segregation may occur in very pure materials : all diffusion experiments must be done on specimens previously stabilized at the same temperature for long times.

Finally, grain boundary diffusion measurements cannot be compared if they are not done on specimens with comparable boundaries (same crystallographic texture, same purity, same grain size).

III.4.2 Experimental techniques (summary)

The most precise diffusion measurements are obtained by radioisotope techniques [59] : after deposition of a thin layer of the radiotracer and diffusion anneal, the average concentration profile is obtained by sectioning or residual activity measurements. The logarithm of the average concentration $\bar{C}$ is plotted against y^2 or $y^{6/5}$ in order to obtain respectively the volume and the grain boundary diffusion parameters (fig. 22). For large grain specimens, the depth ranges of the two phenomena can be separated by simultaneous use of autoradiography.

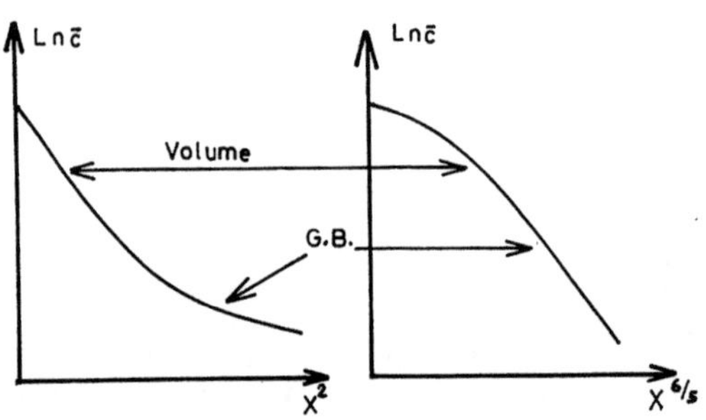

Fig. 22: Determination of the volume and grain boundary diffusion coefficients by an average concentration profiling method

Depth profiling is also obtained by other analysing techniques as SIMS : no radioisotope is needed and small diffusion depth and/or small grain specimens can be investigated [77]; the qualitative control is often impossible.

Another method is to obtain by autoradiography, chemical etching, local analysis or electrical characterization (EBIC) an isoconcentration curve around the boundary, and to measure the angle θ between the curve and the boundary (§ III.2.2). In semiconductors, the chemical etching or the EBIC technique gives in fact the charge carrier isoconcentration curve which may be influenced by other doping impurities.

In thin films with columnar grains (boundaries perpendicular to the surfaces), the accumulation technique can be used : a thin layer of the diffusing species is deposited on one face and, after annealing, the amount accumulated on the other face is measured by radioisotope counting or surface analysis. The mathematical formulation is complex [59] and the boundary conditions may be difficult to define, as the surface diffusion on both faces of the film may be rate controlling.

III.4.3 Examples of experimental results

Grain boundary diffusion of phosphorus in silicon has been measured by several authors. In small grain polycrystalline silicon (e.g. LPCVD recrystallized films), no accelerated diffusion is observed by SIMS measurements [78], whereas other authors have measured, by radioisotopes [60] or EBIC technique [62] an accelerated diffusion with an activation energy (including the segregation energy) of 1.5 to 1.95 eV. It seems that antimon diffuses more rapidly with a lower activation energy (0.83 eV) [60].

In large grain silicon (ingot cast or ribbon grown), grain boundary diffusion depends strongly on the nature of the boundaries : it is not observed in twin boundaries [61] but can be measured in more general tilt and twist boundaries [60]. (Both measurements use radioisotope). Moreover a strong discrepancy has been found [60] between radioisotope and chemical etching measurements; the discrepancy might be attributed to dopant segregation effects.

Aluminum diffusion in silicon has been also measured by various techniques [77] [79] but on different temperature ranges. If one compares the results on an Arrhenius plot [77], they are quantitatively compatible, showing a grain boundary diffusion effect for temperatures below 1300°C, with a small activation energy.

The accelerated diffusion of metalloids, as carbon and hydrogen in silicon is also controversial. Carbon does not diffuse preferentially in twin boundaries [61] but is strongly segregated [24]. It is often assumed [52] that hydrogen diffuses very rapidly in grain boundaries, but a recent estimation of the grain boundary diffusion by an EBIC technique [80] lead to an estimated D_B value (10^{-8} to 10^{-9} cm^2s^{-1} at 350°C) which is not larger than the volume diffusion coefficient at the same temperature. On the other hand, SIMS measurements on small grain silicon films [81] show a strongly accelerated hydrogen diffusion.

III.5 Discussion : influence of segregation on diffusion

The phenomenological description of grain boundary diffusion shows that the experimental parameter P_B is in fact a combination of the diffusion coefficient D_B and the segregation coefficient k. It is thus impossible, from a single diffusion experiment to know the effect of the segregation on the intergranular mobility. Extended studies on metals and alloys [47] [69] show that all kinds of situation can be encountered : a large segregation with a small influence on D_B (hence P_B increases with concentration); a large influence of segregation on D_B which is often lowered (P_B may either increase or decrease with concentration). Such complex effects should be accounted for in the controversial discussion about the diffusion of phosphorus in silicon.

It must also be mentioned that grain boundary diffusion of a given impurity can be affected by the segregation or diffusion of other impurities.

CONCLUSION

Grain boundary diffusion and grain boundary segregation are very complex related phenomena. The phenomenological description of these phenomena is now well established in simple cases. There is a great need of reliable experimental data, especially in semiconductors where the processes may have a critical influence on the properties.

REFERENCES

1. W.C. Winegard, "Fundamentals of solidification", Metall. Review 6 (1961) 57
2. G. Revel, Internat. Symposium "non-crystalline semiconductors", M.R.S. Europe Meeting, June 1984, to be published in J. Physique
3. J. Fally, D. Guignot / ibid
4. D.A. Porter, K.E. Easterling, "Phase transformations in metals and alloys", p. 208
5. P.J. Clemm, J.C. Fisher, Acta Metal;;. 3 (1955) 70
6. G.W. Lorimer, J. Physique 36, (1975), C4-233
7. F.A. Trumbore, Bell Syst. Tech. J. 39 (1968) 205
8. Boltaks, "Diffusion and point defects in semiconductors", Ed. MIR, Moscow (1977)
9. L. Priester, J. Microsc. Spectrosc. Electron., 8 (1983) 211
10. S. Lartigue, L. Priester, Acta Metall. 31 (1983) 1809
11. R. Sharko, A. Gervais, C. Texier-Hervo, J. Physique 43 (1982) C1-129
12. P. Ruterana, A. Bary, G. Nouet, J. Physique 43 (1982) C1-27
13. M. Guttmann, Surface Sci. 53 (1975) 213
14. M. Guttmann, D. Mc Lean, "Interfacial Segregations", Ed. W.C. Johnson J.M. Blakely, A.S.M. Seminars, Metal Park (1979), 251
15. M. Guttmann, "Advances in the Mechanics and Physics of Surfaces", Ed. R.M. Latanision, A.J. Courtel, Nato Editions (1981)
16. E.D. Hondros, J. Physique 36 (1975) C4-117
17. D. Mc Lean, "Grain Boundaries inMetals", Clarendon Press, Oxford (1957)
18. A. Larere, D. Sc. Thesis, Orsay, 1983
19. W.R. Tyson, Acta Metall. 26 (1978) 1471
20. A. Larere, M. Guttmann, P. Dumoulin, C. Roques-Carmes, Acta metall. 30 (1982) 685
21. C. Loier, J.Y. Boos, Met. Trans. 12A (1981) 1223
22. A. Bourret, J. Thibault, D.M. Seidman, J. Appl. Phys. 55 (1984) 825
23. D. Bouchet, L. Beaunier, C. Colliex, P. Trebbia, C. Vignaud, Int. Conf. on "Structure and Properties of Internal Surfaces", Irsee Germany, Sept. 1984, to be published in J. Physique (1984)
24. O. Rallon, M. Aucouturier, C. Texier-Hervo, M. Mautref, C. Belouet, Solar Cell 9 (1983) 149
25. K. Sicrafus, S.L. Sass, Scripto Metall. 18 (1984) 165
26. H.L. Marcus, ASTM Data Ser. STP 479 (1982) 90
27. W.D. Kingery, "Grain boundary Phenomena in Electronic Ceramics", Advance in Ceramics vol.1, ACS (1981) 1
28. W.K. Kingery, Solid States Ionic 12 (1984) 299
29. H.J. Frost, F. Spaepen, J. Physique 43 (1982) C6-73
30. V. Vitek, A.P. Sutton, D.A. Smith, R.C. Pond, "Grain boundary structure and kinetics" A.S.M., Metal Park (1980)
31. R.W. Baluffi, P.D. Bristowe, C.P. Sun, J. Amer. Cer. Soc. 64 (1981) 29
32. D.Wolf, J. Physique 43 (1982 C6-45
33. D.M. Duffy, P.W. Tasker, AERE Harwell, T.P. 1016 (1983) To be published in Phil.Mag.
34. V.Pontikis, J. Physique 43 (1982) C6-65
35. R. Hashimoto, Y. Ishida, R. Yamamoto, D.Doyama , Acta Metall. 32 (1984) 13 and 21

36. C.L. Briant, R.D. Messmer, J. Physique 43 (1982) C6-255
37. O. Khalfallah, D. Bouchet, L.Priester, Men. Et.Scient. Rev. Metall. (1980) 985
38.G.Saindrenan, A.Larere, Scripta Metall., to be published

39.B.Carriere, J.P. Deville, A. Elmaachi, J. Quesada, Internat. Sympo. "non-crystalline semiconductors", MRS Europe Meeting, Strasbourg June 1984, to be published in J. Physique
40.L.L. Kazmerski, P.E.Russel, J. Physique, 43 (1982), C1-171
41. D.R. Clarke, "Grain boundary phenomena in electronic ceramics", Advances in ceramic, vol. 1 A.C.S. (1981), 67
42. P. Doig, P.E.J. Flewitt, J. Microsc. Spectrosc. Electron., 8 (1983) 193
43. A. Bourret, C.Colliex., P. Trebbia, J. Physique Lettres 44 (1983) L-33
44.J.H. Rose, R. Gronsky, Appl. Phys. Lett., 41 (1982) 993
45.J.L. Maurice, J.L. Pastol, J.E. Bourée, J.Y.Laval, G. Revel M. Rodot Internat. Sympos. on "Non-crystalline semiconductors", M.R.S. Europe Meet., Strasbourg, June 1984, to be published in J. Physique
46.F.Moya, G.E.Moya-Gontier, J. Physique 36 (1975) C4-147
47.F.Cabane-Brouty, J.Bernardini , J. Physique 43 (1982) C6-163
48.A.Carabelas, D. Nobili, S.Solmi, J. Physique 43 (1982) C1-187
49.M.M.Nandurah, K.C. Saraswat, C.R. Helms, T.I. Kamins, J. Appl. Phys. 51, (1980) 5755
50.A.Broniatowski, Internat. Sympo. on "Non-crystalline semiconductors", M.R.S. Europe Meet., Strasbourg, June 1984; To be published in J. Physique
51.J.L.Maurice, J.Y. Laval, J. Physique 43 (1982) C1-207
52.C.H.Seager, D.J. Sharp, J.K.G. Panitz, J.I. Hanoka, J. Physique 43 (1982) C1-103
53.H.Amzil, L.Ammor, M.Sebbar, G. Nathian, S. Martinuzzi, Intern. Symp.on "non-crystalline semiconductors". MRS Europe Meet; Strasbourg, June 1984, To be published in J.Physique
54.M.Aucouturier, O. Mallon, M. Mautref, C. Belouet, J. Physique 43 (1982) C1-117
55.M.Mautref, C.Belouet, A. Beunas, M.Aucouturier, M.Groos, Intern. Symp. on "Non-crystalline semiconductors", MRS Europe Meet.; Strasbourg, June 1984, To be published in J. Physique
56.M.Froment, J.Phys. 36 (1975) C4-371
57.L.B.Beaunier, C.Chefi, A. Larere, C.Vignaud, J. Microsc.Spectrosc. Electron., 7 (1982) 171
58.G.Martin, B.Peraillon, J. Physique 36 (1975) C4-165
59. C.Leymonie, "Radioactive tracers in Physical Metallurgy" Dunod, Paris (1960)
60.J.L.Liotard, R. Biberian, J.Cabane, J. Physique 43 (1982) C1-213
61.A.Chari, M.Aucouturier, ECS Sympos. St.Louis, May 1980
62.H.Baumgart, H.S.Leamy, G.K. Celler, L.E. Trimble, J. Physique 43 (1982) C1-363
63.J.C. Fisher, J. Appl. Phys. 22 (1951) 74
64. H.J. Queisser, K.Hubner, W. Shockley, Phys. Rev., 123 (1961) 1245
65.R.T.P. Whipple, Phil. Mag. 45 (1954) 1225
66.T. Suzuoka, Trans. Jap. Inst. Met. 2 (1961) 25 and 176; and J. Phys. Soc. Jap., 19 (1974) 839
67 A.D. Le Claire, Br. J. Appl. Phys. 14 (1963) 351
68 R. Ghez, Surf. Sci 20 (1970) 326
69 J. Bernardini, P. Gas, E.D. Hondros, M.P. Seah, Proc. Roy. Soc. London A 379 (1982) 159
70 P. Benoit, G. Martin, J. Physique 36 (1975) C4-213
71 D. Lemuet, D. Sci. Thesis (1981) Grenoble, France
72 P. Guiraldenq, J. Physique 43 (1982) C6-137

73 P. Guiraldenq, J. Physique 36 (1975) C4-201
74 V.T. Borisov, V.M. Golikov, G.V. Scherbedinski, "Diffusion Processes in metals" Transl. Israel Progr. Scient. Trans. Jerusalem (1970) 122
75 S.R.L. Couling, R. Smoluchowski, J. Appl. Phys. 25 (1954) 1538
76 J. Herbeuval, M. Biscondi, G. Goux, Mem. Sci. Rev.Met. 70 (1977) 39
77 C. Leray, J.E. Bouree, M. Rodot, Symp. on "Laser annealing", M.R.S. Meet., Strasbourg 1983
78 D.L. Losee, J.Q. Lavine, E.A. Trabka, S.T. Lee, C.M. Jarman, J. Appl. Phys. 35 (1984) 1218
79 J.C.M. Hwang, P.S. Ho, J.E. Lewis, D.R. Campbell, J. Appl. Phys. 51 (1980) 1576
80 C. Dube, J.I. Hanoka, D.B. Sandstrom, Appl. Phys. Lett. 44 (1984) 425
81 N.M. Johnson, D.K. Biegelsen, M.D. Moyer, Appl. Phys. Lett. 40 (1982) 882

ACKNOWLEDGMENTS

The author is endebted to J. BERNARDINI, J. CABANE and C. MONTY for valuable informations and discussions.

Part II

Electronic Characterization of Grain Boundaries

Electronic Properties of Grain Boundaries

Jürgen Werner

Max-Planck-Institut für Festkörperforschung, Heisenbergstraße 1
D-7000 Stuttgart 80, Fed. Rep. of Germany

Abstract - This contribution discusses methods for the determination of the density of states at grain boundaries in polycrystalline semiconducting films and bicrystals. All methods are based on measurements of the electronic transport of majority carriers. The doping method reveals the dominance of band tails in the density of states of fine-grained films of Si and GaAs. This is confirmed by photocapacitance measurements at a grain boundary in a Si bicrystal. Ac-measurements on bicrystals are described within the framework of the "trap transistor model". General solutions of this model are given in graphical form. It allows for the evaluation of the density of states and the capture cross section for majority carriers. Moreover, potential fluctuations at grain boundaries are resolved and a correlation of electronic and structural properties can be carried out. The application of this technique to a grain boundary in a Si bicrystal indicates that the interface charge is caused by secondary dislocations.

1. Introduction. The properties of polycrystalline semiconductors are governed by the inherent grain boundaries of these materials. The electrical properties of dislocations and boundaries in semiconductors were summarized in two excellent reviews [1,2]; they will not be reviewed here again. The present paper discusses some experimental methods for the determination of the energetic distribution of grain boundary interface traps, i.e. of the density of states. All these methods are based on measurements of the electronic transport of majority carriers across grain boundaries in either fine-grained films or across single boundaries in bicrystals. Measurements of minority carrier transport have so far not been used for the evaluation of the density of states. Instead, such measurements are used as a powerful tool for the evaluation of the interface recombination velocity. This is discussed in the contribution of C. Donolato.

Section 2 of the present contribution presents the model of thermionic emission and section 3 discusses the "doping method" for the determination of the density of states at grain boundaries in fine-grained polycrystalline films. Methods, which are based on steady state measurements at bicrystals are discussed in section 4. The emphasis is placed on section 5, which discusses the ac-admittance of isolated grain boundaries in bicrystals. Such measurements are interpreted within a recently developed "trap transistor model". General solutions of this model are given in the appendix. Some of the methods discussed are illustrated,with results which arise from an extended investigation of a grain boundary in a p-type silicon bicrystal.

2.Thermionic Emission. The thermionic emission model was originally formulated by Petritz [3] and Mueller [4], and as the result of the pioneering work of the Sandia group [5-8] it is now well established. For a charged grain boundary in a p-type bicrystal as shown in Fig.1 this model predicts a net current j_{th} of the majority carrier holes flowing from left to right,

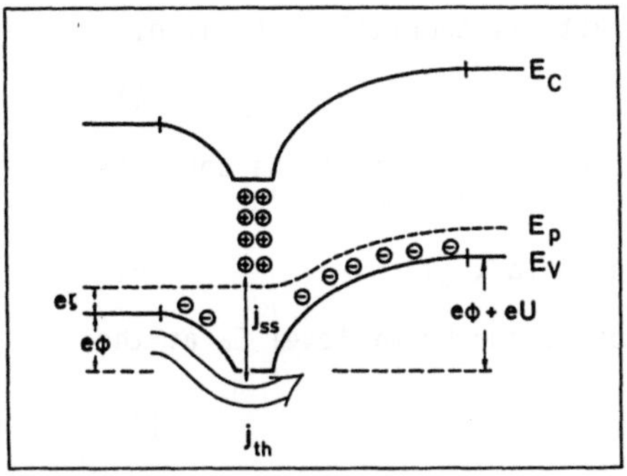

Fig.1.
Band diagram of a charged grain boundary in a p-type bicrystal. Holes are captured in interface states above the Fermi level E_F. This positive charge is compensated by the negative acceptors in the space charge region. Current j_{th} flows from left to right by thermionic emission. j_{ss} is the current due to dynamic equilibrium between the capture and emission of holes

$$j_{th} = A^*T^2\exp(-\beta(\zeta+\phi))\,(1-\exp(-\beta U))\ . \tag{1}$$

Here β=e/kT is the inverse of the thermal voltage, A* the effective Richardson constant and U the applied voltage. The forward biased (left) barrier is termed with eϕ, whereas eζ is the Fermi level in the single crystalline grains, which depends on doping. Apart from the thermionic current, j_{th}, there is also a second current, j_{ss}, at the grain boundary as indicated in Fig.1. This current j_{ss} is caused by the fact that the grain boundary interface states permanently capture holes from the valence band and re-emit holes to the valence band. The current j_{ss} is the difference between the capture and the emission rate of holes and can be described with the help of Shockley-Read-Hall statistics. This will be discussed in section 5. Under conditions near thermodynamical equilibrium (eu$\ll$kT, zero-bias conditions) $j_{ss}\approx 0$ holds because of the principle of detailed balance.

3. The Doping Method. One of the first steps towards a systematic evaluation of a density of states on the basis of Eq.(1) was done by Seager and Castner [5]. They modelled the resistivity of fine-grained polycrystalline silicon. A model with an energy-independent density of states showed the poorest agreement with the data whereas an exponential density of states fitted the data somewhat better than a single level model. A more precise conclusion could not be drawn from those model fits.

This situation changed fundamentally, when de Graaff et al. [9] found a deconvolution method for the determination of the density of states. The method is based on measurements of the temperature-dependent conductivity of thin films and so far it has been applied to fine-grained polycrystalline films of Si [9-11] and GaAs [12] with grain sizes of the order of 1μm. An exponential dependence of the conductivity on the inverse of temperature is usually found, i.e. an Arrhenius behavior is observed. This is easily understood on the basis of Eq.(1).

For a polycrystalline film of length L and average grain size a the current j_{th} is given by:

$$j_{th} = A^*T^2\cdot\exp(-\beta(\zeta+\phi))\,(1-\exp(-\beta Ua/L)) \tag{2}$$

Here Ua/L is the voltage drop per grain boundary. This value is usually small compared to kT/e. The conductivity σ is then obtained by a Taylor expansion of the voltage-dependent term in Eq.(2):

$$\sigma = \frac{j_{th}L}{U} = \frac{e}{k}A^*Ta\,\exp(-\beta(\zeta+\phi)) \tag{3}$$

Experimentally one finds that the conductivity is thermally activated:

$$\sigma = \sigma_\infty \exp(-E_{act}/kT) \tag{4}$$

Here E_{act} is the activation energy in the Arrhenius plot of the conductivity σ, and σ_∞ is the σ-value at 1/T=0. From Eqs.(3), (4) one obtains:

$$E_{act} = e\zeta(T) + e\phi(T) + kT \ln(eA^*Ta/k\sigma_\infty) \tag{5}$$

Figure 1 shows that the sum of $e\zeta$ and $e\phi$ gives the Fermi level E_F at the grain boundary.

$$E_F(T) = e\zeta(T) + e\phi(T) \tag{6}$$

Therefore, from Eqs.(5), (6) one obtains:

$$E_{act} = E_F(T=0) \tag{7}$$

The activation energy of the thermally activated conductivity σ gives a measure for the Fermi level at the grain boundary extrapolated to zero temperature. Experimentally one finds a strong dependence of the activation energy E_{act} upon the doping of the films [5,9-12]. This means that the Fermi level E_F at the grain boundary can be changed by a change of doping! The shift of E_F through the density of states at the grain boundary goes hand-in-hand with a change of the trapped interface charge. This charge Q can be calculated from the barrier height $e\phi$. It is in turn calculated from $e\phi = E_{act} - e\zeta$. The Fermi level in the grains, $e\zeta$, is obtained from the grain doping N, which is either known from the implantation dose or is determined by Hall effect measurements. The density of states is then determined by taking the derivative of the charge Q with respect to the Fermi level E_F, which both depend on doping:

$$N_{ss}(E_F) = \frac{dQ}{dE_F} = \frac{dQ}{dN} \Big/ \frac{dE_F}{dN} \tag{8}$$

The doping of thin films of Si and GaAs can be varied over orders of magnitude. That is why the Fermi level can be shifted through a wide energy range within the forbidden gap of these semiconductors. Results are given in Figs.2,3. Both figures show that the density of states within the forbidden gap increases towards the band edges.

A very interesting feature, measured by Yang et al. [12], can be seen in Fig.3. If one extrapolates the measured density of states towards the band edges one finds a reasonable agreement with the values of the density of

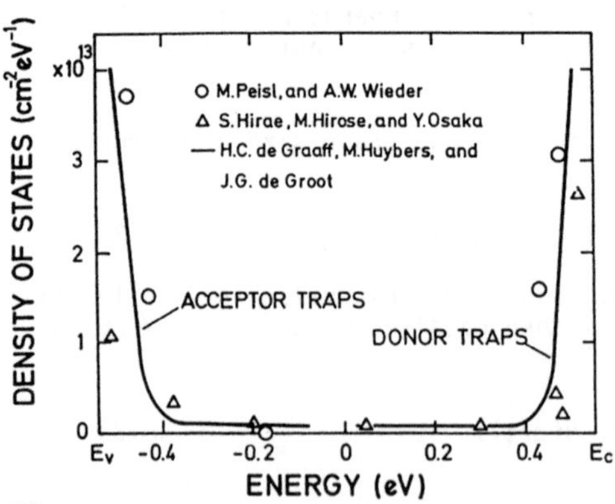

Fig.2. Density of states of polycrystalline silicon films. Peisl and Wieder [10] as well as de Graaff et al. [9] used the doping method, which is based on measurements of the thermally activated dc conductivity. Hirae et al. [13] used an ac-technique (redrawn after [10])

states for a two-dimensional gas of free carriers within the conduction band and the valence band. Conversely, it seems that the density of states of the bands of the free carriers is smeared out into the forbidden gap of the semiconductor. This was recently also reported by Peisl [11] who found an excellent agreement of such extrapolated and directly measured values for polycrystalline silicon. Such a smear-out of the density of states is well-known from disordered semiconductors, as for example highly doped single crystals [14,15] and amorphous silicon [16] where it is termed as "band tailing". As a consequence of the potential fluctuations at grain boundaries it is also expected for polycrystalline semiconductors.

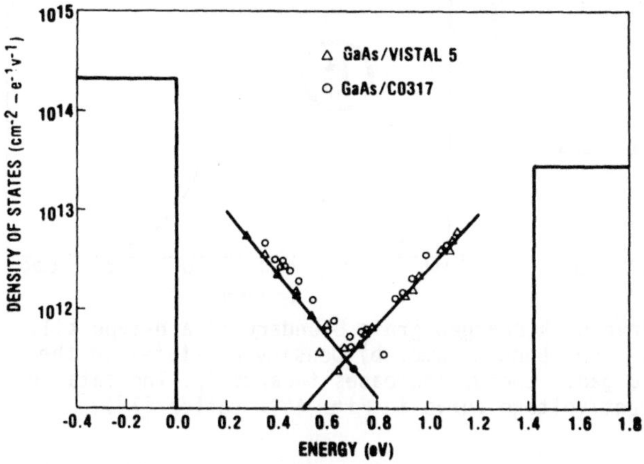

Fig.3. Density of states within the forbidden gap of polycrystalline GaAs films. The density of states for a two-dimensional gas of free electrons and holes within the conduction band and the valence band of singlecrystalline GaAs are shown for comparison. The doping method was used. (After [12])

The advantage of the doping method can be seen in Figs.2,3: Information over a wide energy range is obtained, especially when one uses n-type and p-type samples. The method requires a lot of samples, which are identical apart from the doping. This disadvantage seems not to be a serious drawback for polycrystalline films. In this case one takes average of a lot of grain boundaries, anyhow. For bicrystals, however, the method is inappropriate, because it is extremely difficult to grow a series of completely identical bicrystals which differ only in the amount of doping. Therefore for bicrystals other methods with a wide energy resolution are more favorable as for example the optical method.

4. Steady State Bicrystal Measurement.

4.1. The I/U-Method. Already Taylor, Odell and Fan [17] measured and discussed current(I)/voltage(U) curves of bicrystals in Ge. They had some ideas about the influence of interface traps on these characteristics. Goldman et al. [18,19] were the first who discussed the occupation of grain boundary traps and the influence of different shapes of the density of states on I/U-curves. The development of a deconvolution method for the density of states, which is based on such considerations, is nowadays attributed to Pike and Seager [6,7].

The technique is based on current/voltage curves of bicrystals as shown in Fig.4. The basic ideas are as follows: The imref of the majority carriers (i.e. electrons in Fig.4) is assumed to be flat through the forward biased (left) space charge and is identical to the quasi Fermi level of the inter-

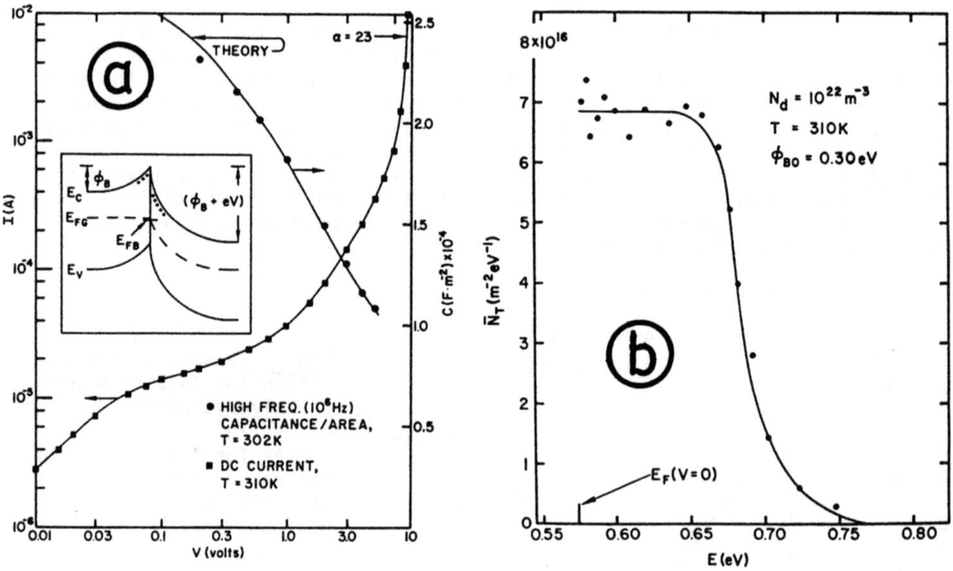

Fig.4 a) Current/voltage curve of a charged grain boundary in a n-type silicon bicrystal. Inset shows band scheme. b) Density of states in the upper part of the band gap. Energy increases towards E_C. The data are deduced from the current/voltage curve in Fig. 4a). (after [7])

face states. The left barrier decreases under bias whereas the right one is increased. The flat Fermi level causes an increase of the boundary charge, i.e. more electrons are trapped, when the voltage is increased. Increasing charge is expected until the Fermi level reaches a decreasing density of states. A rapid (varistor) breakdown of the barrier and a strong increase of the current will then occur. Conversely, from the breakdown of a current/voltage curve a density of states was derived (Fig.4). A continuous distribution dropping rapidly for energies more than 0.65eV above the valence band was found [7]. The measurements required voltages >5V, fields around 10^5V/cm and powers up to 10W/cm^2. The collapse therefore might have been caused not by a decreasing density of states but by sample heating and impact ionisation, which was shown to occur in the varistor regime [20]. De Groot et al. [21] showed that such apparent decreases of the density of states could also be misinterpretations of the fact that the captured electrons are released from traps by thermionic field emission.

4.2. The Optical Method. Any evaluation of a density of states, when based on current measurements at bicrystals, requires verification of sample homogeneity. Such a verification can be given by a comparison of measurements of the dc-current and the high frequency capacitance. With the help of Eq.(1) the barrier ϕ can be deduced from the measured current, j_{th}. This barrier ϕ in turn yields the space charge width and the high frequency capacitance. An agreement of such derived capacitance data with directly-measured values was given earlier [22,23] for measurements in darkness as well as for samples illuminated with sub-band-gap light.

A spectroscopic determination of the density of states has then been carried out on the basis of such photocapacitances [22,23]. In Fig.5 the data for a p-type silicon bicrystal are shown. The capacitance is measured under zero

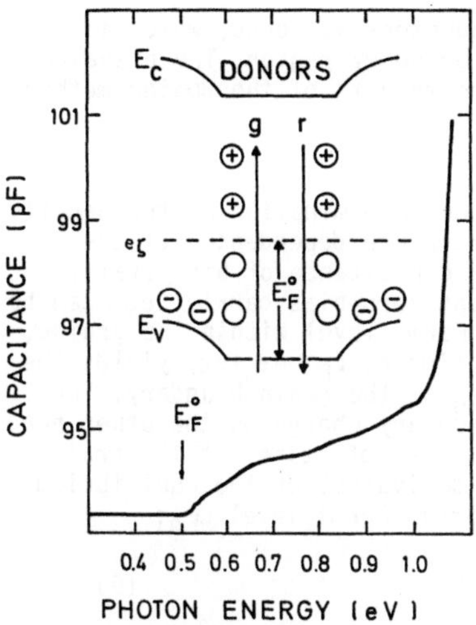

Fig.5. Capacitance of an isolated grain boundary in a p-type bicrystal for sub-band-gap illumination (after [22])

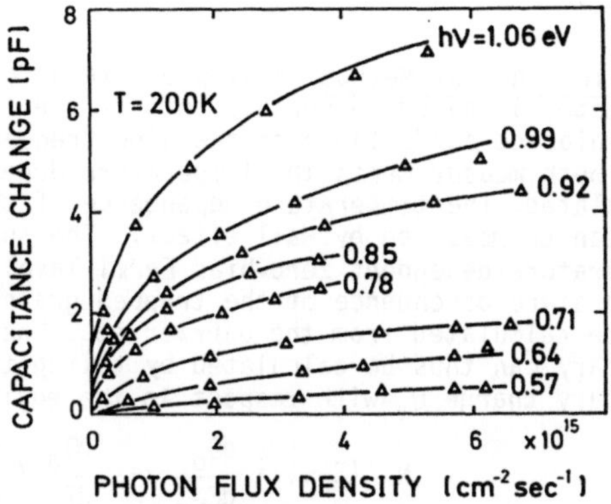

Fig.6. Photocapacitance as a function of light intensity for eight different photon energies (after [22])

bias, avoiding the application of high fields, while the sample is illuminated with infrared light using a grating monochromator/globar source. The onset of the capacitance change agrees with the energy difference between the dark Fermi level E_{Fo} and the edge of the valence band. This indicates a direct optical excitation of trapped holes from the forbidden gap into the valence band. Fig.6 shows the capacitance change tending towards a saturation value when measured as a function of light intensity for a constant photon energy $h\nu$. This saturation value corresponds to a situation where all traps between the valence band edge E_V and an energy $h\nu$ apart are emptied by the optical excitation of trapped holes. The density of states can thus be obtained by differentiating this remaining charge, Q_∞, with respect to photon energy $h\nu$.

The charge Q_∞ is obtained from measurements at finite light intensities with the help of a rate equation and a fit to Fig.6 [22,23]. Results for the density of states are shown in Fig.7 where they are labeled as "optical method".

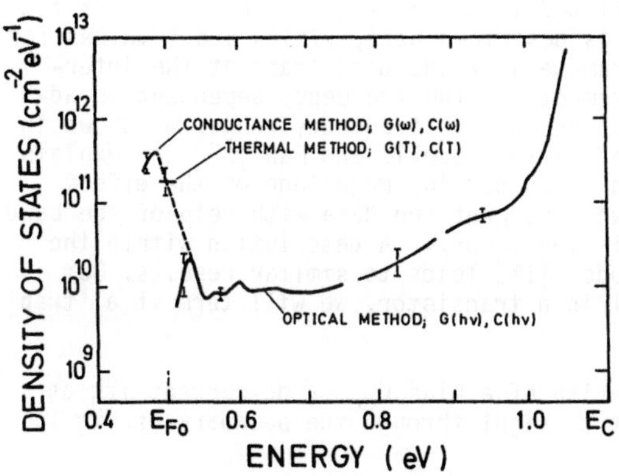

Fig.7. Density of states for a p-type silicon bicrystal boundary as derived from photoconductance and -capacitance ("optical method"), the thermally activated zero-bias conductance ("thermal method"), and the frequency dependence of the ac admittance ("conductance method"). The latter method is discussed in section 5

The typical band tailing of disordered semiconductors is found, which is caused by the loss of long-range order in the boundary region. The prevalence of the bandtails in Fig.7 strongly confirms the results of the doping method as shown in Figs.2,3.

4.3. The Thermal Method. A simple method for the determination of the density of states is based on measurements of the temperature dependence of the zero-bias dc-conductance or the high frequency capacitance of bicrystals. From both measurements the temperature dependent zero-bias barrier $e\phi_0$ can be calculated. The temperature dependence of the Fermi level within the grains, $e\zeta$, can be measured by Hall effect. The sum of both, $E_{Fo}=e\phi_0+e\zeta$, yields the temperature dependent zero-bias Fermi level E_{Fo} at the grain boundary. The temperature dependence of the trapped grain boundary charge on the other hand can be calculated from the barrier $e\phi_0$. The density of states at the grain boundary can thus be calculated by taking the derivative of the equilibrium boundary charge Q_0 with respect to the equilibrium Fermi level E_{Fo}:

$$N_{ss}(E_{Fo}) = \frac{dQ_o}{dE_{Fo}} = \frac{dQ_o}{dT} \Big/ \frac{dE_{Fo}}{dT} \tag{9}$$

Results of this method for a grain boundary in p-type silicon are also shown in Fig.7. The disadvantage of the method can be clearly seen: It has a very limited energy resolution. The reason for this limitation is very simple: The Fermi level at the grain boundary is nearly pinned by the high density of states. By changing the temperature from 180-260K, for example, the Fermi level at the grain boundary can only be shifted by about 30meV [23].

5. AC admittance. Admittance spectroscopy is a powerful tool for the determination of at least three interface parameters: density of states, majority carrier capture cross section, and standard deviation of potential fluctuations. From a combination of such measurements with transmission electron microscopy, moreover, conclusions about the origin of the grain boundary charge were drawn [24].

A frequency-dependent capacitance of germanium bicrystals was already reported by Taylor, Odell, and Fan [17]. They tried to attribute their results to an effect which is nowadays called "diffusion capacitance" [25]. A quantitative analysis was not tried in those days of the early fifties. Seager and Pike [8] were the first who published frequency dependent capacitance/voltage curves of grain boundaries in silicon. They suggested that the effect is caused by an "out-of-phase modelling" of the grain boundary barrier. On the basis of this model the data were fitted in a puzzling manner. In contrast to such fitting procedures the model, as described here, yields exact values for the above-quoted characteristic parameters of the deep traps at the interface. The method is based on measurements of the frequency dependent ac-admittance of single grain boundaries over a wide frequency range, as shown in Fig.8. The minority carrier model of Taylor, Odell, and Fan [17] can explain neither the frequency dependence in Fig.8 nor the magnitude of the effect (J.Werner, unpublished). Instead, we interpret the data with help of the band diagram of Fig.1 and thermionic emission theory. A description within the framework of the drift/diffusion model [17] leads to similar results. Our model resembles the current control in a transistor, we will term it a "trap transistor model", therefore.

5.1. DC currents. Under the application of a bias U_{dc}, a dc-current j_{th}^{dc} as given by Eq.(1) will flow from left to right through the boundary of Fig.1.

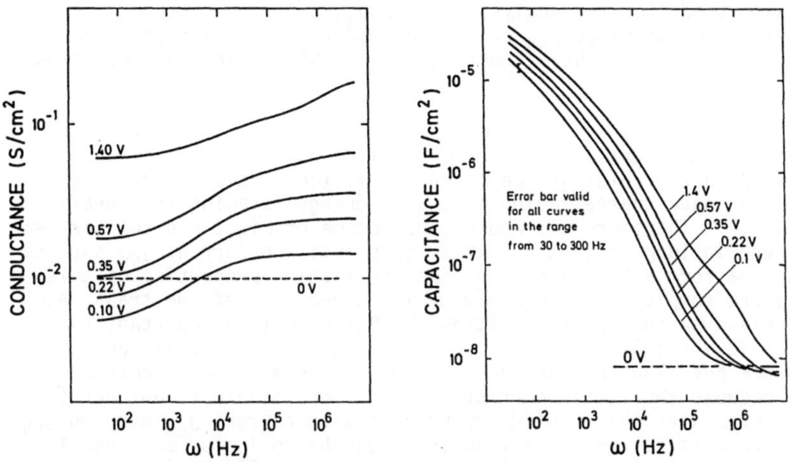

Fig.8. Frequency dependence of the conductance and capacitance of a single boundary in a p-type silicon bicrystal. The bias voltage is given as a parameter. Ac amplitude is 10mV, temperature is 260K

The quasi Fermi level of majority carriers (holes) is assumed to be flat up to the interface. We assume $\sigma_p p_{dc} \gg \sigma_n n_{dc}$ to hold at the interface, where p_{dc} and n_{dc} give the concentrations of free carriers under dc conditions, and σ_p, σ_n are the respective capture cross sections of the interface states. Under this assumption the quasi Fermi level of the traps is equal to the imref of free holes [26]. Also, detailed balance for captured and reemitted holes is then nearly fulfilled: $j_{ss}^{dc} \approx 0$.

5.2. AC currents. The application of a small ac-voltage $\delta U \ll kT/e$ in addition to the dc voltage $U_{dc} \gg kT/e$ leads to a time periodic change of the barrier $\phi = \phi_{dc} - \delta\phi$, $\delta\phi \ll kT/e$ and to three different currents:

a) Thermionic current:

$$j_{th} = A^* T^2 \exp(-\beta(\zeta + \phi)). \tag{10}$$

b) The high frequency capacitance C_{HF} of the space charge yields a displacement current:

$$j_{sc} = i\omega \, C_{HF} \, \delta U \quad , \tag{11}$$

where $i = (-1)^{1/2}$.

c) Due to the periodic barrier change, the hole concentration at the interface also varies periodically in time. This change drives a periodic capture and emission of holes by the interface states. The difference between the capture and emission rate, the current j_{ss} in Fig.1, is phase shifted with respect to $\delta\phi$ because the capture and emission process at the interface is not infinitely fast. For the description of this current j_{ss}, which is measured as a displacement current, we write:

$$j_{ss} = Y_{ss} \, \delta\phi \; . \tag{12}$$

Here Y_{ss} is the characteristic admittance of the traps, which depends on their distribution in energy and space, and on the capture cross section. The definition and methods for the analysis of $Y_{ss}=G_{ss}+i\omega C_{ss}$ in terms of a conductance G_{ss} and a capacitance C_{ss} can be adopted from the physics of MOS-interfaces, especially from the pioneering work of Nicollian and Goetzberger [27].

The current j_{ss} of Eq.(12) is identical to the time derivative of the trapped interface charge and is driven by $\delta\phi$. The puzzling fact for a grain boundary is as follows: Whenever holes are captured or emitted at the interface then the width of the space charge region must also change to maintain overall charge neutrality. Therefore the whole band scheme of Fig.1 and the barrier variation $\delta\phi$ are influenced in turn. This means that the current j_{ss} and the barrier change $\delta\phi$ form a feedback loop. The oscillation properties of this feedback loop are completely determined by the properties of the traps, and the loop is driven by the applied voltage δU. But even more important for a grain boundary is the following fact: Not only the current j_{ss} is controlled by this feedback but also the current j_{th} that is thermionically emitted over the barrier, because the latter current responds sensitively to any barrier change. This means that the coupled capture/emission current j_{ss} acts on j_{th} similarly as the base current in a transistor on the collector current. The stated action leads us to the name "trap transistor model". It is this transistor action which causes the "out-of-phase modelling" of the barrier ϕ. The high capacitances in Fig.8 are, indeed, caused by capacitive thermionic currents as suspected by Seager and Pike [8] although their formalism is inappropriate for a systematic evaluation of the measurements.

5.3. AC admittance. The measured small-signal admittance $Y_{tot}=G+i\omega C$ of Fig.8 is given by the derivative of the total current $j_{th}+j_{ss}+j_{sc}$ with respect to the applied voltage δU.

$$Y_{tot} = \gamma\ \delta\phi/\delta U + Y_{ss}\ \delta\phi/\delta U + i\omega C_{HF} \quad (13)$$

$$\gamma = \frac{e}{k} A^{*}T \exp(-\beta(\zeta+\phi_{dc})) = \beta\ j_{th}^{dc} \quad (14)$$

The total ac current $j_{tot}=Y_{tot}\delta U$ is given by:

$$\begin{aligned} j_{tot} &= \gamma\delta\phi + Y_{ss}\delta\phi + i\omega C_{HF}\delta U \\ &= j_{th}^{ac} + j_{ss} + j_{sc} \end{aligned} \quad (15)$$

Equation (15) shows that the thermally emitted ac current j_{th}^{ac} is determined by amplitude and phase of $\delta\phi$ which in turn depend on j_{ss}. Therefore j_{ss} enters twofold in the measured ac current j_{tot}. Firstly it is measured directly as a displacement current and secondly it influences j_{th}^{ac} via the above stated transistor action.

The cornerstone of our method is to extract the admittance Y_{ss} of the "base current" j_{ss} from the measured admittance Y_{tot}. The well-known methods of the admittance spectroscopy of MOS interfaces [26,27] are then applied to the extracted Y_{ss}. For this purpose the relationship between $\delta\phi$ und δU has to be found first. This is achieved in the following manner: Charge neutrality requires the current j_{ss} at the interface to be equal to the displacement current dQ^-/dt which is caused by the change of the width of the space charge region whenever holes are captured or emitted at the interface. This yields:

$$j_{ss} = Y_{ss}\ \delta\phi = e\ dQ^-/dt \quad (16)$$

where Q^- is the sum of acceptor charges in the left and right space charge and can be expressed in terms of barrier $\phi=\phi_{dc}-\delta\phi$ and voltage $U=U_{dc}+\delta U$ [6,26]. Then we finally get:

$$\delta\phi = i\omega C_R/(Y_{ss}+i\omega C_\Sigma)\,\delta U \tag{17}$$

Here C_R and C_L are the high-frequency capacitances of the left and right space charge region:

$$C_L = (e\varepsilon\varepsilon_0 N_A/2\phi_{dc})^{1/2}; \qquad C_R = (e\varepsilon\varepsilon_0 N_A/2(\phi_{dc}+U_{dc}))^{1/2} \quad , \tag{18}$$

where $\varepsilon\varepsilon_0$ is the permittivity and N_A is the doping of the grains. The high-frequency capacitance C_{HF} is given by: $1/C_{HF}=1/C_L+1/C_R$, and $C_\Sigma=C_L+C_R$. Equation (17) clearly reveals that the barrier response $\delta\phi$ to the applied voltage δU is determined by the properties of the interface states, which are contained in Y_{ss}. It shows also that for $|\delta U|<kT/e$ also $|\delta\phi|<kT/e$ holds for the ac-change of the potential barrier, $\delta\phi$.

5.4. Transistor Current Amplification. The capture/emission current of holes, j_{ss}, which flows between the grain boundary traps and the valence band is the "base current" in our trap transistor model. The thermionically emitted ac-current j_{th}^{ac} corresponds to an "emitter-collector current". As in an usual transistor we can express the transistor action of j_{ss} on j_{th}^{ac} in terms of a current amplification factor. The thermionically emitted ac current is given by:

$$j_{th}^{ac} = \gamma\,\delta\phi \tag{19}$$

On the other hand, j_{ss} is given by:

$$j_{ss} = Y_{ss}\,\delta\phi \tag{20}$$

From Eqs.(19), (20) it then follows:

$$j_{th}^{ac} = \frac{\gamma}{Y_{ss}}\,j_{ss} \tag{21}$$

Equation (21) is the central equation of the trap transistor model. It shows that the current j_{th}^{ac} is controlled by the capture/emission-current j_{ss}. The current amplification factor, γ/Y_{ss}, is controlled by the properties of the interface states, which are represented by the admittance Y_{ss}. The admittance Y_{ss} has real and imaginary parts. That is why, for the grain boundary trap transistor, there is a phase shift between the "base current" j_{ss} and the emitted current j_{th}^{ac}.

5.5. Equivalent Circuits. The total measured ac current j_{tot} divides into three parts, as given by Eq.(15) and shown in Fig.9. The conductance G_{ss}^x and capacitance C_{ss}^x are the equivalent circuit elements of the current j_{ss} with respect to the applied δU. They are different from G_{ss} and C_{ss} of the admittance $Y_{ss}=G_{ss}+i\omega C_{ss}$, which describe the current j_{ss} with respect to $\delta\phi$! The mutual relationships of these quantities are given by Eqs.(12), and (22):

$$j_{ss}= Y_{ss}\,\delta\phi=(G_{ss}+i\omega C_{ss})\;\delta\phi = (G_{ss}^x+i\omega C_{ss}^x)\;\delta U = Y_{ss}^x\,\delta U \tag{22}$$

The three parts of the ac current in Fig.9 are assumed to be driven by the applied ac-voltage δU. This voltage is responsible for the barrier change $\delta\phi$. Equation (15) shows that we can also give an equivalent circuit of the total

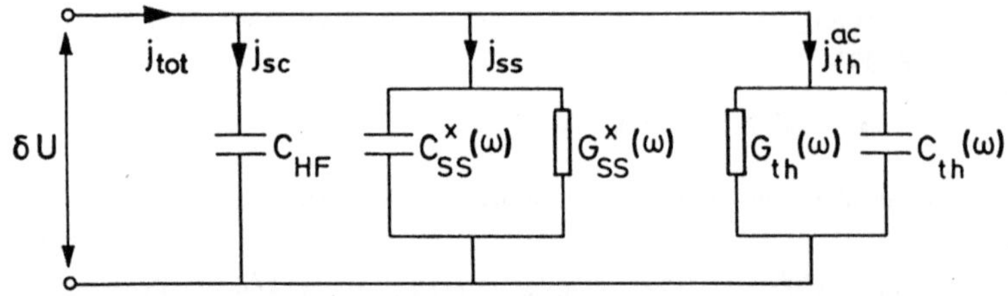

Fig.9. Equivalent circuit for ac measurements at a single grain boundary. The total ac-current j_{tot} divides into three parts. j_{sc} is the displacement current of the high frequency capacitance C_{HF} of the space charge width. j_{ss} is the current in Fig.1, which is exchanged between the traps and the valence band. C^x_{ss} and G^x_{ss} are its capacitance and conductance with respect to δU. j^{ac}_{th} is the ac-part of j_{th} in Fig.1. The conductance G_{th} and the capacitance C_{th} depend on frequency because of the transistor action of j_{ss} on j^{ac}_{th}

ac current with respect to $\delta\phi$, if we relate the displacement current j_{sc} of the high frequency capacitance C_{HF} to $\delta\phi$ by:

$$j_{sc} = i\omega C_{HF}\ \delta U = (G_\phi + i\omega C_\phi)\,\delta\phi \tag{23}$$

where G_ϕ, C_ϕ describe the real and imaginary part of j_{sc} with respect to $\delta\phi$. They are calculated from Eqs.(17), (23) with the result:

$$G_\phi = \frac{C_\Sigma}{C_L}\, G_{ss} \tag{24}$$

$$C_\phi = \frac{C_\Sigma}{C_L}\,(C_{ss}+C_\Sigma). \tag{25}$$

The total ac current is then given by:

$$\begin{aligned} j_{tot} &= \gamma\delta\phi + (G_{ss}+i\omega C_{ss})\,\delta\phi + (G_\phi + i\omega C_\phi)\,\delta\phi \\ &= j^{ac}_{th} + \qquad j_{ss} \qquad + \qquad j_{sc} \end{aligned} \tag{26}$$

The equivalent circuit is shown in Fig.10. The thermionically emitted current j^{ac}_{th} has no capacitive part in this representation and the conductance γ is independent of frequency. All other currents have capacitive parts and the driving "voltage" $\delta\phi$ depends on frequency according to Eq.(17).

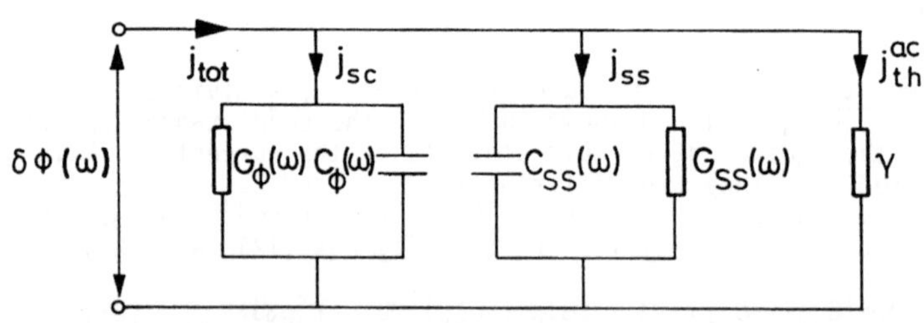

Fig.10. Equivalent circuit for the total ac current j_{tot} with respect to the ac-change of the barrier, $\delta\phi$. The thermionically emitted current j^{ac}_{th} has only a conductance γ. The other currents have also capacitive parts. The driving force $\delta\phi$ depends on frequency

Figures 9 and 10 show the equivalent circuits for the total measured ac current, j_{tot}. As the main part it contains the thermally emitted current j_{th}^{ac}, the "emitter-collector current" of our trap transistor model. The equivalent circuit for the "base current", the capture/emission current j_{ss}, is simple: It consists of a parallel connection of the conductance G_{ss} and the capacitance C_{ss}, and it is driven by $\delta\phi$.

5.6. Extraction of the Base Current. The aim of our analysis is to extract the admittance Y_{ss} of the "base current" j_{ss} from the measured admittance Y_{tot} of the total ac current j_{tot}. The currents j_{ss}, j_{sc} are usually small if compared to j_{th}^{ac} [26]. In this case the measured G and C are identical to G_{th} and C_{th} of the thermionically emitted current j_{th}^{ac}. Equations (15), (17) then yield:

$$G = \gamma \cdot \frac{C_R(C_{ss}+C_\Sigma)}{(C_{ss}+C_\Sigma)^2+(G_{ss}/\omega)^2} \tag{27}$$

$$C = G \cdot \frac{G_{ss}/\omega^2}{C_{ss}+C_\Sigma} \tag{28}$$

The information about the traps is contained in G_{ss}, C_{ss} which are in turn obtained from the measured G, C by:

$$G_{ss} = \gamma \cdot \frac{C_R \cdot C}{(G/\omega)^2+C^2} \tag{29}$$

$$C_{ss} = G_{ss}G/(\omega^2 C) - C_\Sigma \tag{30}$$

The quantities G_{ss}, C_{ss} from Eqs.(29), (30) can now be subjected to an analysis. Both quantities contain the same information about the traps because they fulfill the Kramers-Kronig relations. But it is more convenient to analyse G_{ss} because G_{ss}/ω goes through a maximum when plotted against frequency. That is why we call this method the "conductance method".

5.7. Potential Fluctuations Model. Fig.11 shows values for G_{ss}/ω evaluated with the help of Eq.(29) from the data of Fig.8. The measured curve is compared to three different models for the energetic and spatial distribution of the traps at the grain boundary. Curve (a) has a Lorentzian shape and is expected for a monoenergetic trap [27]. Curve (b) is expected for a continuous density of states N_{ss}, with [27]:

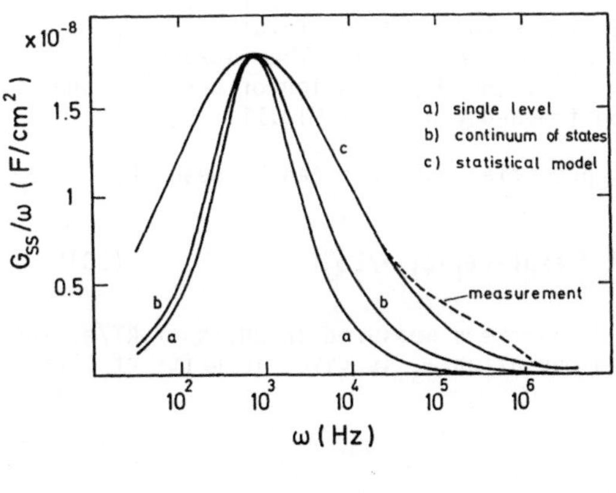

Fig.11. G_{ss}/ω for the 0.35V curve of Fig.8. The measurements are compared to three different models for the energetic and spatial distribution of the traps. The width of the measured curve can only be explained by a statistical model of interface potential fluctuations due to the spatial distribution of the traps

$$G_{ss}/\omega = \frac{e^2 N_{ss}}{2\omega\tau_m} \ln(1+\omega^2\tau_m^2) \qquad (31)$$

$$\tau_m = \exp(\beta\phi_{dc})/(\langle v_p\rangle \sigma_p p_g) \qquad (32)$$

where p_g is the hole concentration in the grains and $\langle v_p \rangle$ their thermal velocity.

Neither the single level model nor the model of the continuous density of states explain the width of the measured curve in Fig.11. This width can only be explained if we take into account that the dc barrier ϕ_{dc} at the grain boundary is not homogeneous but fluctuates due to the random spatial distribution of the traps. This model is very similar to the model of Nicollian and Goetzberger [27] for the MOS interface. Figure 12 illustrates our model for a charged grain boundary in p-type material. The grain boundary is not a homogeneously charged plane but the holes are captured in localized states. The sites of these states are statistically distributed over the boundary. This results in a spatial distribution of potential barriers ϕ as shown in Fig.12 for zero-bias conditions.

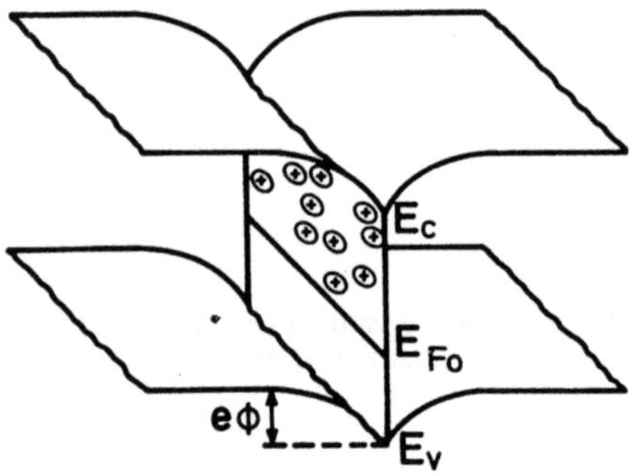

Fig.12. Potential fluctuations for a charged grain boundary in p-type material. The positively charged traps are randomly distributed. This results in spatial fluctuations of the potential barrier ϕ

The important influence of such fluctuations on the current transport through grain boundaries was also indicated by Thomson and Card [28] and Mahan [29], recently. It should be stated therefore that Fig.11 gives an indirect depiction of such fluctuations: For a continuous density of states the maximum of G_{ss}/ω from Eq.(31) is expected at $\omega\tau_m$=1.98. Equation (32) shows that the time constant τ_m depends sensitively on the potential barrier ϕ_{dc}. Any spatial fluctuation of barriers thus results in a spatial distribution of time constants. This spatial time constant dispersion in turn yields a spatial dispersion of the frequency for which G_{ss}/ω is a maximum. The integral G_{ss}/ω is thus widened over the frequency axis and the distribution of barriers and time constants is delineated in the frequency axis of Fig.11.

We assume that the distribution of potential barriers can be described by a Gaussian distribution $P(\phi)$:

$$P(\phi) = (2\pi\sigma_s^2)^{-1/2} \exp(-(\phi_T-\bar{\phi}_T)^2/2\sigma_s^2) \qquad (33)$$

where $\bar{\phi}_T=\beta\bar{\phi}_{dc}$ and $\phi_T=\beta\phi_{dc}$ are the dc barriers measured in units of kT/e, and $\bar{\phi}_{dc}$ is the mean value. The standard deviation σ_s is given in units of kT/e. The distribution is normalized:

$$\int_{-\infty}^{+\infty} P(\phi)\, dU_s = 1 \qquad (34)$$

$$U_s = \phi_T - \bar{\phi}_T \qquad (35)$$

The overall curve of G_{ss}/ω for the total grain boundary area is obtained by multiplying (31) with the probability function (33) and integrating over the deviations (35):

$$G_{ss}/\omega = e^2 N_{ss}/2 \int_{-\infty}^{+\infty} (\omega\tau_m)^{-1} \ln(1+\omega^2\tau_m^2)\, P(\phi)\, dU_s \qquad (36)$$

If we adopt the notation of Simonne [30] by introducing the quantity

$$\Omega = (\omega/\langle v_p\rangle \sigma_p p_g)\, \exp(\bar{\phi}_T) \qquad , \qquad (37)$$

we finally arrive at the following equation:

$$G_{ss}/\omega = e^2 N_{ss}/2 \quad (2\pi\sigma_s^2)^{-1/2}\Omega^{-1}\int_{-\infty}^{+\infty} \exp(-U_s^2/2\sigma_s^2 - U_s)\cdot \ln(1+\Omega^2\exp(2U_s))\, dU_s \qquad (38)$$

We are interested in the density of states N_{ss}, the capture cross section σ_p for majority carriers, and the standard deviation σ_s of the potential distribution. These values have to be determined from the measured curve as shown in Fig.11. Instead of fitting Eq.(38) to the measured curve we can use exact solutions of the integral Eq.(38). Such solutions were published by Simonne [30]. We have extended his solution procedure. It is based on the value at the maximum of the G_{ss}/ω-curve and an additional value at any multiple of the frequency where the curve goes through the maximum. This is described in the appendix. From the measured curve of Fig.11 we find: $N_{ss}=3.9\cdot10^{11}/cm^2eV$, $\sigma_p=1.1\cdot10^{-14}cm^2$, $\sigma_s=1.75kT/e$. These data are determined from the values at the maximum at about 1kHz and the value at 5kHz. For a test of the applicability of the model of potential fluctuations we then integrate Eq.(38) over the whole frequency range. This integration yields curve (c) in Fig.11. The agreement with the measured curve over a wide frequency range demonstrates the internal consistency of the solution procedure.

For the determination of the values for N_{ss}, σ_p, σ_s we use the data of G_{ss}/ω. The data of the capacitance C_{ss}, which can also be calculated from the measured G,C with the help of Eq.(30) are not used. But the frequency dependence of C_{ss} should also be predictable if our model is correct, because the three parameters of the model are already fixed.

For spatially constant potential barriers the capacitance C_{ss} is given by [27]:

$$C_{ss} = \frac{e^2 N_{ss}}{\omega\tau_m} \arctan(\omega\tau_m) \qquad (39)$$

Within the model of potential fluctuations we have to integrate over the potential distribution. This finally leads to an equation similar to Eq.(38):

$$C_{ss} = e^2 N_{ss} (2\pi\sigma_s^2)^{-1/2}\ \omega/\Omega \int_{-\infty}^{+\infty} \exp(-U_s^2/2\sigma_s^2 - U_s)\cdot \arctan(\Omega\exp(U_s))\, dU_s . \qquad (40)$$

We integrate Eq.(40) with the values of N_{ss}, σ_p, σ_p, which are determined from the G_{ss}/ω curve. This integration yields the solid C_{ss} curve in Fig.13. The agreement with the broken C_{ss}-curve, which is calculated from the mea-

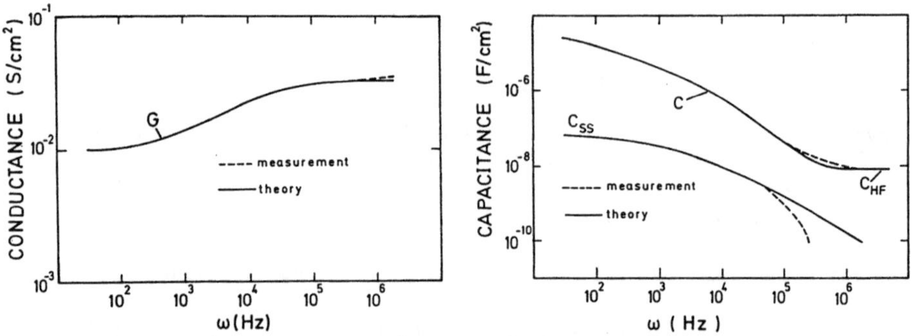

Fig.13. Comparison of measured and recalculated values for the measured conductance G, the measured capacitance C and the capacitance C_{SS}. The curves "theory" arise from the integration of Eqs.(38), (40) and the use of Eqs.(27), (28) after the determination of the values for N_{SS}, σ_p, σ_S from the G_{SS}/ω curve in Fig.11

sured G, C on basis of Eq.(30) demonstrates the validity of the model of potential fluctuations.

The analysis and the determination of N_{SS}, σ_p, σ_S with help of the procedure of Simonne is based on two values of the experimental G_{SS}/ω curve, which is calculated from the measured G, C of Fig.8 with the help of Eq.(29) and is given by the broken line in Fig.11. The integration of Eqs.(38), (40) with the values for N_{SS}, σ_p, σ_S results in the solid curves for G_{SS} and C_{SS} in Figs.11,13. These theoretical data for G_{SS}, C_{SS} are are then taken to predict values for the measured G and C by using Eqs.(27), (28). The comparison of such measured and re-calculated data is also shown in Fig.13. The agreement demonstrates the validity and selfconsistency of the trap transistor/potential fluctuations model.

5.8. Comparison to other methods. Admittance spectroscopy yields the density of states, capture cross section, and the standard deviation of the distribution of potentials. No other method yields more information about majority carrier properties of grain boundaries. The disadvantage of the method at present is its limited energy resolution as shown in Fig.7. We measure N_{SS} around the Fermi level of free holes. If we shift this energy towards the valence band edge, i.e. to lower energies in Fig.7, we find an apparent dropping of the density of states. This putative drop is calculated from a decrease of the low frequency maximum in Fig.11 for $U_{dc}>0.5V$, which goes hand-in-hand with the appearance of a second maximum at about 100kHz. This second maximum most probably is caused by the injection of minority carriers and their interaction with the grain boundary traps [26,31]. Our present model cannot quantitatively account for this effect. Therefore the results of the method are only given for the limited energy range in Fig.7.

Figure 7 compares also the results from photocapacitance to the results of admittance spectroscopy. The energy ranges of the two methods do not overlap. This is because photocapacitance measures only states above E_{F0} whereas admittance spectroscopy gives results at lower energies. The peak near 0.55eV which was formerly attributed to vacancy complexes [22,23], tends towards the results from the conductance method. A careful analysis of occupation statistics [26] elucidates that photocapacitance underestimates the density of states from 0.50 eV to about 0.55eV by at least a factor of two due to the thermal width of the Fermi function. The density of states increases as indicated by the dashed line in Fig.7. The increase is confirmed by the thermal method.

5.9. Correlation with transmission electron microscopy. One of the most interesting features of the conductance method is that it allows for a correlation of electronic and structural grain boundary properties. The spatial fluctuations of the dc-barriers at the grain boundary are caused by a spatial distribution of the charges. A reasonable agreement of average potential barriers, as derived from measurements of current and capacitance, indicates that no gross nonuniformities as charge clusters occur at the interface. We can then assume that the distances of charges follow a Poisson distribution.

Brews [32] discussed the influence of such a Poisson distribution of point charges on the potential fluctuations at a MOS-interface. He showed how to derive a minimum charge distance λ_{min} from a measured standard deviation σ_s of the Gaussian distribution of potentials. With some modifications, his results are transferable to grain boundaries. We get a minimum charge distance of $\lambda_{min}=(7\pm2)$nm from such a calculation. From the total charge of the grain boundary we get moreover an average charge distance of $\lambda=(27\pm1)$nm. These two values are compared to the crystallographic features of the grain boundary [24].

The analysis in the transmission electron microscope reveals a periodical contrast structure with a repetition distance of 1.8nm. This contrast is caused by a basic dislocation network, which can be imagined to be arranged in the interface similar to a periodic grid with a width of mesh of 1.8nm. In addition to this basic dislocation network secondary dislocations are also found, which can be imagined as faults in the "grid". Figure 14 shows a histogram of the distances between these secondary dislocations. It can be seen that they are statistically distributed as it is assumed for the charges. The minimum and average distance of charges fit well into the distribution.

The correlation of the electronic and structural properties of the grain boundary leads to the following conclusions: The charges and the statistical

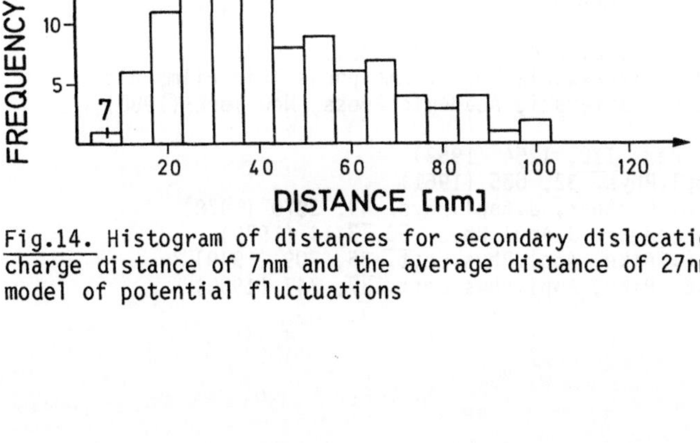

Fig.14. Histogram of distances for secondary dislocations. The minimum charge distance of 7nm and the average distance of 27nm arise from the model of potential fluctuations

potential fluctuations cannot be caused by the periodic arrangement of the basic dislocation network. On the contrary, the secondary dislocations are compatible with the model of potential fluctuations. They are, as the charges, irregularly distributed at the interface and the characteristic distances of charges and secondary dislocations show reasonable agreement. This indicates that the charges are correlated with secondary dislocations. The electrical activity of the regular basic dislocation network is of inferior significance [1].

6. Discussion and conclusions. Five different methods for the determination of the density of states at grain boundaries in polycrystalline semiconductors have been presented and discussed in this contribution. The doping method and the optical method reveal the dominance of band tails. Both methods yield information about a wide energy range. The energy resolution of the thermal method is poor, but on the other hand it is very simple. The I/U method yields a density of states which drops rapidly towards the majority carrier band. This drop most probably is an apparent one [7,21,33] because the evaluation is carried out up to high voltages, where the influence of thermionic field emission and the injection of minority carriers seems not to be neglectable.

Information about the density of states, the capture cross section for majority carriers, and about the potential fluctuations at the interface is obtained by the conductance method. It is based on measurements of the ac admittance and is the only method, which resolves inhomogeneities at the interface. The ac measurements have been analyzed using the trap transistor model, which describes the control of the thermionically emitted current by the capture/emission current at the grain boundary. It can be shown [26] that this model describes not only ac measurements, but also dc measurements. The I/U-method of Pike and Seager [6,7] can be shown to be a special case of this model, for example. The analysis of the ac measurements is carried out with the help of the methods of admittance spectroscopy, which are well-known from the MOS-interface [27,30]. On the basis of such ac measurements information about the origin of the charge can be obtained. A correlation of the results of the electrical measurements with an analysis in the transmission electron microscope has been carried out. This comparison has shown that the charge at a special Si grain boundary is correlated with secondary dislocations. Ac measurements as discussed in this contribution are even more powerful: Apart from localized states at the grain boundary, traps within the space charge region can also be analyzed [34]. This is achieved on the basis of measurements, which show a strong frequency dispersion at zero-bias.

Acknowledgments. The author is grateful to Prof. H.J. Queisser for his continuous support and encouragement during the course of the work. Many thanks are due to Prof. H. Strunk for many discussions and the TEM work as well as A. Vierhaus for typing the manuscript.

1. H.J. Queisser, Mat.Res.Soc.Proc. 14, 323 (1983)
2. L.L. Kazmerski, in "Polycrystalline and Amorphous Thin Films and Devices", ed. by L.L. Kazmerski, Academic Press, New York (1980), p.59
3. R.L. Petritz, Phys.Rev. 110, 1254 (1958)
4. R.K. Mueller, J.Appl.Phys. 32, 635 (1961)
5. C.H. Seager and T.G. Castner, J.Appl.Phys. 49, 3879 (1978)
6. G.E. Pike and C.H. Seager, J.Appl.Phys. 50, 3414 (1979)
7. C.H. Seager and G.E. Pike, Appl.Phys.Lett. 35, 709 (1979)
8. C.H. Seager and G.E. Pike, Appl.Phys.Lett. 37, 747 (1980)

9. H.C. de Graaff, M. Huybers, and J.G. de Groot, Solid State Electron. 25, 67 (1982)
10. M. Peisl and A.W. Wieder, IEEE Trans. Electron Devices ED-30, 1792 (1983)
11. M. Peisl, to be published in J. de Physique
12. J.J.J. Yang, P.D. Dapkus, R.D. Dupuis, and R.D. Yingling, J.Appl.Phys. 51, 3794 (1980)
13. S. Hirae, M. Hirose, and Y. Osaka, J.Appl.Phys. 51, 1043 (1980)
14. E.O. Kane, Phys.Rev. 131, 79 (1963)
15. B.I. Halperin and M. Lax, Phys.Rev. 148, 722 (1966)
16. M.H. Cohen, H. Fritzsche, and S.R. Ovshinsky, Phys.Rev.Lett. 22, 1065 (1969)
17. W.E. Taylor, N.H. Odell, and H.Y. Fan, Phys.Rev. 88, 867 (1952)
18. E.I. Gol'dman and A.G. Zhdan, Fiz.Tekh.Poluprovodn. 10, 1839 (1976) [Sov.Phys.Semicond. 10, 1098 (1976)]
19. E.I. Gol'dman, A.G. Zhdan, and V.N. Nemenushchiĭ, Fiz.Tekh.Poluprovodn. 12, 833 (1978) [Sov.Phys.Semicond. 12, 491 (1978)]
20. G.E. Pike, P.L. Gourley, and S.R. Kurtz, Appl.Phys.Lett. 43,939 (1983)
21. A.W. de Groot, G.C. McGonigal, D.J. Thomson, and H.C. Card, J.Appl.Phys. 55, 312 (1984)
22. J. Werner, W. Jantsch, and H.J. Queisser, Sol.St.Comm. 42, 415 (1982)
23. J. Werner, W. Jantsch, K.H. Fröhner, and H.J. Queisser, Mat.Res.Soc.Proc. 5, 99 (1982)
24. J. Werner, and H. Strunk, J. de Physique C1, 99 (1982)
25. S.M. Sze, Physics of Semiconductor Devices, John Wiley, New York 1981, p.94
26. J. Werner, Thesis, University of Stuttgart 1983
27. E.H. Nicollian and A. Goetzberger, Bell Syst.Techn.J. 46, 1055 (1967)
28. D.J. Thomson and H.C. Card, J. Appl.Phys. 54, 1976 (1983)
29. G.D. Mahan, J.Appl.Phys. 55, 980 (1984)
30. J.J. Simonne, Solid State Electron. 16, 121 (1973)
31. J.A. Cooper, and J.R. Schwartz, Solid State Electron. 17, 641 (1974)
32. J.R. Brews, J.Appl.Phys. 43, 2306 (1972)
33. G.C. McGonigal, D.J. Thomson, J.G. Shaw, and H.C. Card, Phys.Rev. B28, 5908 (1983)
34. F. Greuter and G. Blatter, 1st workshop on "Polycrystalline Semiconductors", Stuttgart, February 22/23, 1984 and to be published.

Appendix

General Solutions of the Potential Fluctuations Model. The evaluation is based on the value of the G_{ss}/ω-curve, Eq.(38), at its maximum and an additional value at a frequency $\omega = n\omega_{max}$. Here $n>0$ and ω_{max} is the frequency of the maximum of the curve. This procedure was first described by Simonne [30].

A one-to-one relation between σ_s and Ω_{max} can be derived from the condition that G_{ss}/ω goes through a maximum. Here Ω_{max} is the Ω of Eq.(37) at the frequency ω_{max}. This first relation is given by the lower left quadrant of Fig.15. With the known relation $\Omega_{max}(\sigma_s)$ we calculate the ratio:

$$R(n) = \frac{(G_{ss}/\omega)_{max}}{G_{ss}/n\omega_{max}} \qquad (A1)$$

This ratio R(n) gives a measure for the width of the G_{ss}/ω-curve and therefore for σ_s, too. The relation between R(n) and σ_s is shown in the lower right quadrant of Fig.15 for different values of n.

Fig.15. Determination of the density of states N_{SS}, the capture cross section σ_p for majority carriers, and the standard deviation σ_S from the ratio

$$R(n) = \frac{(G_{ss}/\omega)_{max}}{G_{ss}/n\omega_{max}}$$

for some values of n.

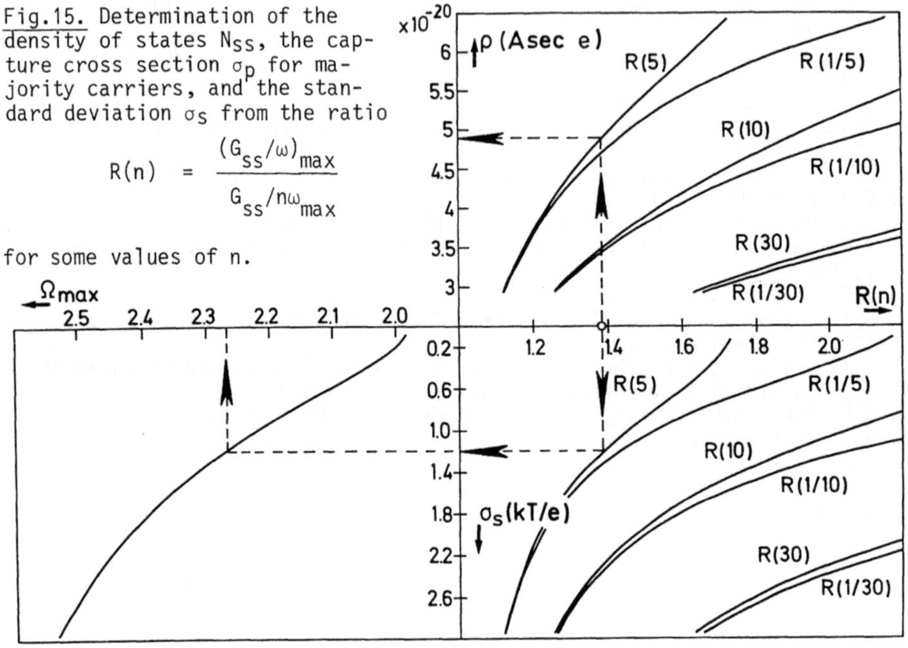

Finally, the upper right quadrant of the figure gives the relationship between R(n) and ρ, where ρ gives the maximum of the G_{ss}/ω-curve, when normalized to the density of states:

$$\rho = \frac{(G_{ss}/\omega)_{max}}{N_{ss}} \tag{A2}$$

Determination of N_{ss}, σ_p, σ_s. The evaluation is based on R(n) which is taken from the measured curve. In Fig.15 we have assumed that n=5 was taken and R(5)=1.39 was found. The evaluation is then carried out in three steps.

1) From the relation between R(5) and σ_s a value σ_s=1.2 kT/e is found for the standard deviation of the potentials.

2) From the $\Omega_{max}(\sigma_s)$-curve we find Ω_{max}=2.27. This value yields the capture cross section σ_p by:

$$\sigma_p = (\omega_{max}/\langle v_p\rangle \Omega_{max} p_g) \exp(\bar{\Phi}_T) \tag{A3}$$

3) The upper right quadrant of the figure yields $\rho=4.89\cdot10^{-20}$Asec e. The density of states is then obtained by:

$$N_{ss} = \frac{(G_{ss}/\omega)_{max}}{\rho} \tag{A4}$$

Here G_{ss}/ω is given in units of F/cm² and N_{ss} in 1/eVcm².

The procedure as described here for n=5 can be carried out with any multiple n of the frequency where the G_{ss}/ω-curve goes through a maximum. In Fig.15 values for n=1/30, 1/10, 1/5, 5, 10, 30 are given. Exact values in tabular form can be obtained from the author.

Electronic States at Grain Boundaries in Semiconductors

A. Broniatowski

Groupe de Physique des Solides de l'Ecole Normale Supérieure
Université Paris VII, Tour 23, 2 place Jussieu, F-75251 Paris Cedex 05, France

1. Introduction

It has long been suspected that dangling bond states were involved in the barrier effects observed at grain boundaries in covalent semiconductors. The recent discovery by electron spin resonance experiments, that this is indeed the case for polycrystalline silicon [1,2], raises a renewed interest in the electrical measurement of the electron states at grain boundaries. Among the various measurement methods, transient capacitance spectroscopy is potentially one of the most powerful, and its application to the grain boundary states in bicrystals forms the major part of this chapter.

Section 2 presents a brief review of recent data on the grain boundary states in various semiconductors, with an emphasis on the case of silicon. Section 3 recalls the basic features of the electrostatic potential barrier model for charged grain boundaries. Section 4 is concerned with transient capacitance measurements of the grain boundary states. After recalling the principle of Deep Level Transient Spectroscopy (Section 4.2), we shall discuss some basic experiments on bicrystals, starting with the question of distinguishing the grain boundary states from carrier traps located in the vicinity of the boundary (Section 4.3). Some implications for the detection of impurity segregation at grain boundaries will be discussed in Section 4.4. A procedure to describe the density of the grain boundary states will be presented in Section 4.5. Lastly, the kinetics of carrier capture at grain boundaries with an application to capture cross-section measurements will be discussed in Section 4.6.

2. Electron States at Grain Boundaries

2.1 Introduction

As discussed in the previous chapters, the creation of a grain boundary generally involves local lattice distortions in the vicinity of the crystal interface. These local distortions result in bond stretching, with the eventual formation of unsaturated dangling bonds. Impurities may also precipitate at the grain interface, thus modifying the electronic properties of the boundary : such effects are indeed demonstrated by the effectiveness of hydrogen passivation of the boundary traps in polycrystalline silicon.

2.2 Dangling Bonds at Grain Boundaries in Silicon

A significant advance has been made recently, with the discovery of paramagnetic centres associated with dangling bonds at grain boundaries in polycrystalline silicon by electron spin resonance experiments [1,2]. In small grain polycrystalline films (average grain size of ~ 80 nm), the dangling bond density is estimated to approximately 10^{12} per cm^2 of grain boundary. The electron spin centres can be removed by atomic hydrogen passivation through the formation of Si-H bonds [34], the fraction of removed centres being in direct relation to the amount of hydrogen diffused within the silicon film [1]. Further experiments performed on poly- and bicrystalline samples have shown that these spin centres could behave as majority carrier traps [2], thus providing the first conclusive evidence that dangling bonds are involved in the barrier effects observed at grain boundaries in silicon.

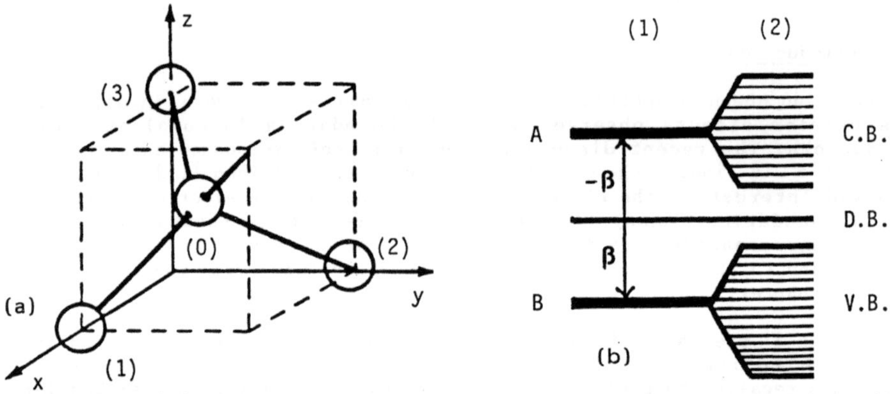

Figure 1 - (a) Atomic configuration at a dangling bond site. The atom (0) has a three-fold coordination, while the three back atoms (1), (2) and (3) are tetra-coordinated. (b) Energy levels in the molecular model (1) and when broadening effects are taken into account (2). A : antibonding state; B : bonding state; D.B. : dangling bond state; V.B. : valence band; C.B. : conduction band.

The existence of electron states localized at dangling bond sites can be explained in simple terms, using a tight-binding approximation [35]. Consider Figure 1a, representing the atomic configuration at a dangling bond site. The atom (0) has a three-fold coordination, while its three nearest neighbours (1), (2) and (3) are tetra-coordinated. Let ϕ_{ij} be the sp^3 hybridized orbital, centered at the atom i and pointing towards j. The dangling orbital is denoted by ϕ_o. Within the tight binding approximation, all the overlap integrals are taken equal to 0, except $\langle \phi_{ij} | \phi_{ij} \rangle = 1$. Let H be the one-electron Hamiltonian of the system. The matrix elements of H may be classified in decreasing order of magnitude, the leading terms being $\langle \phi_{ij} | H | \phi_{ij} \rangle = \alpha$ and $\langle \phi_{ij} | H | \phi_{ji} \rangle = \beta$. As a zero-order approximation (known as the molecular model), we shall neglect all the other matrix elements of H. With these simplifying assumptions, the electron states fall into three different classes, the bonding (B), antibonding (A) and dangling bond (DB) states, with the respective energies (Fig. 1b) :

$$E_B = \alpha - \beta$$
$$E_A = \alpha + \beta \qquad (1)$$
$$E_{DB} = \alpha$$

and wave functions :

$$\psi_B = (\phi_{ij} + \phi_{ji})/\sqrt{2}$$
$$\psi_A = (\phi_{ij} - \phi_{ji})/\sqrt{2} \qquad (2)$$
$$\psi_{DB} = \phi_o$$

Including the intra-atomic matrix elements $\Delta = \langle \phi_{ij}|H|\phi_{ij'} \rangle$ leads to a broadening of the bonding and antibonding states into the valence and the conduction band, respectively, while leaving the dangling bond state unaffected. Only the inclusion of second nearest neighbour matrix elements leads to a shift in the dangling bond energy [35]. According to this simple picture, the dangling bond site might exist in three different charge states, depending upon the relative positions of the Fermi level E_F and the neutral dangling bond energy E_{DB} : + 1 (elementary charge) for $E_F < E_{DB}$, 0 for $E_F = E_{DB}$ and - 1 for $E_F > E_{DB}$, respectively. Thus, the dangling bond site is expected to have an amphoteric character, with both donor and acceptor properties.

2.3 Comparison with Experimental Data from Electrical Measurements on Bicrystals

The data available on bicrystals relate to widely different boundary orientations and impurity contents, and therefore, only major trends will be stated with an emphasis on the case of silicon. Barrier effects have been detected in both n- and p-type, Czochralski grown silicon bicrystals [36] showing the existence of donor and acceptor-like states at the interface. The magnitude of the barrier effects appears strongly dependent upon annealing treatments, at temperatures where the diffusion and precipitation of impurities (especially at the grain boundary) cannot be ignored [37,38]. There is still, therefore, a measure of doubt concerning the intrinsic character of the boundary states detected in these experiments.

Admittance spectroscopy measurements indicate a maximum in the density of states about midgap [39], while photo-capacitance and photoconductance experiments reveal an increase of the interface states density towards the edge of the conduction band [40]. By analogy with the case of amorphous semiconductors, this band tailing is attributed to the loss of the long-range order in the boundary region.

Transient capacitance measurements have been performed on silicon, germanium and GaAs bicrystals. In the case of silicon, they suggest a broad spread of the boundary states across the energy gap, rather than well-defined levels [41]. On the other hand, in germanium [22] and GaAs [21] bicrystals, discrete energy levels were detected, although no definite conclusion could be drawn in terms of the dangling bond model yet (see section 4.5 for a discussion in the case of a low angle tilt boundary in germanium).

3 The Electrostatic Potential Barrier Model for Grain Boundaries in Semiconductors

3.1 Introduction

The purpose of this section is to describe the electrostatic potential barrier model, proposed by TAYLOR, ODELL and FAN [4] to explain the non-ohmic character of the electrical conduction in polycrystalline semiconductors. Consider Fig. 2a, showing the bending of the energy bands in the vicinity of an electrically charged boundary in n-type material. (The effect of the boundary in p-type material is described similarly.) As the interface states are filled with electrons, a depletion layer forms on both sides of the boundary, resulting in an electrostatic potential barrier. At equilibrium, the electron capture and emission rates adjust, so that the electrons localized at the interface form a Fermi distribution. As a voltage bias is applied across the boundary (Fig. 2b), the capture rate increases, resulting in a change in the occupancy of the interface states. For large values of the applied bias, the barrier breaks. Breakdown voltages as large as a hundred volts have been recorded [4].

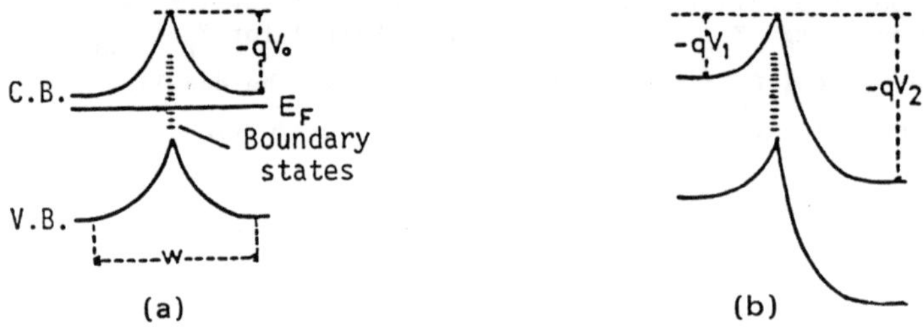

Figure 2 - (a) Equilibrium band structure of a grain boundary in n-type material, showing the barrier height V_o and the barrier width w; (b) band scheme under applied bias $V = V_1 - V_2$.

Section 3.2 discusses the mechanisms of electrical conduction across the boundary. Two different models apply, depending on the relative values of the barrier width w and the carrier mean free path $\bar{l}$: the diffusion model for $\bar{l} \ll w$ and the thermoelectronic emission model for $\bar{l} \gtrsim w$. In both cases, the current density is found to be thermally activated, with an activation energy depending upon the value of the barrier potential. An extra contribution to the current due to tunnelling across the boundary must also be considered in highly doped materials [5,6]. We shall restrict ourselves, however, to the simpler case of low or medium range doping ($\lesssim 10^{17}$ cm^{-3}) where this effect can be neglected.

Section 3.3 discusses the processes of electron capture and emission at the interface. It will be shown that, under steady state conditions, the boundary electrons form a system in quasi-equilibrium with the conduction electrons on the negatively biased side of the boundary. This property is used to define the quasi-Fermi level of the boundary electrons, the latter being used, in turn, to derive the grain boundary density of states from the current-voltage characteristics. Lastly, in section 3.4, we shall consider the response of the grain boundary to a small, alternative applied voltage, from which the a.c. impedance of the boundary is shown to consist of a capacitance in parallel with a leakage resistance.

The barrier model presented in this section assumes the semiconductor material to be homogeneously doped. This amounts to neglecting any possible effect of impurity segregation at the boundary. Some implications of impurity segregation on the electrical properties of the barrier will be discussed in section 4.4.

3.2 Electrical Conduction across the Boundary [4,7]

3.2.1 *The Diffusion Theory*

If the barrier width w is much larger than the mean free path $\bar{l}$ of the carriers, then the diffusion theory applies. The current density j across the boundary is given by :

$$j = \sigma(- dV/dx) + qD(dn/dx) \tag{3}$$

where V represents the electrostatic potential and n the electron density at the distance x from the interface. $\sigma = nq\mu$ is the local conductivity and $D = kT\mu/q$ is the carrier diffusion coefficient, μ being the electron mobility and q the (positive) elementary charge. After integration (see for instance [8]), the following expression is obtained for j :

$$j = qN_d\mu \frac{E_1 E_2}{E_2 - E_1} \left(\exp (qV_2/kT) - \exp (qV_1/kT)\right) \tag{4}$$

where N_d represents the free carrier density in the bulk of the grains, V_1 and V_2 being the barrier heights on the negatively ($x < 0$) and positively ($x > 0$) biased sides of the boundary (Fig. 3b), and E_1 and E_2 the electric field strengths for $x = 0^-$ and $x = 0^+$, respectively.

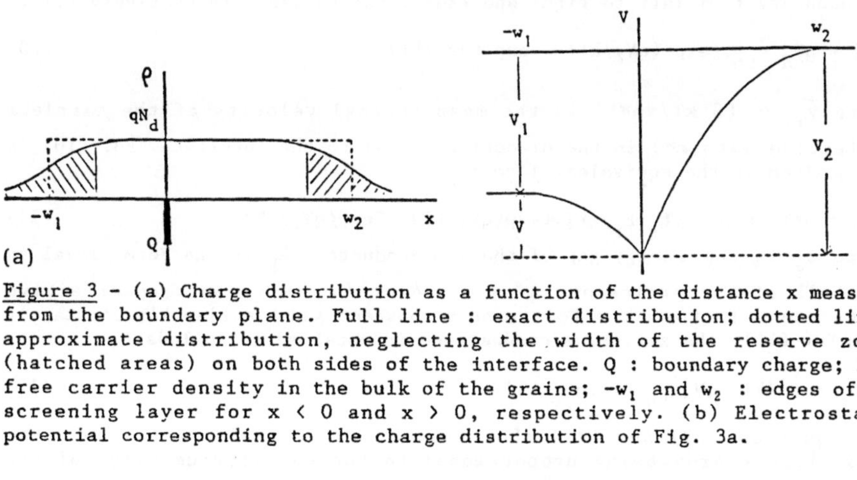

Figure 3 - (a) Charge distribution as a function of the distance x measured from the boundary plane. Full line : exact distribution; dotted line : approximate distribution, neglecting the width of the reserve zones (hatched areas) on both sides of the interface. Q : boundary charge; N_d : free carrier density in the bulk of the grains; $-w_1$ and w_2 : edges of the screening layer for $x < 0$ and $x > 0$, respectively. (b) Electrostatic potential corresponding to the charge distribution of Fig. 3a.

V_1, V_2, E_1 and E_2 are obtained by integrating Poisson's equation for the charge distribution represented by Fig. 3a. Neglecting the reserve zones (hatched areas in Fig. 3a), the width of which is on the order of the Debye screening length $\lambda_D = (\varepsilon kT/q^2 N_d)^{\frac{1}{2}}$, the following approximate expressions are found :

$$V_1 = - \frac{\varepsilon}{2qN_d} \left(\frac{Q}{2\varepsilon} - \frac{qN_d V}{Q}\right)^2 \tag{5}$$

$$V_2 = - \frac{\varepsilon}{2qN_d} \left(\frac{Q}{2\varepsilon} + \frac{qN_d V}{Q}\right)^2 \tag{6}$$

$$E_1 = -\left(\frac{Q}{2\varepsilon} + \frac{qN_dV}{Q}\right) \quad (7)$$

$$E_2 = \frac{Q}{2\varepsilon} - \frac{qN_dV}{Q} \quad (8)$$

where Q represents the boundary charge per unit area and ε the (static) dielectric constant. These expressions will be used in section 3.3.3 to derive the grain boundary density of states from the shape of the current-voltage characteristics of bicrystals. The zero-bias conductance G_o is deduced from Eq. (4) as follows :

$$G_o = (q/kT).qN_d\mu\ (|E_o|/2)\ \exp\ (qV_o/kT) \quad (9)$$

where $|E_o| = (2\ qN_d|V_o|/\varepsilon)^{\frac{1}{2}}$ is the field strength for $x = 0$ and V_o is the equilibrium barrier height.

3.2.2 *The Thermoelectronic Emission Theory*

The thermoelectronic emission theory applies for $\bar{l} \gtrsim w$. According to this model, only those among the carriers which possess a kinetic energy larger than the barrier height can move across the boundary. The net current is proportional to the difference of the electron fluxes, crossing the boundary from left to right and from right to left, respectively [7] :

$$j = qN_d\ \bar{v}_{th}.(\exp\ (qV_2/kT) - \exp\ (qV_1/kT)) \quad (10)$$

where $\bar{v}_{th} = (8\ kT/\pi m^*)^{\frac{1}{2}}$ is the mean thermal velocity of the carriers (effective mass m^*) in the direction normal to the interface. Eq. (10) is often given in the equivalent form :

$$j = A^*T^2\ \exp - (E_g-E_F)/kT.(\exp(qV_2/kT) - \exp(qV_1/kT)) \quad (11)$$

where E_g is the energy gap of the semiconductor, E_F is the Fermi level in the bulk of the grains and $A^* = 4\ \pi m^*\ qk^2/h^3$ represents an effective Richardson factor (k = Boltzmann's constant and h = Planck's constant). From Eq. (11), the zero-bias conductance is obtained as follows :

$$G_o = (q/kT).qN_d\bar{v}_{th}.\exp(qV_o/kT) \quad (12)$$

The equations (9) and (12) show that G_o is thermally activated, the activation energy being proportional to the equilibrium value of the barrier height.

3.3 Emission and Capture Rates, and Quasi-Fermi Level of the Electrons Localized at the Interface

3.3.1 *Equilibrium Values of the Capture and the Emission Rates*

Consider the interface state, characterized by the binding energy E (measured from the top of the valence band) and the electron capture cross-section σ_T (Fig. 4). Let n_q be the density of the conduction electrons at the boundary plane. At equilibrium, the capture (c_n) and emission

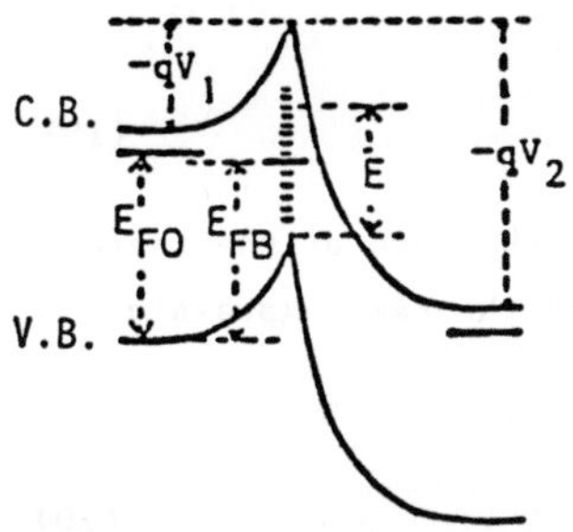

Figure 4 - Band scheme of the boundary under applied bias, showing the quasi-Fermi level of the boundary electrons E_{FB} and the Fermi level E_{FO} on the negatively biased side of the interface. Energy references : E_{FO} and E_{FB} are measured from the top of the valence band within the negatively biased grain; the boundary level E is measured from the top of the valence band at the interface.

(e_n) rates adjust so that the fractional occupancy f_B^o of the level E obeys the Fermi statistics. Thus :

$$c_n \, (1 - f_B^o) = e_n \, f_B^o \tag{13}$$

where $f_B^o = [1 + \exp{(E - qV_o - E_F)/kT}]^{-1}$ (14)

The capture rate is given by :

$$c_n = 2 \, n_o \, \sigma_T \, \bar{v}_{th} \tag{15}$$

(the factor 2 accounts for the capture processes for electrons coming from both sides of the interface), n_o being given by the expression :

$$n_o = N_c \exp - (E_g - E_F - qV_o)/kT \tag{16}$$

where N_c represents the effective number of states in the conduction band. Using the last four equations, one finds for the emission rate from the boundary state the well-known expression [9] :

$$e_n = 2 \, N_c \, \sigma_T \, \bar{v}_{th} . \exp - (E_g - E)/kT \tag{17}$$

3.3.2 *Grain Boundary under bias : the Steady State case and the quasi-Fermi Level of the Boundary Electrons*

As a voltage bias is applied across the boundary, the capture rate is increased and so is the magnitude of the boundary charge. Under steady state conditions, the occupancy f_B of the interface state (E) is still given by the expression :

$$c_n \, (1 - f_B) = e_n \, f_B \tag{18}$$

where c_n differs, however, from Eq. (15). By analogy with the equilibrium case, we shall define the quasi-Fermi level E_{FB} of the boundary electrons by the expression :

$$f_B = [1 + \exp (E - qV_1 - E_{FB})/kT]^{-1} \tag{19}$$

(fig. 4). In order to obtain E_{FB}, we shall treat the case of the thermo-electronic emission theory (section 3.2.2). Almost the same expression is obtained from the diffusion model [7].

The capture rate is given by :

$$c_n = \sigma_T N_d \bar{v}_{th} [\exp (qV_1/kT) + \exp (qV_2/kT)] \tag{20}$$

the two terms on the right hand side giving the capture probabilities, per unit time, for electrons coming from the left and the right hand side of the boundary respectively. From Eq. (17) to (20), one finds for E_{FB} the expression :

$$E_{FB} = E_{FO} + kT \text{ Log } \frac{1}{2}\left(1 + \exp^{-\frac{q|V|}{kT}}\right) \tag{21}$$

where $E_{FO} = E_g - kT \text{ Log}(N_c/N_d)$ represents the position of the Fermi level on the negatively biased side of the boundary. Eq. (21) shows that, whatever the amplitude of the applied bias, E_{FB} and E_{FO} remain practically coincident. Thus, as long as a steady state is maintained, the electrons localized at the interface remain in quasi-equilibrium with the electrons on the negatively biased side of the boundary.

3.3.3 *Determination of the Grain Boundary Density of States from the Current-Voltage Characteristics of Bicrystals*

Eq. (21) forms the basis for a practical procedure to derive the grain boundary density of states from the current-voltage characteristics of bicrystals [10]. Let $N_B(E)$ be the density of the boundary states. If one assumes $N_B(E)$ to form a continuum across the energy gap, then the quasi-Fermi level may be identified with the occupation limit E_1 of the boundary states (Fig. 4). It follows from Eq. (21) that :

$$E_1 \sim E_{FO} + qV_1 \tag{22}$$

If the applied bias is increased by dV, then the boundary charge varies by dQ and the occupation limit is shifted by $dE_1 = qdV_1$. Thus, the interface density of states is given by :

$$N_B(E_1) \sim \frac{1}{q^2}\left|\frac{dQ}{dV}\cdot\frac{dV}{dV_1}\right| \text{ (in } J^{-1} \text{ m}^{-2}) \tag{23}$$

The procedure proposed by SEAGER and PIKE [10] consists in determining V_1 and dV_1/dV from the current-voltage characteristics of the bicrystal using Eq. (10). The boundary charge and its derivative dQ/dV are obtained next by means of Eqs. (5) and (6). Figure 5 shows an example of grain boundary density of states in silicon, obtained by this technique and reproduced from ref. [10].

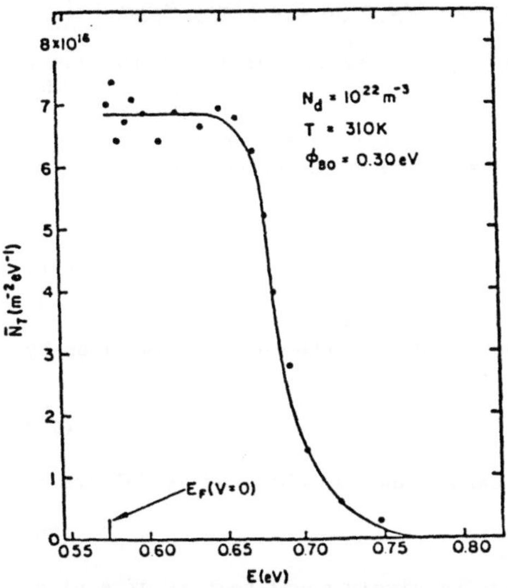

Figure 5 - Density of states for a grain boundary in silicon, computed from the current-voltage characteristics (after [10]).

3.4 Frequency Dependence of the Boundary Impedance

Only a brief outline of the frequency effects will be given, as the subject is treated in more detail in the lectures by J. WERNER.

3.4.1 *Frequency Ranges*

A.c. measurements performed on bicrystals have shown a strong dependence of the grain boundary impedance upon the frequency of the measuring signal : thus, the boundary capacitance may be increased by more than one order of magnitude as the measuring frequency f is decreased from the megahertz down to the kilohertz range [11,12]. A discussion of these effects involves, as major parameters, the characteristic frequencies for carrier capture (c_n) and emission (e_n) at the occupation limit of the interface states. Qualitatively, three different frequency ranges may be distinguished :

a) a high frequency range ($f \gg (e_n, c_n)$) where the boundary charge remains constant during one cycle of the modulation signal;

b) a low frequency range ($f \ll (c_n, e_n)$) where the boundary charge varies in phase with the a.c. applied voltage;

c) an intermediate range ($f \sim (c_n, e_n)$) where the boundary charge exhibits a time dependence at the frequency f, although not in phase with the modulation signal.

For a measuring signal of a given frequency, the transition from one type of behaviour to another depends upon both the temperature (due to the temperature variations of e_n and c_n, Eqs. (17) and (20)), and the magnitude of the d.c. applied bias (due to the voltage dependence of the occupation limit of the interface states, section 3.3.3). The grain boundary density of states can be obtained from a detailed discussion of these frequency effects, using the technique of admittance spectroscopy [12]. In the following section we shall restrict ourselves to the simple case where the boundary is submitted to a small amplitude, high frequency voltage superimposed on a constant d.c. bias.

3.4.2 *The Grain Boundary Capacitance in the high Frequency Range* [4]

The current density across the boundary is the sum of two components, namely the conduction current j_c (given by Eq. (4) or (10)) and the displacement current $j_D = \partial D/\partial t$, where D is the electric displacement. Since the boundary charge is assumed to remain constant through one cycle of the modulation signal, the displacement current is deduced from Eq. (7) or (8) as follows :

$$j_D = - \frac{q\varepsilon N_d}{Q} \frac{\partial V}{\partial t} \tag{24}$$

The width of the depleted layer on both sides of the boundary is given by the condition of electrical neutrality :

$$w = - Q/qN_d \tag{25}$$

From Eqs. (24) and (25), the boundary capacitance is obtained as follows :

$$C = \varepsilon/w = - q\varepsilon N_d/Q \tag{26}$$

Thus, the high-frequency capacitance is simply equal to that of a plane capacitor, whose plates are set at both edges of the depletion layer. Eq. (26) also shows that C can be used to monitor the variations of the boundary charge. Advantage will be taken of this fact to study the transient electrical properties of the boundary in the next section.

4. Transient Electrical Properties of the Grain Boundary Levels

4.1 Introduction

In this section we shall be concerned with the transient electrical effects associated with carrier capture and emission at the interface levels. Deep Level Transient Spectroscopy (D.L.T.S.) has emerged as a major tool to study these effects, and its principle will be recalled briefly in section 4.2. In studies of polycrystalline materials, the question arises to distinguish the grain boundary states from point defect or impurity levels located within the space charge region of the boundaries. Means to answer this question will be discussed in section 4.3, in the simple case of a single boundary in a bicrystal. Some implications for the study of impurity segregation at grain boundaries will be considered in section 4.4. Section 4.5 presents a procedure to describe the grain boundary density of states. Lastly, some basic aspects of the kinetics of carrier capture at grain boundaries will be discussed in section 4.6.

4.2 Deep Level Transient Spectroscopy [13]

The prime purpose of D.L.T.S. is to measure the activation energies for carrier emission from deep levels in semiconductors. For this purpose, a p-n junction or a Schottky barrier is built on the surface of the material (Figs. 6a and b). The traps within the space-charge region of the barrier are filled periodically by means of voltage pulses applied across the junction, and the carrier emission is detected by the concomitant variation of the junction capacitance. The capacitance transients are processed by means of a dual-gate correlator (boxcar averager or lock-in amplifier) and the resulting signal $C(t_1) - C(t_2)$ (Fig. 6c) is recorded as a function of

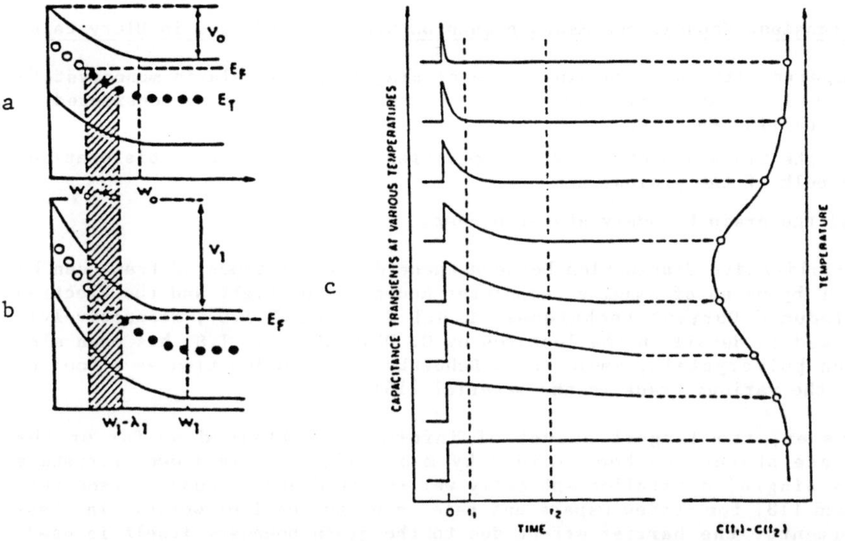

Figure 6 - Band scheme of a Schottky barrier during the trap filling pulse (a) and during the emission transient (b). The full traps are represented as closed circles (•) and the empty traps as open circles (o). V_o and V_1 are the barrier heights during the phases (a) and (b) respectively. w_o and w_1 are the corresponding barrier widths. The hatched area represents the barrier region from which carrier emission is detected (after [3]). (c) : Formation of the D.L.T.S. signal from the capacitance transient. Left, from bottom to top : shape of the emission transients for increasing temperatures, also showing the gate setting t_1 and t_2 of the correlator. Right : variation of the D.L.T.S. signal as a function of temperature (after [13]).

the sample temperature. Consider the contribution to this signal, due to an electron trap characterized by the binding energy E and the capture cross-section σ_T. The emission rate e_n varies with temperature in accordance with Eq. (17). Moreover, it can be shown that the D.L.T.S. signal $C(t_1) - C(t_2)$ goes through a maximum when e_n takes the value $[(t_2 - t_1)/\mathrm{Log}(t_2/t_1)]^{-1}$ [13] determined by the gate settings of the correlator. Thus, every single type of trap contributes a well defined peak to the spectrum. Furthermore, the contributions due to traps with different activation energies are separated in the temperature scale, as they give rise to various maxima in the spectrum.

The temperature dependence of the emission rate is obtained by measuring the temperature shift of the maximum of the D.L.T.S. signal, for various gate settings (t_1, t_2). The activation energy is deduced from an Arrhenius plot of $\mathrm{Log}(T^2/e_n)$ vs. 1/T, named the signature of the trap. (The factor T^2 accounts for the temperature dependences of N_c and $\bar{v}_{th}$ in Eq. (17)). This energy is a characteristic parameter for every single trap level. However, it may differ from the energy $E_c - E$ in Eq. (17), in particular when capture cross-section is temperature dependent. For further details, the reader is referred to ref. [13] and the review papers [14] and [15].

4.3 Transient Capacitance Measurements on Single Boundaries in Bicrystals

As compared with the more usual case of single trap levels in monocrystalline samples, two different types of carrier traps must be considered in polycrystalline materials :

(i) the traps associated with impurities, point defects or dislocations in the bulk of the grains, and

(ii) the grain boundary states proper.

A qualitative distinction between these different types of traps can be achieved by means of imaging techniques known as the Light and the Electron Beam Induced Current techniques (L.B.I.C. and E.B.I.C., respectively), discussed in detail in the lectures by C. DONOLATO. D.L.T.S. has been also used on polycrystals, employing a Schottky or a p-n junction as a tool to detect the various traps in the material [16].

However, the largest amount of information obtained so far on the interface states has been gained by means of specially grown bicrystals with a single, crystallographically well-defined grain boundary (see refs [17] and [18] for review papers and references to earlier works). In these experiments, the barrier effect due to the grain boundary itself is used, rather than that of a Schottky junction as described in the previous section. Some basic experiments on bicrystals will be described next, starting with the problem of differentiating the boundary states from the bulk traps located in the vicinity of the interface.

4.3.1. *Detection of Trap Segregation at Grain Boundaries using Deep Level Transient Spectroscopy* [19].

The purpose of this experiment is to detect the carrier traps located in the bulk regions adjacent to the grain boundary : such traps may be associated with point defects or impurities segregated during the growth of the crystal or subsequent heat treatments.

Consider Figs. 7a and b, showing the bending of the energy bands at a grain boundary in n-type material. In addition to the interface levels, traps have been also represented within the space charge region of the boundary. Fig. 7c shows the voltage cycle applied across the boundary : the charge carriers are captured (from time 0 to t_p) and released (from t_p to t_r) in turn by the traps within the region (zz') on the positively biased side of the interface. Thus, to a first approximation, the D.L.T.S. spectrum obtained under these conditions is essentially the same as one would expect from a Schottky diode built along the grain interface. More precisely, however, one must also take into account the effects of carrier capture and emission by the interface states, as well as the traps in the region (yy') on the negatively biased side of the boundary. In fact, it is readily seen from Fig. 7 that carrier emission from the interface states is coincident with capture by the traps at (zz'), whereas emission from the latter is coincident with capture at the interface. The capture and emission processes give rise to opposite variations in the grain boundary capacitance, and their kinetics are also strongly different. Thus, they are expected to produce opposite effects in the D.L.T.S. spectrum : a positive D.L.T.S. signal for carrier capture at the interface, and a negative signal for emission from the traps located on the positively biased side of the barrier. An example of the results obtained using this procedure will be given in section 4.4.2.

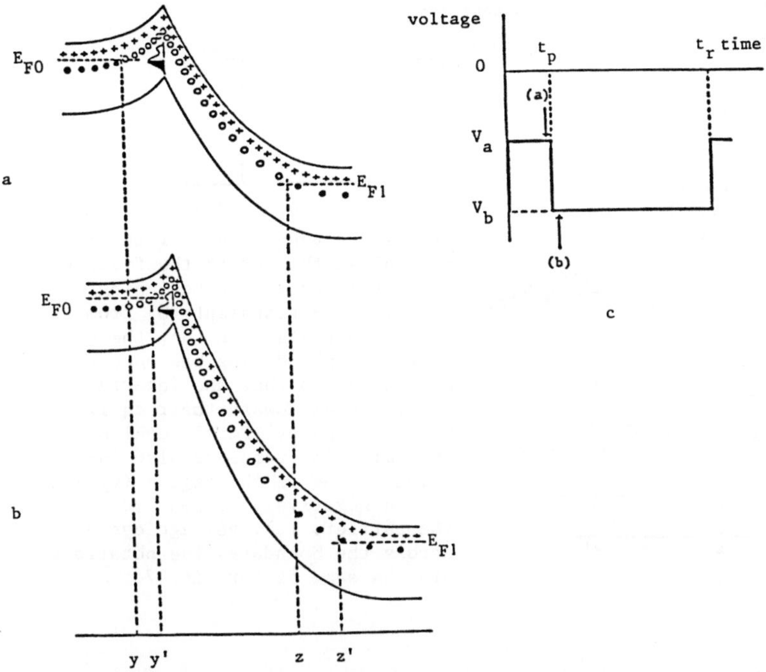

Figure 7 - Band scheme of the bicrystal at the end of the filling pulse (a) and at the beginning of the emission transient (b) from the traps located in the layer (zz') of the space charge region. E_{F0} and E_{F1} are the quasi-Fermi levels of the electrons on the negatively and positively biased sides of the interface, respectively. (c) : Voltage cycle applied across the grain boundary : from $t = 0$ to $t = t_p$, trap filling pulse (amplitude V_a); from t_p to t_r, emission transient under bias V_b.

4.3.2 *Detection of the Grain Boundary States* [20-22]

The complementary experiment consists in applying voltage pulses across the grain boundary, in order to fill the interface levels, and detecting the carrier emission therefrom under reduced bias (Fig. 8), again by means of the concomitant change of the boundary capacitance. For reasons similar to those discussed in section 4.3.1, a contribution from the traps located at (yy') on the negatively biased side of the barrier, should be also expected in the spectrum so obtained.

4.4 Illustrative Example

4.4.1 *Sample Description*

As an example, we shall consider the case of a low angle (3.5 degrees) tilt boundary in an n-type germanium bicrystal, phosphorus doped to $\sim 2.10^{13}$ cm^{-3}. The boundary plane is $(\underline{1}11)$ and the tilt axis (also the growth direction of the bicrystal) is $[1\bar{1}0]$. The dislocation structure of this boundary has been studied by high resolution transmission electron

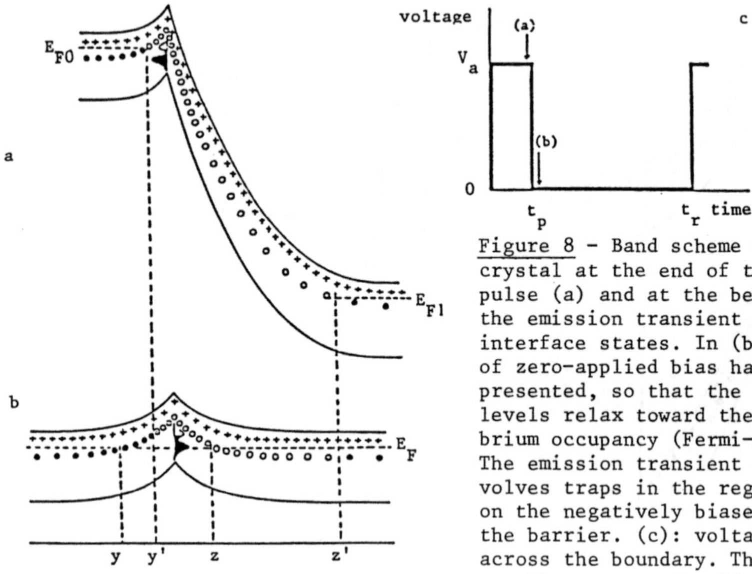

Figure 8 - Band scheme of the bicrystal at the end of the filling pulse (a) and at the beginning of the emission transient (b) from the interface states. In (b), the case of zero-applied bias has been represented, so that the interface levels relax toward their equilibrium occupancy (Fermi-level E_F). The emission transient also involves traps in the region (yy') on the negatively biased side of the barrier. (c): voltage cycle across the boundary. The notations are the same as for Fig. 7c.

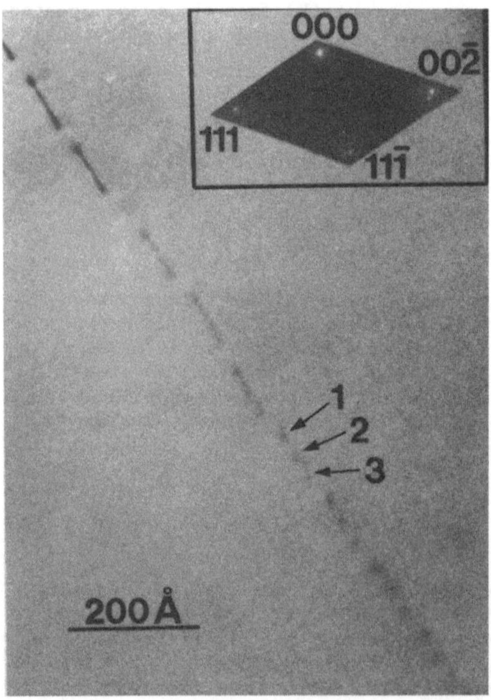

Figure 9 - Bright field electron micrograph of the boundary dislocation structure. The zone axis $[1\bar{1}0]$ is oriented parallel to the beam, so that the dislocations are seen end on. The Frank partials are marked 1,2,3. The stacking faults between the partials are visible as dark lines in the upper left part of the micrograph. Insert: diffraction pattern showing the relative rotation (3.5 degrees) of the crystal lattices on both sides of the boundary.

microscopy [24], and shown to consist mainly of [111] edge dislocations, split into 1/3 [111] Frank partials (Fig. 9). The average spacing of the dislocations is in the order of 60 Å, the value expected from Frank's formula [25].

4.4.2 *Trap Segregation at the Interface*

Fig. 10a shows a typical spectrum of the traps located in the vicinity of the boundary, obtained as explained in section 4.3.1. Carrier capture at the interface is not apparent from this spectrum, as the D.L.T.S. signal remains negative throughout the temperature range investigated (from 77 K up to room temperature). Four different electron traps are detected, B'1 to B'4, with the respective activation energies of 0.22, 0.25, 0.36 and 0.39 eV (Fig. 10b). For the purposes of comparison, the signature of the traps B1 (0.22 eV) and B2 (0.38 eV) in the bulk of the sample have been also represented, as obtained from a standard D.L.T.S. analysis using a Schottky junction on one of the grains of the bicrystal. B1 and B'1 are probably one and the same, as their signatures are almost identical. On the other hand, B'2, B'3 and B'4 are observed in the vicinity of the boundary exclusively, and should therefore be associated with impurity or point defect segregation.

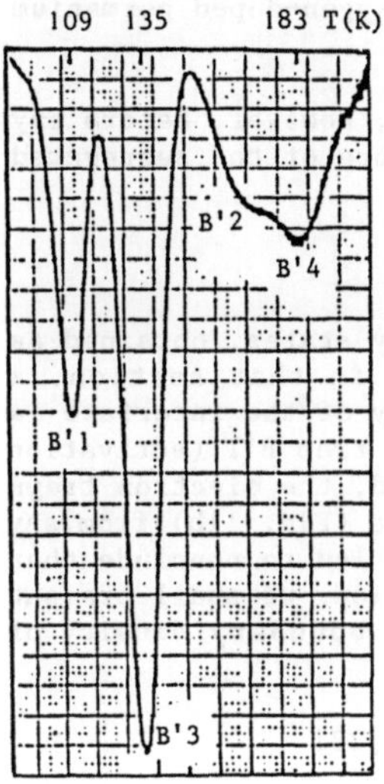

(a)

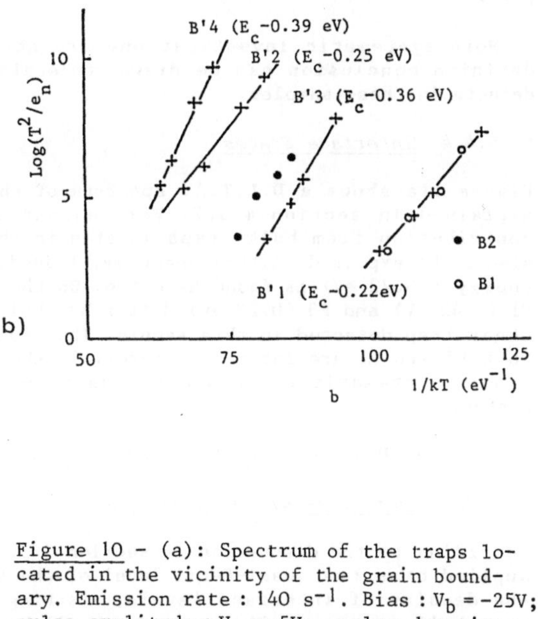

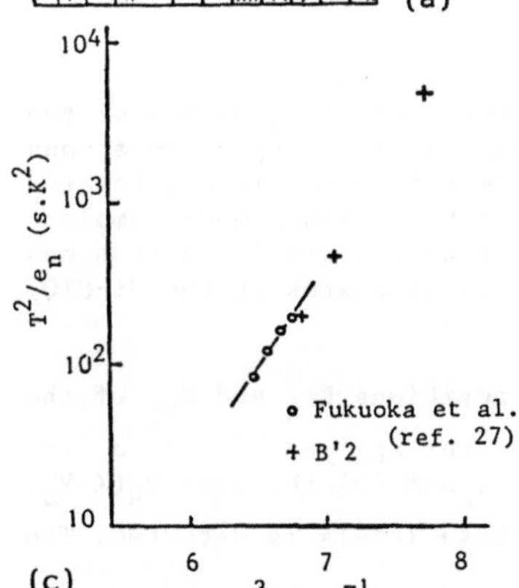

(c)

Figure 10 - (a): Spectrum of the traps located in the vicinity of the grain boundary. Emission rate : 140 s^{-1}. Bias : $V_b = -25V$; pulse amplitude : $V_a = -5V$;, pulse duration : 4 ms. (b): Signature of the traps in the spectrum (a). The trap energies are given as measured from the conduction band edge (E_c). The signature of the bulk traps B1 and B2 are also represented for comparison (see text). (c): signature of the trap (B'2) and comparison with the oxygen-related electron trap at $E_c - 0.25$ eV, from ref. 27.

The amount of segregated defects can be estimated from the amplitude of the emission transient as compared with the overall magnitude of the grain boundary capacitance (see ref. 13 for a discussion in the case of the Schottky and the p-n junctions). In the present case, the density of segregated defects is found comparable with the doping level of the bicrystal (a few 10^{13} cm^{-3}), over a region extending several tens of microns on either side of the interface.

4.4.3 *Nature of the Segregated Defects*

The following observations have suggested a possible connection between one of the traps (B'2) and an oxygen-related defect segregated in the vicinity of the boundary :

(i) Despite the low oxygen content of the bicrystal (below the detection limit for infrared absorption measurements at liquid helium temperature, ~ 10^{13} cm^{-3}), an oxygen precipitation has been detected along the core of the boundary (Frank) dislocations by electron energy loss spectroscopy [26]. Therefore, one can assume that some amount of oxygen is also present in the vicinity of the interface.

(ii) An oxygen-related trap with the same signature as B'2 (Fig. 10c) has been found recently in electron-irradiated, oxygen-doped germanium [27].

More systematic investigations are still needed, however, before any definite conclusion can be drawn concerning the nature of the segregated defects in these samples.

4.4.4 *Interface States*

Figure 11a shows a D.L.T.S. spectrum of the boundary states, obtained as explained in section 4.3.2. For reasons discussed in that section, a contribution from bulk traps located in the vicinity of the interface is also to be expected in this spectrum. Indeed, the bulk trap B'1 (activation energy : 0.22 eV) is found here too. On the other hand, the electron traps E1 (0.42 eV) and E3 (0.29 eV) differ by their signature (Fig. 11b) from any other trap detected in this sample. Therefore, we are led to conclude that both E1 and E3 are interface states. In the next section, we shall present a more systematic procedure to describe the grain boundary density of states.

4.5 Energy Distribution of the Grain Boundary Levels [23]

4.5.1 *Principle of the measurement method*

According to the discussion in section 3.3.3, for increasing values of the applied bias the quasi-Fermi level of the boundary electrons moves across the density of interface levels, starting from the deeper levels towards the shallower ones. This property will be used to obtain a more complete description of the grain boundary density of states. A similar method has been used by JOHNSON [28] to measure the density of states at the Si-SiO_2 interface in MOS structures.

Figures 12a and b represent the respective positions E_{Fa} and E_{Fb} of the quasi-Fermi level, for two different values of the applied bias : (a) the bias V_a of the interface states filling pulses, and (b) the bias V_b ($< V_a$) under which the emission of carriers from these levels is detected. The

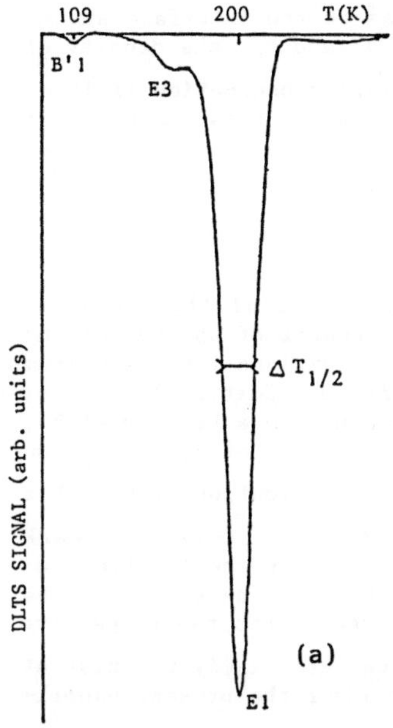

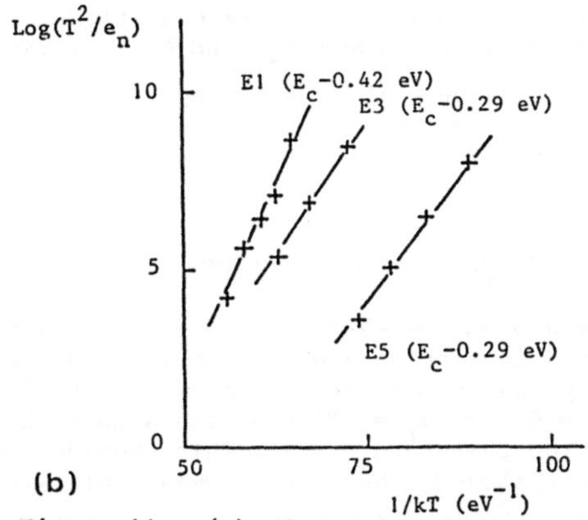

Figure 11 - (a): Spectrum of the grain boundary states E1 and E3. The spectrum also reveals the bulk trap (B'1) (see Fig. 10a). Emission rate: 140 s^{-1}. Bias : $V_b = 0V$; pulse amplitude : $V_a = 0.5$ V; pulse duration : 4 ms. (b): signature of the grain boundary states E1, E3 and E5 (see also Fig. 12).

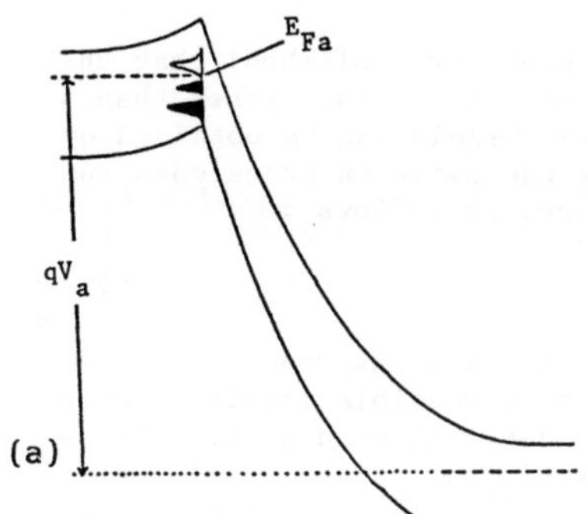

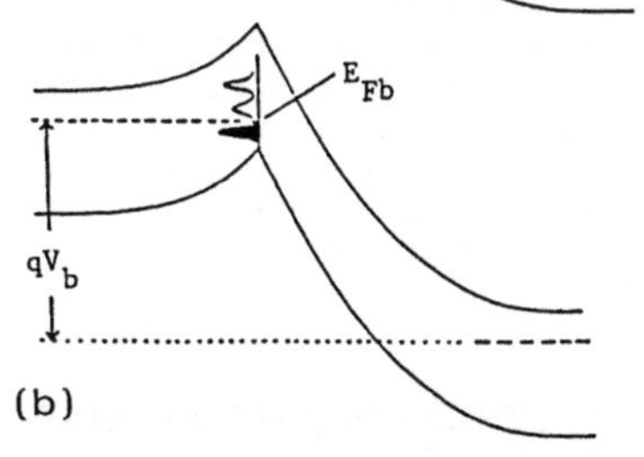

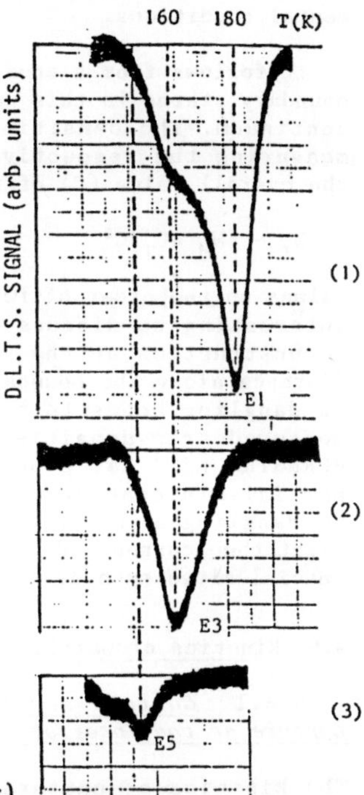

Figure 12 - (a): Band scheme of the bicrystal during the interface states filling pulse (bias V_a) and (b) during the emission transient (bias V_b). The quasi-Fermi levels E_{Fa} and E_{Fb} are also represented. (c): Spectra of the grain boundary states obtained for various biasing conditions: (1) $V_a = 0.2$ V, $V_b = 0V$; (2) $V_a = 0.6$ V, $V_b = 0.4$ V; (3) $V_a = 0.8$ V, $V_b = 0.6$ V. Emission rate: 5.7 s^{-1}; pulse duration : 40 ms.

spectrum obtained under these conditions, relates to the interface states whose energy lies between E_{Fb} and E_{Fa}. By varying V_a and V_b, the density of states can be described, starting from the equilibrium occupation limit of the boundary states ($V_b = 0$). The energy resolution of this technique is limited by the width of the Fermi distribution, i.e. 2-3 kT (~ 0.05 eV at 200 K).

4.5.2 *Example of Application*

As an example, we shall consider once again the case of the bicrystal discussed in section 4.4. Figure 12c represents a number of spectra of the boundary states obtained for different biasing conditions : the spectrum (1) ($V_a = 0.2$ V, $V_b = 0$V) exhibits a peak (E1), with an additional shoulder on the low temperature side. This shoulder corresponds to a trap level E3, shallower than E1 but hardly measurable with this choice of V_a and V_b. On the spectrum (2), ($V_a = 0.6$ V, $V_b = 0.4$ V), the contribution from E1 has been cancelled, and the shoulder is replaced by a well-defined peak corresponding to E3. Here again a shoulder is observed on the low temperature side of (E3); this shoulder is resolved into a new peak (E5) in the spectrum (3) ($V_a = 0.8$ V, $V_b = 0.6$ V). Figure 11b represents the respective signatures of these three levels. Shallower states are likely to exist at the interface, however they were not detected under the present experimental conditions.

It follows from a study of the shape of the peaks (unpublished) that the boundary states in this sample form a discrete set of levels, rather than a continuum. The densities (N_T) of the various levels can be obtained by measuring the respective amplitudes (δC_T) of the emission transients and the overall value (C) of the boundary capacitance, as follows :

$$N_T \sim \varepsilon N_d \, \delta C_T / C^2 \tag{27}$$

Values on the order of 10^{11} cm^{-2} are obtained using this expression. It is interesting to discuss this result in relation to possible models of bond reconstruction at the interface. Assuming one dangling bond every lattice distance along the boundary dislocations, one would expect the total number of dangling bonds to be on the order of 10^{13} cm^{-2}, i.e. larger than the measured trap densities by two orders of magnitude. On the other hand, HORNSTRA [29] has proposed a model of core reconstruction of the Frank partial dislocations, consisting of two adjacent 5- and 7-atom rings with no dangling bond (Fig. 13). The low trap densities detected in this sample might then correspond to a few remaining, unreconstructed bonds, located at special sites such as kinks or jogs along the boundary dislocations.

4.6 Kinetics of Carrier Capture at the Interface

4.6.1 *Current and Capacitance Transients Associated with Carrier Capture at the Interface*

The kinetics of capture at grain boundaries has been studied using both current and capacitance transient measurements [30,31]. Large transient effects are observed, especially in the current mode (Fig. 14a). The explanation of these effects is basically similar to that of carrier capture by charged dislocations in plastically deformed semiconductors [32], the capture rate being limited in both cases by the electrostatic

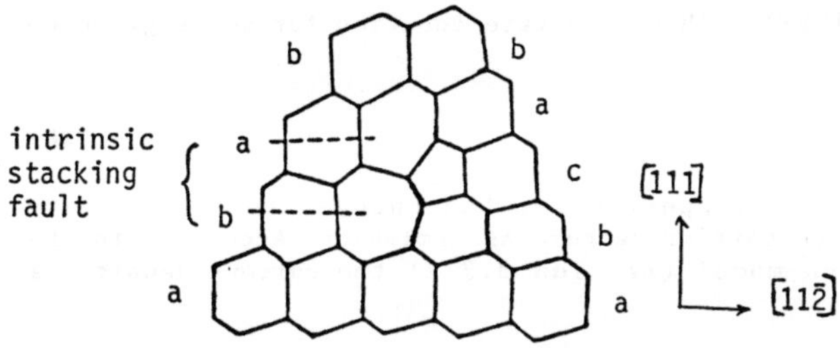

Figure 13 - Model of core reconstruction of the Frank partial dislocation in the diamond cubic lattice, according to HORNSTRA [29]. The core consists of adjacent 5- and 7-atom rings with no dangling bond.

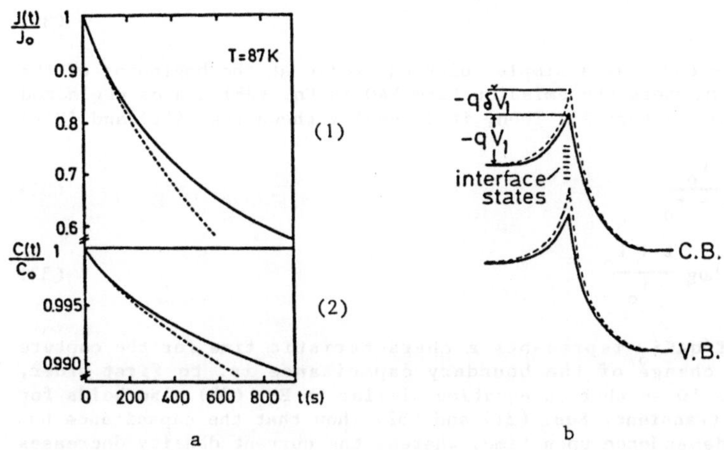

Figure 14 - (a) : Current (1) and capacitance (2) transients following the application of a voltage step (9 V) across the grain boundary. Temperature : 87 K. J_o and C_o represent the initial values of the current and the capacitance for t = 0. Full lines : experiment; dotted lines : empirical fit using Eq. (31) and (32) with t_o = 1100 s. (b) Bending of the energy bands near the boundary under applied voltage : band bending at t = 0 (full line) and at t > 0 (dashed line) showing the increase in the barrier height δV_1.

repulsion of the free carriers by the charge of the defects. In addition to changing the magnitude of the boundary charge, carrier capture also affects the height of the potential barrier and therefore, transient electrical effects are expected both on the boundary capacitance and the current density across the interface.

In order to discuss these effects, we shall consider the case where a voltage step is applied across the boundary at t = 0. Let Q(0) be the boundary charge, j(0) the current density and $V_1(0)$ the barrier height at $t = 0^+$. j(t) represents the current density at time t, and $\delta V_1(t)$ and $\delta Q(t)$ the respective changes of the barrier height and the boundary charge from their initial values for t = 0 (Fig. 14b). According to Eqs. (10) and (20), the capture rate c_n is proportional to j, at least for large values of the applied bias compared with kT/q. Thus, the rate equation for δQ is given to a good approximation by :

$$\frac{d}{dt}(\delta Q) = Aj - B\delta Q \tag{28}$$

where Aj and $B\delta Q$ represent, respectively, the changes in the boundary charge per unit time, due to carrier capture and emission. According to the thermoelectronic emission model (section 3.2.2) the current density is given by :

$$j(t) = j(0)\exp(q\delta V_1/kT) \tag{29}$$

Lastly, for small variations of the boundary charge, δV_1 is proportional to δQ as follows :

$$\delta V_1 = K\delta Q \tag{30}$$

The system (28) - (30) has a simple solution, valid at the beginning of the capture transient, where the emission term $B\delta Q$ in Eq. (28) can be neglected against the capture term Aj. Then, it is easily shown that j(t) and $\delta Q(t)$ are given by :

$$j(t) = j(0)\frac{t_o}{t + t_o} \tag{31}$$

$$\delta Q(t) = -\frac{kT}{qK}\,\mathrm{Log}\,\frac{t + t_o}{t_o} \tag{32}$$

where $t_o = -kT/qAKj_o$ represents a characteristic time for the capture transient. The change of the boundary capacitance is, to first order, proportional to δQ so that an equation similar to Eq. (32) also holds for the capacitance transient. Eqs. (31) and (32) show that the capacitance has a logarithmic dependence upon time, whereas the current density decreases as $(t + t_o)^{-1}$. A similar behaviour is obtained for the relaxation of the photoconductivity in plastically deformed semiconductors, also logarithmic vs. time [32]. As discussed above, the reason for this peculiar capture kinetics is electrostatic in nature, the probability of an electron being captured at a dislocation or at the grain boundary being a function of the charge state of the defect.

The dotted lines in Fig. 14a represent an empirical fit of the curves j(t) and C(t) to Eqs. (31) and (32), obtained with $t_o = 1100$ s. The deviation from the computed curves for large values of t is explained by the approach to the steady state, where carrier emission can no longer be neglected against capture processes.

4.6.2 *Capture Cross-Section Measurement*

The capture cross-section of single trap levels is commonly deduced from the dependence of the emission transient amplitude upon the filling pulse duration t_p, as discussed in ref. 33. In the case of grain boundary states, however, a different procedure must be used for reasons pertaining to the peculiar capture kinetics of these traps discussed in the previous section [31]. Let us consider the boundary state, characterized by the energy E, the density N_T and the capture cross-section σ_T. Assuming the amplitude of the filling pulses to be large compared with kT/q, the capture rate of the trap is given according to Eqs. (10) and (20) as follows :

$$c_n = \sigma_T \, j/q \tag{33}$$

Thus, the capture cross-section can be deduced from the simultaneous knowledge of both j and c_n.

Let f be the fractional occupation ratio of the trap. Then, for a given pulse amplitude (V), f varies as a function of t_p in accordance with the equation :

$$df/dt_p = c_n (1 - f) - e_n f \tag{34}$$

In order to obtain c_n, we shall consider the particular case where the pulse duration t_p is large enough to obtain the steady state occupancy of the trap. Then $df/dt_p = 0$ and therefore :

$$c_n = e_n f/(1 - f) \tag{35}$$

The emission rate is readily obtained from the signature of the interface state (section 4.5). In order to obtain f, let us consider the dependence of the emission transient amplitude δC upon the amplitude V of the filling pulses (Fig. 15). For increasing values of V, δC increases towards the limiting value δC_T corresponding to f = 1 (all states filled). For small trap densities ($\delta C_T \ll C$), the fraction of occupied states is to a first approximation, proportional to δC, and therefore :

$$f = \delta C/\delta C_T \tag{36}$$

On the other hand, for large trap densities ($\delta C_T \sim C$), the boundary charge-capacitance relationship must be used explicitly. According to section 3.4.2, C is proportional to 1/Q and therefore :

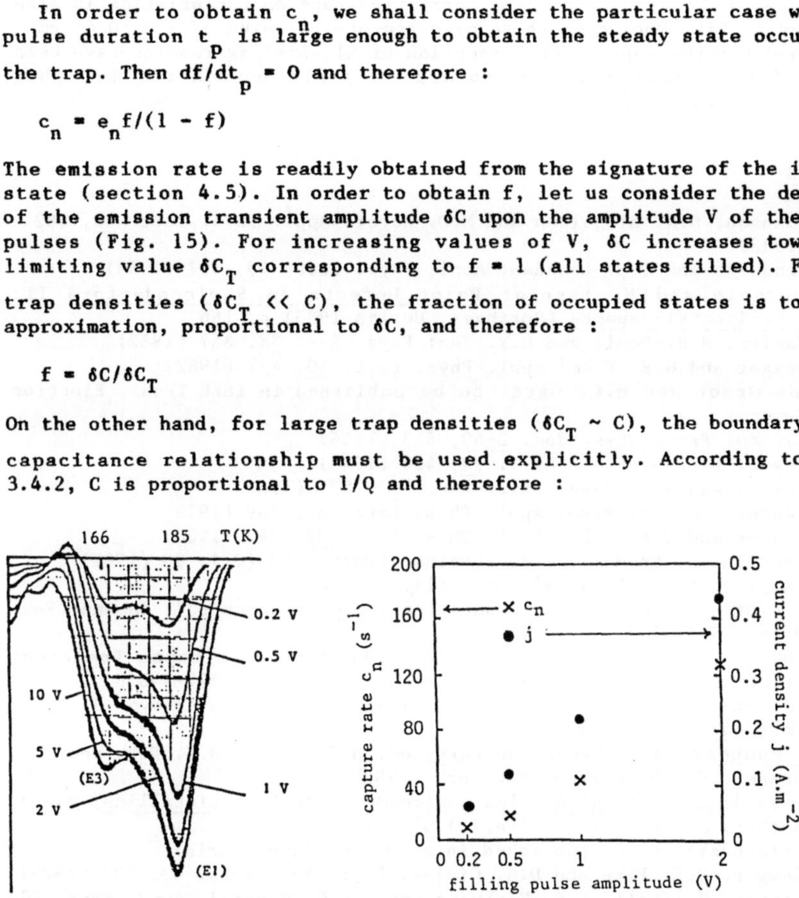

Figure 15 - (a) Spectra of the boundary states E1 and E3 for various values of the filling pulse height V_a, between 0.2 V and 10 V. The emission transients are detected under bias V_b = 0 V. The amplitude of the peak (E1) saturates for pulses larger than 5 V. Emission rate : 14 s^{-1}; pulse duration : 40 ms. (b) Capture rate of the trap E1(x) and current density (•) across the boundary, as a function of the filling pulse amplitude.

$$f = \frac{\delta C/C}{\delta C_T/C_T} \qquad (37)$$

where C_T represents the capacitance of the boundary when all N_T traps are filled with electrons, and C the boundary capacitance when only a fraction f among these states are occupied. These measurements being completed by a determination of the (steady state) current-voltage characteristics j(V), the capture cross-section σ_T of the trap is obtained by means of Eq. (33).

As an example, let us consider the case of the boundary state El (section 4.5.2). Fig. 15a shows a number of spectra obtained for various filling pulse heights between 0.2 V and 10 V. The amplitude of the peak (El) saturates for pulses larger than 5 V. The trap occupancy f(V) has been estimated using Eq. (36), whereas the capture rate $c_n(V)$ has been deduced by means of Eq. (35) (Fig. 15b). The current-voltage characteristics is also reproduced in Fig. 15b. From these measurements, a value of ~ 2.10^{-13} cm^2 is obtained for the capture cross-section of El. Similar results have been obtained from measurements performed on grain boundaries in other materials, e.g. GaAs [30].

REFERENCES

1 N.M. Johnson, D.K. Biegelsen and M.D. Moyer: Appl. Phys. Lett. 40, 882 (1982).
2 W.K. Schubert and P.M. Lenahan: Appl. Phys. Lett. 43, 497 (1983).
3 J. Bourgoin and M. Lannoo: Point Defects in Semiconductors II; Experimental aspects (Springer, Berlin 1983) p. 166
4 W.E. Taylor, N.H. Odell and H.Y. Fan: Phys. Rev. 88, 867 (1952)
5 C.H. Seager and G.E. Pike: Appl. Phys. Lett. 40, 471 (1982)
6 A.W. de Groot and H.C. Card: to be published in IEEE Trans. Electron Devices
7 R. Stratton: Proc. Phys. Soc. B-69, 513 (1956)
8 J. Bardeen: Bell Syst. Tech. J. 28, 428 (1949)
9 W. Shockley and W.T. Read: Phys. Rev.. 87, 835 (1952)
10 C.H. Seager and G.E. Pike: Appl. Phys. Lett. 35, 709 (1979)
11 C.H. Seager and G.E. Pike: Appl. Phys. Lett. 37, 747 (1980)
12 J. Werner and H. Strunk: J. de Physique (Paris) 43 (Cl), 89 (1982)
13 D.V. Lang: J. Appl. Phys. 45, 3023 (1974)
14 H.G. Grimmeis: "Deep Level Impurities in Semiconductors", in Ann. Rev. Mater. Sci. 1977, pp. 341-376
15 G.L. Miller, D.V. Lang and L.C. Kimerling: "Capacitance Transient Spectroscopy", in Ann. Rev. Mater. Sci. 1977, pp 377-448
16 P.C. Srivastava, J. Bourgoin, F. Rabajo and J. Mimila-Arroyo: J. Appl. Phys. 53, 8633 (1982)
17 Grain Boundaries in Semiconductors, ed. H.J. Leamy, G.E. Pike and C.H. Seager, North Holland, New York (1982)
18 Colloque International sur les Semiconducteurs Polycristallins, ed. A. Rocher, J. de Physique (Paris) 43, Cl (1982)
19 A. Broniatowski: to be published in J. de Physique (Paris)
20 C.H. Seager, G.E. Pike and D.S. Ginley: Phys. Rev. Lett. 43, 532 (1979)
21 M. Spencer, R. Stall, L.F. Eastmann and C.E.C. Wood: J. Appl. Phys. 50, 8006 (1979)
22 A. Broniatowski and J. Bourgoin: Phys. Rev. Lett. 48, 424 (1982)
23 A. Broniatowski: J. de Physique (Paris) 44 (C4), 339 (1983)
24 A. Bourret and J. Desseaux: Phil. Mag. 39, 405 (1979)
25 F.C. Frank in Conference on Plastic Deformation of Crystalline Solids (Carnegie Institute of Tech. and Office of Naval Research, 1950) p. 150

26 A. Bourret, C. Colliex and P. Trebbia: J. de Physique (Paris) Lettres 44, L33 (1983)
27 N. Fukuoka, H. Saito and Y. Kambe: Jap. J. Appl. Phys. 22, L353 (1983)
28 N.M. Johnson: Appl. Phys. Lett. 34, 802 (1979)
29 J. Hornstra: J. Phys. Chem. Solids 5, 129 (1958)
30 M.G. Spencer, W.J. Schaff and D. Ken Wagner: J. Appl. Phys. 54, 1429 (1983)
31 A. Broniatowski: J. de Physique (Paris) 43 (Cl), 63 (1982)
32 T. Figielski: Solid State Electron. 21, 1403 (1978)
33 C.H. Henry, H. Kukimoto, G.L. Miller and F.R. Merritt: Phys. Rev. B7, 2499 (1973)
34 D.S. Ginley and D.M. Haaland: Appl. Phys. Lett. 39, 271 (1981)
35 M. Lannoo and J. Bourgoin: Point Defects in Semiconductor I; Theoretical aspects (Springer, Berlin 1981), chap. 3
36 Y. Matukura: Jap. J. Appl. Phys. 2, 91 (1963)
37 L.L. Kaszmerski and P.E. Russel: "Chemical and Electrical Characterization of Polycrystalline Semiconductors", in J. de Physique (Paris) 43 (Cl), 171 (1982)
38 C. Dianteill, Thesis (Toulouse, 1983)
39 J. Werner: to be published in J. de Physique (Paris)
40 J. Werner, W. Jantsch and H.J. Queisser: Solid State Comm. 42, 415 (1982)
41 C.M. Shyu and L.J. Cheng: "Electronic states associated with grain boundaries in silicon", in Grain Boundaries in Semiconductors, ed. H.J. Leamy, G.E. Pike and C.H. Seager, North Holland, New York (1982) p. 131

Electrical Properties of Grain Boundaries in the Presence of Deep Bulk Traps

G. Blatter and F. Greuter

Brown Boveri Research Center, CH-5405 Baden, Switzerland

The static and dynamic electrical properties of a grain boundary are calculated within linear response theory. In particular, the admittance of a double-Schottky-type potential barrier is determined self-consistently, including the contribution of deep bulk traps. Relaxation effects are found due to the finite response time of both, the interface and the bulk defect states. Contrary to the interface, the time constants of the deep traps do not depend on bias and therefore give rise to a dispersion in the conductance and capacitance under zero bias conditions. Such a behavior is experimentally observed for the electrically active grain boundaries in ZnO varistors.

1. Introduction

The electrical properties of many polycrystalline semiconductors and semiconducting ceramics are dominated by the formation of potential barriers at the grain boundaries. A large variety of such materials has been studied in the past, among which poly-Si and ZnO-ceramics probably provide the best known examples due to their immediate technical applications [1]. Since their electrical properties are controlled by double Schottky-type potential barriers forming at grain boundaries, the microscopic understanding of the underlying grain boundary phenomena is essential to any technical improvements of devices like solar cells or ZnO-varistors. Furthermore, the physics of interfaces, especially grain boundaries, has attracted increasing interest over the past few years.

Among others, the admittance spectroscopy [2,3] is a valuable experimental tool for studying the electrical properties of grain boundaries and interfaces. The method provides information on the density of states, the trapping cross-section and the relaxation time for the interface states. Furthermore the corresponding quantities can be obtained for the donor and acceptor states within the depletion region. There is a close relation to the well known technique of deep level transient spectroscopy (DLTS), which can provide similar information [1].

In this paper, we give a self-consistent description of the time-dependent properties of a double Schottky barrier (DSB), defined by a given set of interface states and an arbitrary number of deep donor or acceptor levels within the depletion region adjacent to the grain boundary. In particular we will treat two different models for the description of the dynamics of the interface states. For strongly localized states, the interface is assumed to be in quasi-equilibrium with the bulk of the grains and no relaxation within the grain boundary exists. The second model considers the possibility of perfect relaxation within the interface itself. The physical situation in general then will be somewhere between these two extremes.

Models for the admittance of a grain boundary have been developed by SEAGER and PIKE [2] as well as by WERNER [3], however the influence of deep bulk trap levels has not been included in their analysis. On the other hand, several model calculations for the time dependent properties of a p-n junction or a single Schottky contact in the presence of deep volume traps have been reported in the literature [4-6].

Experimentally, the need to include deep bulk traps in the DSB-model for the grain boundary admittance has come from polycrystalline ZnO-ceramics (varistors). There a large zero bias dispersion in the small signal properties is found both for the conductance and capacitance. Such a dispersion however cannot be explained by a DSB-model without deep trap levels, as will be shown below.

The outline of the present paper is as follows. To establish the notation, we briefly review in section 2 the steady state properties of a simple double Schottky barrier neglecting the deep trap levels in the bulk. In section 3, the time dependent properties of this simple model are discussed for: A) a single interface level, B) for a continuous density of interface states in the strongly localized limit and C) in the ideally fast relaxing model. In section 4 we develop the theory for a double Schottky barrier with one deep bulk trap. The general case, including an arbitrary number of deep donor and acceptor states, will be published elsewhere. The results are illustrated in section 5 by some numerical examples, using parameter values typical for a ZnO varistor, and comparison is made to experimental data. Section 6 finally contains a summary and some concluding remarks.

2. Steady State DSB without Deep Traps

The geometry of the bands around a charged interface is easily calculated in the Schottky approximation. We have to solve Poisson's equation

$$\Delta\Phi(x) = \rho(x)/\varepsilon_o\varepsilon \quad , \qquad (1)$$

for the potential $\Phi(x)$ with a given charge distribution $\rho(x)$,

$$\rho(x) = eN_o[\theta(x+x_{lo})-\theta(x-x_{ro})]-Q_i\;\delta(x) \quad ,$$

taking into account the boundary conditions

$$\Phi(-\infty) = \Phi(-x_{lo}) = 0 \qquad \text{and} \qquad \Phi(\infty) = \Phi(x_{ro}) = -V$$

Here V is the applied bias. N_o is the uniform density of shallow donors, Q_i is the interface charge, ε is the bulk dielectric constant of the semiconductor, ε_o the permittivity of the vacuum, and the electronic charge is denoted by e.

The solution of (1) is parabolic, as illustrated in Fig.1, with

$$\begin{aligned} \Phi(x) &= \frac{\gamma_o}{2}(x+x_{lo})^2 \quad , \qquad -x_{lo} \le x \le 0 \quad , \\ \Phi(x) &= \frac{\gamma_o}{2}(x-x_{ro})^2-V \quad , \qquad 0 \le x \le x_{ro} \quad , \end{aligned} \qquad (2)$$

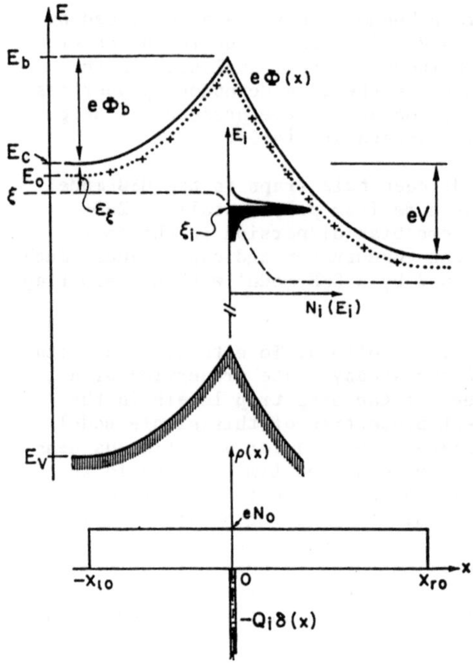

Fig. 1: Energy band diagram and charge distribution for a double Schottky potential barrier at a grain boundary. The shallow donor states are typically 50 meV below the conduction band

where we have introduced the abbreviation

$$\gamma_o = eN_o/\varepsilon_o\varepsilon \quad .$$

The barrier height Φ_b is

$$\Phi_b = \Phi(0) = \frac{V_c}{4}\left(1 - \frac{V}{V_c}\right)^2 \quad , \text{ with} \tag{3}$$

$$V_c = \frac{Q_i^2}{2\varepsilon_o\varepsilon\, eN_o} \quad .$$

In general Q_i depends on the applied bias V. The positions x_{lo} and x_{ro} are:

$$x_{lo} = \frac{Q_i}{2eN_o} - \frac{\varepsilon_o\varepsilon V}{Q_i} = \sqrt{2\Phi_b/\gamma}$$

$$x_{ro} = \frac{Q_i}{2eN_o} + \frac{\varepsilon_o\varepsilon V}{Q_i} = \sqrt{2(\Phi_b+V)/\gamma} \quad .$$

The remaining free parameters for the neutral junction are the interface charge Q_i and the applied bias V. For zero bias conditions, the interface charge is determined by the spatially uniform Fermi level ξ and can be calculated by integrating up the density of interface states $N_i(E)$,

$$Q_i = e \int_{-\infty}^{\infty} dE\ N_i(E) f_i(E) \quad , \quad \text{with} \tag{4}$$

$$f_i(E) = \frac{1}{1+e^{(E-\xi_i)/k_BT}} \quad , \quad \xi_i(V=0) = \xi \quad . \tag{5}$$

Here T is the temperature and k_B the Boltzmann constant. We assume that the interface is neutral before charge trapping, otherwise we have to subtract a constant charge from the right hand side of (4).

When we apply a bias V across the junction, the interface states are only in a quasi-equilibrium. Their occupation statistics however can still be described by a Fermi distribution function and the value of the quasi Fermi level ξ_i at the interface is determined by the detailed balance condition:

$$j_t = j_{em} \quad \text{where}$$

$$j_t = j(1+e^{-eV/k_BT}) \int dE\ c(E) N_i(E)[1-f_i(E)] \quad , \tag{6}$$

denotes the electronic current trapped by the interface states ($j = A \exp[-(e\Phi_b+\varepsilon_\xi)/k_BT]$, $\varepsilon_\xi = E_c-\xi$) and

$$j_{em} = \int dE\ b(E) e^{-(E_b-E)/k_BT} N_i(E) f_i(E) \tag{7}$$

is the emission current out of the interface assuming a thermionic emission model for the currents. The capture cross-section c(E) and the charge emission rate b(E) are approximated by constants and for zero bias we find the relation b = 2Ac. The energy E_b - E is the activation energy for thermionic emission out of the interface (see Fig.1) and $A = A^*T^2$ with A^* an effective Richardson constant (A^* may include any prefactor caused by scattering effects). The result of the detailed balance condition for the quasi Fermi level at the interface, ξ_i, is

$$\xi - \xi_i = k_BT \ln \frac{2}{1+e^{-eV/k_BT}} \quad .$$

Note that for the strongly localized interface model (see section 3B), the interface occupation probability f_i for steady state condition can be shown to be a Fermi function; a result which might not be obvious from the beginning. For the ideally relaxing interface, f_i of course is a priori the Fermi function.

Finally we have to give an expression for the thermionic emission current over the barrier, which is:

$$j_{lb} = j(1-e^{-eV/k_BT}) - j_{em}/2 + j_{rt} \tag{8}$$

when calculated to the left of the barrier or

$$j_{rb} = j(1-e^{-eV/k_BT}) + j_{em}/2 - j_{lt}$$

when evaluated on the right hand side. Here j_{rt} and j_{lt} are the currents trapped at the interface coming from the right and the left, respectively (j_{rt} + $j_{lt} = j_t$). Both formulae give the same final expression for the net over barrier current measured in the circuit,

$$j_{lb} = j_{rb} = j(1-\hat{c}/2)(1-e^{-eV/k_BT}) \quad ,$$

where we have used the abbreviation

$$\hat{c} = c \int dE\, N_i(E)[1-f_i(E)]$$

for the trapping probability. Note that $j_{lb} - j_{rb} = j_t - j_{em} = j_i$ is the interface current and for steady state conditions this is zero. The emission out of the interface has been assumed to be symmetric.

The simple thermionic emission model for the currents applies sufficiently well to a large variety of materials, including poly-Si and ZnO, but care has to be taken in the case of heavily doped samples [7], where tunnelling through the barrier cannot be neglected. A few possible restrictions for the thermionic emission model are discussed in the literature [8,9].

3. Time Dependent Properties

A. Single Interface Level

When a time dependent bias

$$V(t) = V_o + \tilde{V} \exp(i\omega t) \quad , \qquad e\tilde{V} \ll k_BT \quad ,$$

is applied to the junction, we expect that relaxation effects due to the time-dependent trapping of electrons by the interface states become important. The result of their finite response time is a nonvanishing interface current $j_i = \tilde{j}_i \exp(i\omega t)$. By expansion we find:

$$\tilde{j}_i = j_o \frac{e}{k_BT} \cdot \hat{c}_o(1+e^{-eV_o/k_BT}) \{[\tilde{\Phi}_b - \frac{\tilde{V}}{1+e^{eV_o/k_BT}}] - \frac{k_BT}{e} \frac{1}{f_{io}(1-f_{io})} \tilde{f}_i\}$$

for the case of a single interface level, $N_i(E) = N_i\, \delta(E-E_i)$. We have used here the expression $\Phi_b(t) = \Phi_{bo} - \tilde{\Phi}_b \exp(i\omega t)$ for the time-dependent barrier height, $f_i(t) = f_{io} + \tilde{f}_i \exp(i\omega t)$ for the interface occupation function and $\hat{c}_o = cN_i(1-f_{io})$. Note that in general $\tilde{j}_i$, $\tilde{\Phi}_b$ and $\tilde{f}_i$ are complex. The current j_o is given by its dc-value $j_o = A \exp[-(e\Phi_{bo}+\varepsilon_\xi)/k_BT]$.

The equation of continuity determines the dynamics of the interface. Using (4) we find:

$$Q_i(t) = eN_i f_i(t) = eN_i f_{io} + eN_i \tilde{f}_i \exp(i\omega t)$$

and with $j_i = \dot{Q}_i = i\omega \tilde{Q}_i \exp(i\omega t)$ we get

$$\tilde{Q}_i = C_i(\omega)\,[\tilde{\Phi}_b - \frac{\tilde{V}}{1+e^{eV_o/k_BT}}] \quad , \text{ with} \tag{9}$$

$$C_i(\omega) = \frac{e^2N_i}{k_BT}\, f_{io}(1-f_{io})\, \frac{1}{1+i\omega\tau_i} \tag{10}$$

and the <u>interface relaxation time</u>

$$\tau_i = \frac{ef_{io}}{Ac}\, \frac{1}{1+e^{-eV_o/k_BT}}\, e^{(e\Phi_{bo}+\varepsilon_\xi)/k_BT} = f_{io}\tau_i' \quad . \tag{11}$$

Expanding (8) in the small quantities $\tilde{\Phi}_b$, $\tilde{V}$ and $\tilde{f}_i$ and using the above results, we find for the over barrier current $j_{1b} = j_{1bo} + \tilde{j}_{1b}\exp(i\omega t)$ the following expressions,

$$j_{1bo} = j_o(1-\hat{c}_o/2)(1-e^{-eV_o/k_BT}) \quad \text{and} \tag{12}$$

$$\tilde{j}_{1b} = j_o\,\frac{e}{k_BT}\,\{(1-\hat{c}_o/2)[(1-e^{-eV_o/k_BT})\tilde{\phi}_b + e^{-eV_o/k_BT}\,\tilde{V}]$$

$$+ \frac{ck_BT}{2e^2}\, C_i(\omega)[(1+i\omega\tau_i')-e^{-eV_o/k_BT}(1-i\omega\tau_i')][\tilde{\Phi}_b - \frac{\tilde{V}}{1+e^{eV_o/k_BT}}]\} \quad . \tag{13}$$

The total current measured in the circuit is the sum of the over barrier current j_{1b} and the displacement current j_{1o}, which is due to the filling and emptying of donor states at the edge of the depletion region:

$$j_c = j_{1b} + j_{1o} = j_{1b} - \dot{Q}_{1o} \quad , \text{ with} \tag{14}$$

$$Q_{1o}(t) = eN_o\, x_{1o}(t) \quad .$$

Since we know both, the varying bias and the interface charge, we can use (2) to calculate the dynamics of the barrier, i.e. of $x_{1o}(t)$, $x_{ro}(t)$ and $\Phi_b(t)$. By expansion of (2) we find

$$-\tilde{\Phi}_b = \gamma_o\, x_{1oo}\,\tilde{x}_{1o} = \gamma_o\, x_{roo}\,\tilde{x}_{ro} - \tilde{V} \quad ,$$

with $x_{1o}(t) = x_{1oo} + \tilde{x}_{1o}\exp(i\omega t)$. Hence the displacement current j_{1o} becomes

$$j_{1o} = i\omega\, C_1\,\tilde{\Phi}_b\,\exp(i\omega t) \quad , \tag{15}$$

with $C_1 = \varepsilon_o\varepsilon/x_{1oo}$.

Using the neutrality condition

$$\dot{Q}_i = \dot{Q}_{ro} + \dot{Q}_{1o}$$

we finally get the expression for the modulation of the barrier height, $\tilde{\phi}_b$,

$$\frac{\tilde{\phi}_b}{\tilde{V}} = \frac{C_r + \dfrac{C_i(\omega)}{1+e^{eV_o/k_BT}}}{C_r + C_1 + C_i(\omega)} \quad , \tag{16}$$

where we have used the definition $C_r = \varepsilon_o\varepsilon/x_{roo}$ for the capacitance of the right hand side depletion region.

With the equations (12)-(16) we now have a complete set of relations which allows the calculation of the admittance σ,

$$\sigma = \tilde{j}/\tilde{V}$$

$$= [\sigma_{1bo} + \sigma_{rel} + i\omega C_1] \frac{C_r + \dfrac{C_i(\omega)}{1+e^{eV_o/k_BT}}}{C_r + C_1 + C_i(\omega)} - [\frac{\sigma_{1bo}}{1-e^{eV_o/k_BT}} + \frac{\sigma_{rel}}{1+e^{eV_o/k_BT}}] \ . \tag{17}$$

The terms proportional to σ_{1bo} describe the modulation of the over barrier current for steady state trapping and emission $\hat{c}_o$,

$$\frac{d}{dV} j_{1b}\Big|_{\hat{c}=\hat{c}_o} = j_o \frac{e}{k_BT} (1 - \frac{\hat{c}_o}{2})(1-e^{-eV_o/k_BT}) [\frac{\partial\phi_b}{\partial V} - \frac{1}{1+e^{eV_o/k_BT}}] \tag{18}$$

$$= \sigma_{1bo} [\frac{\partial\phi_b}{\partial V} - \frac{1}{1+e^{eV_o/k_BT}}] \quad .$$

The trapping and emission dynamics of the interface is taken into account by the terms proportional to σ_{rel}, with the definition:

$$\sigma_{rel} = j_o \frac{c}{2e} C_i(\omega)[(1+i\omega\tau_i') - e^{-eV_o/k_BT}(1-i\omega\tau_i')] \quad .$$

This term therefore accounts for the correction due to the time dependence of $\hat{c}$ which is not included in (18).

Finally the term $i\omega C_1\ \partial\phi/\partial V$ accounts for the additional admittance due to the displacement current $\dot{Q}_{1o}$.

The zero bias limit of our final expression (17) is specially simple, since $\tilde{\phi}_b/\tilde{V} = 1/2$. The conductance G and the capacitance C of the junction defined by the real and imaginary part of σ become:

$$G(\omega,0) = \mathrm{Re}\ \sigma = \sigma_o = j_o \frac{e}{k_BT} (1-\hat{c}_o/2),$$

$$C(\omega,0) = 1/\omega\ \mathrm{Im}\ \sigma = C_{HF} = \frac{\varepsilon_o\varepsilon}{x_{1oo} + x_{roo}} \quad ,$$

and there is no dispersion, as mentioned already in the introduction.

For a large bias $eV_o >> k_BT$ the result also simplifies considerably and takes the form

$$\sigma = (\sigma_o + \sigma^v_{rel} + i\omega C_l) \frac{C_r}{C_r + C_l + C_i(\omega)} \quad \text{with}$$

$$\sigma^v_{rel} = j_o \frac{c}{2e} C_i(\omega)\ (1+i\omega\tau_i') \quad .$$

Note that the results do not depend on which side of the barrier the current j_c has been calculated, since

$$j_{lb} - \dot{Q}_{lo} - (j_{rb} + \dot{Q}_{ro}) = j_i - (\dot{Q}_{lo} + \dot{Q}_{ro}) = 0 \quad ,$$

where $j_{ro} + \dot{Q}_{ro}$ is the total circuit current j_c, when evaluated at the right hand side of the junction.

B. Continuous Density of States, Strongly Localized Model

We assume that the continuous density of states (DOS) is made up by a homogeneous spatial distribution of single levels, which have vanishing matrix elements for hopping, such that the interface is in thermal (quasi-) equilibrium with the bulk, but not with itself. As a result, each state acts independently of all others. Therefore we simply have to integrate up the contributions from all single levels weighted by $N_i(E)$. Equation (9) still applies for each level separately. The total interface charge variation becomes

$$\tilde{Q}_i = \int_{-\infty}^{\infty} dE\ e^2\ N_i(E)\ \frac{f_{io}(E)[1-f_{io}(E)]}{k_BT}\ \frac{1}{1+i\omega\tau_i'\ f_{io}(E)} \cdot [.....]$$

$$\cong e^2\ N_i(\xi_{io}) \int_0^1 df_{io}\ \frac{1}{1+i\omega\ f_{io}\ \tau_i'}\ [.....]$$

$$= C_i^1(\omega)\ [\tilde{\phi}_b - \frac{\tilde{V}}{1+e^{eV_o/k_BT}}] \quad ,$$

and $C_i^1(\omega)$ is modified with respect to (10):

$$C_i^1(\omega) = e^2\ N_{io}(\xi_o)\ \frac{\ln(1+i\omega\ \tau_i')}{i\omega\ \tau_i'} \quad .$$

The expression for τ_i' is again given by (11). Here ξ_{io} is the dc-position of the interface quasi Fermi level. In this model the occupation statistics of the interface for the time dependent properties deviates from a Fermi function, since the variation of the quasi Fermi level becomes energy dependent,

$$\tilde{\xi}_i(E) \cong \frac{1}{1+i\omega\ \tau_i(E)}\ [e\tilde{\phi}_b - \frac{e\tilde{V}}{1+e^{eV_o/k_BT}}] \quad .$$

All results of section 3A can now be adapted to this model by simply replacing $C_i(\omega)$ by $C_i^1(\omega)$ and using the integrated value for $\hat{c}_o$ (see section 2).

C. Continuous Density of States, Ideally Relaxing Model

In this subsection we assume that the hopping matrix element between the interface states is large enough to guarantee at any time a quasi-equilibrium of the interface, such that the occupation statistics of these states is always described by a Fermi function. This model is applicable if the time constant for restoring thermal equilibrium within the interface is much smaller than $1/\nu$, where ν is the applied ac-frequency.

We have to calculate the following integrals (see (6) and (7)):

$$I_t = c \int dE\, N_i(E)\, [1-f_i(E)]$$

$$I_{em} = c \int dE\, N_i(E)\, \exp[-(E_b-E)/k_BT] f_i(E) \quad .$$

From the expansions:

$$f_i(E) \cong f_{io}(E) + \tilde{\xi}_i\, \delta(E-\xi_{io})$$

$$N_i(E) \cong N_{io}(E) + \frac{dN_{io}}{dE}\, e\tilde{\phi}_b$$

we find the result

$$I_t = \hat{c}_o - cN_{io}(\xi_{io})\, (e\tilde{\phi}_b + \tilde{\xi}_i)$$

$$I_{em} = v[\hat{c}_o + cN_{io}(\xi_{io})\, (e\tilde{\phi}_b + \tilde{\xi}_i)] \quad \text{with}$$

$$v = \exp[-(e\phi_{bo}+\varepsilon_\xi)/k_BT)] \cdot [1+\exp(-eV_o/k_BT)]/2 \quad .$$

Using the expression

$$\tilde{Q}_i = e \int dE\, N_i(E) f_i(E) \cong eN_{io}(\xi_{io})[e\tilde{\phi}_b + \tilde{\xi}_i]$$

for the interface charge variation, the equation of continuity $j_i = \dot{Q}_i$ leads to the following result for the dynamics,

$$\tilde{Q}_i = C_i^e(\omega)\, [\tilde{\phi}_b - \frac{\tilde{V}}{1+e^{eV_o/k_BT}}] \quad , \quad \text{with}$$

$$C_i^e(\omega) = \frac{e^2\hat{c}_o}{2ck_BT}\, \frac{1}{1+i\omega\tau_i^e} \quad ,$$

$$\tau_i^e = \frac{e}{2Ac}\, \frac{1}{1+e^{-eV_o/k_BT}}\, e^{(e\phi_{bo}+\varepsilon_\xi)/k_BT} \quad .$$

Again all results of section 3A can be adapted to this model by taking first the proper definition of $\hat{c}_o$, i.e., $\hat{c}_o$ (single level) = $cN_i(1-f_{io})$

$$\rightarrow \hat{c}_o \text{ (continuous DOS)} = c \int dE\, N_i(E)\, [1-f_i(E)] \quad ,$$

and replacing afterwards all remaining Fermi functions f_{io} by their value at the quasi Fermi level, $f_{io}(\xi_{io}) = 1/2$.

4. DSB with one Deep Trap Level

When we take into account one deep bulk trap level of density N_1, the form of the potential barrier (see (2)) will change to:

$$\begin{aligned}
\phi(x) &= \frac{\gamma_o}{2}(x+x_{lo})^2 && , -x_{lo} \le x \le -x_{l1} \quad , \\
\phi(x) &= \frac{\gamma_o}{2}(x+x_{lo})^2 + \frac{\gamma_1}{2}(x+x_{l1})^2 && , -x_{l1} \le x \le 0 \quad , \\
\phi(x) &= \frac{\gamma_o}{2}(x-x_{ro})^2 + \frac{\gamma_1}{2}(x-x_{r1})^2 - V && , 0 \le x \le x_{r1} \quad , \\
\phi(x) &= \frac{\gamma_o}{2}(x-x_{ro})^2 - V && , x_{r1} \le x \le x_{ro},\ \gamma_1 = eN_1/\varepsilon_o\varepsilon \ .
\end{aligned} \tag{19}$$

The four unknown positions in (19) are determined by continuity and neutrality,

$$\phi\,(0^+) = \phi\,(0^-) = \phi_b \quad ,$$

$$\phi'(0^-) - \phi'(0^+) = Q_i/\varepsilon_o\varepsilon \quad ,$$

and the condition for ionization of the deep traps,

$$e\phi(-x_{l1}) = \varepsilon_1' \qquad \text{and} \qquad e\phi(x_{r1}) = \varepsilon_1' - eV \quad .$$

Here the energy ε_1' is determined by

$$\varepsilon_1' = \varepsilon_1 - \varepsilon_\xi \qquad \text{with} \qquad \varepsilon_1 = E_c - E_1 \qquad \text{(see Fig. 2)} \quad .$$

With these conditions we obtain:

$$x_{lo} = \frac{1}{2}\,\frac{Q_i + 2N_1\sqrt{2\varepsilon_o\varepsilon\ \varepsilon_1'/N_o}}{e(N_o+N_1)} - \frac{\varepsilon_o\varepsilon V}{Q_i} \quad ,$$

$$x_{l1} = \frac{1}{2}\,\frac{Q_i - 2N_o\sqrt{2\varepsilon_o\varepsilon\ \varepsilon_1'/N_o}}{e(N_o+N_1)} - \frac{\varepsilon_o\varepsilon V}{Q_i} \quad ,$$

$$x_{ro} = x_{lo} + 2\varepsilon_o\varepsilon V/Q_i \quad , \qquad x_{r1} = x_{l1} + 2\varepsilon_o\varepsilon V/Q_i \quad ,$$

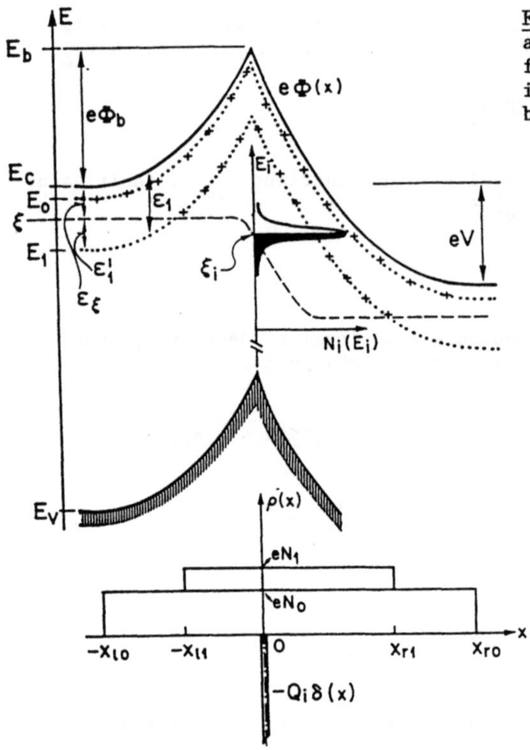

Fig. 2: Energy band diagram and charge distribution for the potential barrier in the presence of one deep bulk trap level

and the barrier height is modified with respect to (3),

$$\phi_b = \frac{V_c}{4}\left(1 - \frac{V}{V_c}\right)^2 + \frac{N_1}{N_o+N_1}\,\frac{\varepsilon_1'}{e} \quad , \quad \text{with} \tag{20}$$

$$V_c = \frac{Q_i^2}{2\varepsilon_o\varepsilon e(N_o+N_1)} \quad .$$

Note that the relative positions of x_{lo} and x_{l1} and of x_{ro} and x_{r1} are equal,

$$x_{lo} - x_{l1} = x_{ro} - x_{r1} = \sqrt{2\varepsilon_o\varepsilon\;\varepsilon_1'/e^2N_o} \quad .$$

All considerations concerning the dynamics of the interface states can be taken over from sections 2 and 3 and we immediately proceed to the calculation of the deep trap dynamics in the depletion region.

We define the total space charge due to the deep levels by:

$$Q_{l1} = e\int_{-\infty}^{o} N_1^+(x)dx \quad \text{and} \quad Q_{r1} = e\int_{o}^{\infty} N_1^+(x)dx \quad , \quad \text{with}$$

$$N_1^+(x) = N_1[1-f_1(x)] \quad , \quad f_1(x) = \frac{1}{1+g_1 e^{[E_1(x)-\xi(x)]/k_BT}} \quad .$$

The degeneracy of the trap states, which can be singly occupied, is denoted by $1/g_1$(usually $g_1 = 1/2$). The filling and emptying of the traps is then governed by the following rate equation:

$$e\dot{N}_1^+(x,t) = i\omega e\tilde{N}_1^+(x)\exp(i\omega t)$$

$$= -Ac_1N_1[1-f_1(x,t)]e^{-[E_c(x)-\xi]/k_BT} + b_1N_1f_1(x,t)e^{-\varepsilon_1/k_BT} \quad , \qquad (21)$$

where the first term describes capture (c_1 = capture cross-section) and the second term emission of electrons by the deep traps (b_1 = charge emission rate). We use again the expansion $N_1^+(x,t) = N_{1o}^+(x) + \tilde{N}_1^+(x)\exp(i\omega t)$ and a similar one for $f_1(x,t)$. Evaluation of (21) for steady state condition $\tilde{V} = 0$ leads to the relation $b_1 = g_1\ A\ c_1$. For the general case $\tilde{V} \neq 0$ we get the result

$$\tilde{N}_1^+(x) = \frac{f_{1o}(x)[1-f_{1o}(x)]}{k_BT}\ \frac{N_1}{1+i\omega\tau_1(x)}\ e\tilde{\phi}(x) \quad . \qquad (22)$$

The <u>relaxation time</u> $\tau_1(x)$ is

$$\tau_1(x) = \frac{ef_{1o}(x)}{Ac_1}\ e^{[E_{co}(x)-\xi_o(x)]/k_BT} \quad .$$

Using the Schottky approximation we can describe $N_1^+(x)$ in the vicinity of $x = -x_{11}$ by

$$N_1^+(x) = N_1\ \theta(x+x_{11})$$

$$= N_1[\theta(x+x_{11o}) + \tilde{x}_{11}\ \delta(x+x_{11o})e^{i\omega t}]$$

and therefore

$$\tilde{N}_1^+(x) = \delta(x+x_{11o})\ N_1\ \tilde{x}_{11} \quad . \qquad (23)$$

Similarly, equation (22) can be reduced to the Schottky approximation and the result is

$$N_1^+(x) = \delta(x+x_{11o})\ \frac{1}{|e\phi_o'(-x_{11o})|}\ \frac{N_1}{1+i\omega\tau_1(-x_{11o})}\ e\tilde{\phi}(-x_{11o}) \quad . \qquad (24)$$

Combining (23) and (24) we find

$$\tilde{x}_{11} = \frac{1}{|\phi_o'(-x_{11o})|}\ \frac{1}{1+i\omega\tau_1(-x_{11o})}\ \tilde{\phi}(-x_{11o}) \quad . \qquad (25)$$

Using the solution of the Poisson equation (19) for a given charge distribution $\rho(x)$, we can relate $\tilde{x}_{11}$ to $\tilde{x}_{1o}$. Expansion of $\phi(x)$ for $x = -x_{11o}$ and $x_{1v} = x_{1vo} + \tilde{x}_{1v}\exp(i\omega t)$, $v = 0,1$, leads to

$$\tilde{\phi}(-x_{11o}) = \gamma_o(x_{1oo}-x_{11o})\tilde{x}_{1o} + O(\tilde{x}_{11}^2)$$

and therefore we get, using (25),

$$\tilde{x}_{11} = \frac{1}{1+i\omega\tau_1}\,\tilde{x}_{1o} \quad , \qquad \text{with} \tag{26}$$

$$\tau_1 = \tau_1(-x_{11o}) = \frac{e}{Ac_1(1+g_1)}\, e^{\varepsilon_1/k_BT} \quad .$$

The dynamics for the trap charges therefore is:

$$Q_{11} = Q_{11o} + \tilde{Q}_{11}\, e^{i\omega t} \quad \text{and} \quad Q_{r1} = Q_{r1o} + \tilde{Q}_{r1}\, e^{i\omega t} \quad \text{with}$$

$$\tilde{Q}_{11} = \frac{eN_1}{1+i\omega\tau_1}\,\tilde{x}_{1o} \quad \text{and} \quad \tilde{Q}_{r1} = \frac{eN_1}{1+i\omega\tau_1}\,\tilde{x}_{ro} \quad .$$

Note that the result for the right hand side depletion region is identical, since $x_{1oo} - x_{11o} = x_{roo} - x_{r1o}$.

The final step is the elimination of $\tilde{x}_{1o}$ and $\tilde{x}_{ro}$. Expanding (19) for x=0 leads to:

$$-\tilde{\phi}_b = \gamma_o\, x_{1oo}\, \tilde{x}_{1o} + \gamma_1\, x_{11o}\, \tilde{x}_{11}$$

$$= \gamma_o\, x_{roo}\, \tilde{x}_{ro} + \gamma_1\, x_{r1o}\, \tilde{x}_{r1} - \tilde{V}$$

and by using (26) we get

$$\tilde{Q}_{1o} = -C_{1o}\, \tilde{\phi}_b \quad \text{and} \quad \tilde{Q}_{ro} = C_{ro}(\tilde{V}-\tilde{\phi}_b) \quad \text{with}$$

$$C_{1o} = \frac{\varepsilon_o\varepsilon}{x_{1oo}\left(1 + \frac{N_1}{N_o}\,\frac{x_{11o}}{x_{1oo}}\,\frac{1}{1+i\omega\tau_1}\right)} \quad \text{and}$$

$$C_{ro} = \frac{\varepsilon_o\varepsilon}{x_{roo}\left(1 + \frac{N_1}{N_o}\,\frac{x_{r1o}}{x_{roo}}\,\frac{1}{1+i\omega\tau_1}\right)} \quad .$$

Next we relate $\tilde{\phi}_b$ to $\tilde{V}$ with the help of the neutrality condition

$$\tilde{Q}_i = \tilde{Q}_{1o} + \tilde{Q}_{ro} + \tilde{Q}_{11} + \tilde{Q}_{r1}$$

and the result is in analogy to (16):

$$\frac{\tilde{\phi}_b}{\tilde{V}} = \frac{C_r + \dfrac{C_i(\omega)}{1+e^{eV_o/k_BT}}}{C_r + C_1 + C_i(\omega)} \quad ,$$

where however C_r and C_1 are now modified to:

$$C_r = C_{ro}(1 + \frac{N_1/N_o}{1+i\omega\tau_1}) \quad , \quad C_1 = C_{1o}(1 + \frac{N_1/N_o}{1+i\omega\tau_1}) \quad .$$

Again we have used (9) for the definition of $\tilde{Q}_i$.

The total measured current j_c now contains an additional contribution from the trapped and released charges

$$j_c = j_{1b} - \dot{Q}_{1o} - \dot{Q}_{11} \quad .$$

The result of section 3, equation (17), is easily adapted to this case by taking the adequate expressions for the capacitances C_r, C_1 and $C_i(\omega)$. As before we quote the simple expression for the admittance in the <u>zero bias</u> case. Again $\tilde{\Phi}_b/\tilde{V} = 1/2$ and we find:

$$G(\omega,0) = \sigma_o + C_{HF}\,\omega^2 \frac{n_1\,\hat{\tau}_1}{1+(\omega\hat{\tau}_1)^2} \quad ,$$

$$C(\omega,0) = C_{HF}\,[1 + \frac{n_1}{1+(\omega\hat{\tau}_1)^2}] \quad , \quad \text{with}$$

$$\hat{\tau}_1 = \tau_1 \frac{1}{1 + \frac{N_1}{N_o}\frac{x_{11o}}{x_{1oo}}} \quad \text{and} \quad n_1 = \frac{N_1}{N_o}\frac{1 - \frac{x_{11o}}{x_{1oo}}}{1 + \frac{N_1}{N_o}\frac{x_{11o}}{x_{1oo}}} \quad .$$

The main result now is a <u>finite dispersion</u> also for the zero bias case. This dispersion is due to the deep trap relaxation. Note that the resonance frequency ω^1_{res}, defined by $\omega^1_{res}\,\tau_1 = 1$, does not depend on bias. This is different for the dispersion due to the interface states at finite bias, since their resonance frequency ω^i_{res} depends strongly on bias due to the changing barrier height $\phi_{bo}(V_o)$.

Thus in practice, the different resonant phenomena can be distinguished and characterized separately. From the frequency- and temperature-dependence of the zero bias admittance, the deep volume traps can be completely analyzed, as will be illustrated elsewhere.

5. Numerical Results and Comparison to Experiments

In the present section a comparison of the above theoretical model to experimental data on commercial ZnO varistors is given. All the experiments have been carried out on the same polycrystalline sample, which was processed by conventional doping and sintering techniques. This provides a consistent set of data for the barrier height, the small signal capacitance and the conductance as a function of applied dc-bias, frequency, and temperature.

The electrical properties of the macroscopic sample are determined by the network of all active grain boundaries within the ceramics. For the present comparison, we reduce this network to the properties of a single grain boundary by assuming identical grains (effective grain size D = 14 μm).

The experimental data are well described using the strongly localized limit for the interface (see section 3B) and including four deep trap levels in the depletion region. From the zero bias dispersion of the capacitance and conductance at various temperatures these bulk trap states have been identified at energies of 0.18 eV, 0.36 eV, 0.5 eV, and ~ 0.7 eV below the conduction band [10]. The following results are given for a temperature T = 440 K (ε_ξ =0.14 eV). In this temperature range the interface becomes resonant for reasonable measuring frequencies at voltages in the prebreakdown regime, i.e., before the generation of minority carriers [2].

In Fig. 3a we show the bias dependence of the experimental barrier height $e\phi_b$, as determined from equation (12) ($A^* = 31$ A cm^{-2} K^{-2}, [11]). The theoretical curve $e\phi_b(V)$ in Fig. 3b on the other hand is obtained

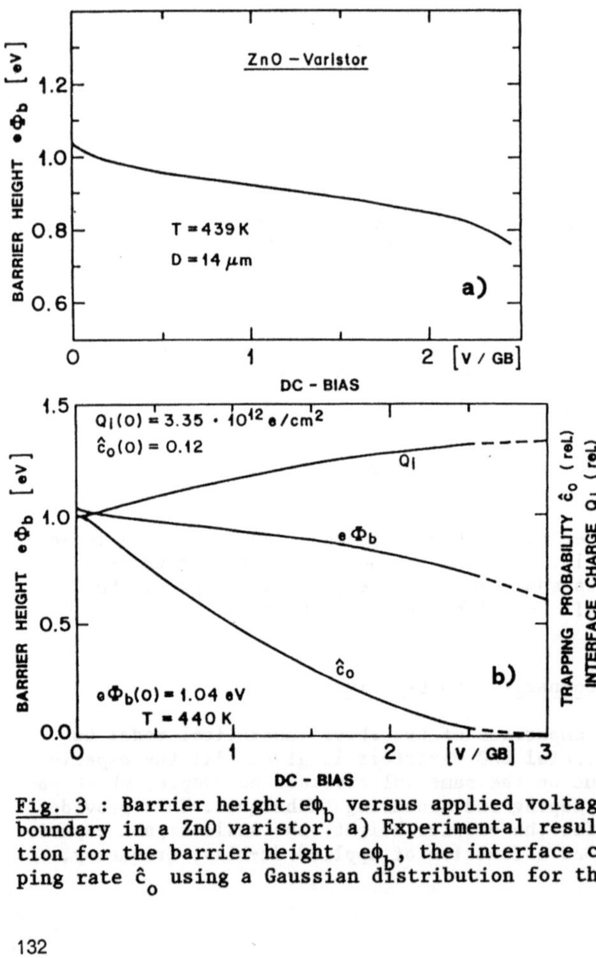

Fig. 3 : Barrier height $e\phi_b$ versus applied voltage of a single grain boundary in a ZnO varistor. a) Experimental results, b) Model calculation for the barrier height $e\phi_b$, the interface charge Q_i, and the trapping rate $\hat{c}_o$ using a Gaussian distribution for the interface states

by : i) Evaluation of the densities of bulk traps from the IR-reflectivity [12] and the zero bias admittance [10]; and ii) Adjustement of the interface density of states $N_i(E)$. The interface charge Q_i is determined self-consistently from equation (4) for $Q_i(\phi_b)$ and a generalized version of (20) for $\phi_b(Q_i)$. The proper Fermi distribution function was used in (4). Agreement with experiment within 0.02 eV over the prebreakdown range is obtained by using a Gaussian for $N_i(E)$:

$$N_i(E) = \frac{N^*}{\sqrt{2\pi}\Delta E}\, e^{-\frac{1}{2}\left(\frac{E-E^*}{\Delta E}\right)^2} ,$$

with $N^* = 4.45 \cdot 10^{12}\ cm^{-2}$, $\Delta E = 0.15$ eV and $E^* = 1.91$ eV. E^* is measured with respect to the valence band edge at x = 0 ($E_c - E_v = 3.2$ eV). This high density of interface states allows for the observed pinning of the barrier height at lower voltages (V < 2.5 V/GB). The faster decay at higher voltages then is caused by the filling of the interface states for the present model density ($\hat{c}_o \rightarrow 0$, $Q_i \rightarrow eN^*$). The situation is different for the experiment, where minority carriers become important above ~ 2.5 V [2,10].

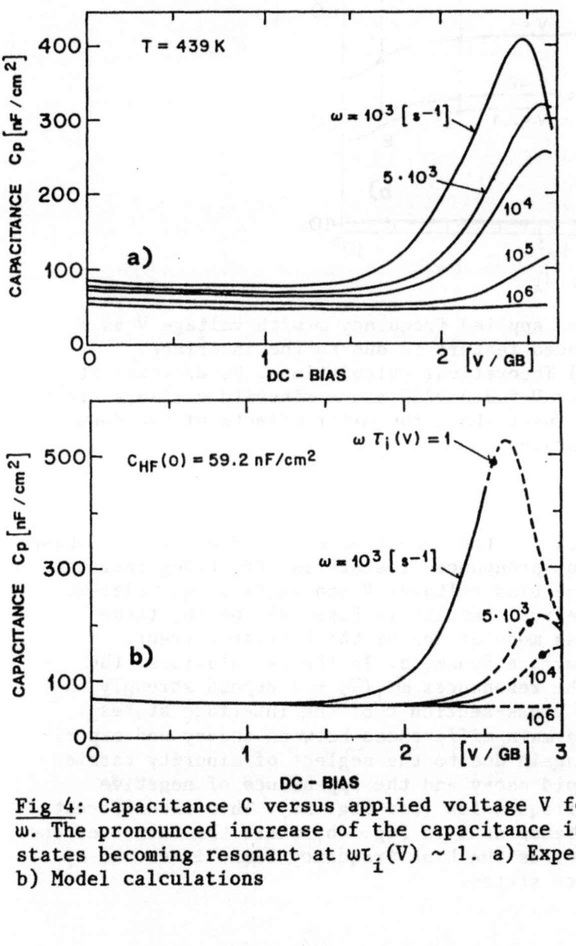

Fig 4: Capacitance C versus applied voltage V for several frequencies ω. The pronounced increase of the capacitance is due to the interface states becoming resonant at $\omega\tau_i(V) \sim 1$. a) Experimental results. b) Model calculations

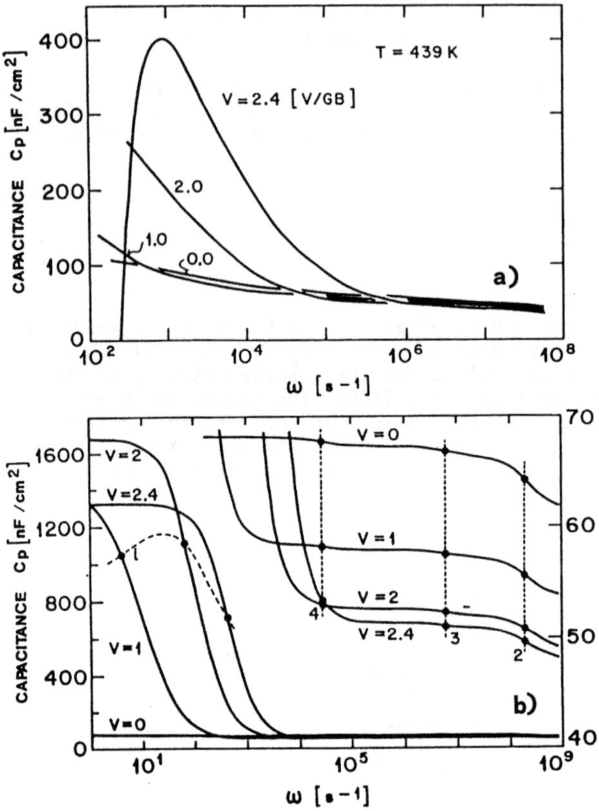

Fig. 5: Capacitance C versus applied frequency ω with voltage V as a parameter. The most pronounced feature is due to the interface. a) Experimental results. b) Theoretical calculations. No decrease at low frequencies is found for V = 2.4 V/GB since minority carriers are not included in the model. The inset shows the minor effects of the deep bulk traps on an enlarged scale

Typical experimental data for the capacitance C(V) and C(ω) are shown in Figs. 4a and 5a. The most pronounced feature is the strong increase in the capacitance for large bias voltages V and small frequencies ω. By comparison to the theoretical results in Figs. 4b and 5b, these resonances are understood as modulations of the barrier current, caused by the retarded interface dynamics. In the calculations, the positions and heights of the resonances $\omega\tau_i(V) \sim 1$ depend strongly on the choice of the trapping cross-section c of the interface states ($c = 1.1 \cdot 10^{-13}$ cm^{-2}). The main differences between theory and experiment at high voltages are again due to the neglect of minority carrier effects which lead to a rapid decay and the appearance of negative capacitance values at low frequencies (see Fig. 5a). This is different for the theoretical capacitance data in Fig. 4b, where the turn-over and subsequent smooth decay towards the high frequency limit is caused by the filling of the interface states.

The influence of the deep bulk traps is best illustrated in Fig. 5b. Their resonances are independent of bias. Some discrepancies still remain at low voltages in the dispersion seen in Figs. 4-5. They can be attributed to : i) the still insufficient characterization of the deep bulk traps at ~.5 eV and ~.7 eV, which dominate the present temperature and frequency range and ii) dispersive effects in the possibly inhomogenous network [2].

The frequency and voltage dependence of the conductance is shown in Figs. 6a - 7b. Here the dominant features are produced by the relaxation effects of the deep bulk traps, which are enhanced by a factor ω^2 compared to their contribution to the capacitance. The zero bias dispersion therefore is very sensitive to the presence of any deep bulk traps and provides the most accurate value of their densities. Measurements for T = 80 - 440 K have been used for the characterization of the defect states at 0.18 eV, 0.36 eV, and 0.5 eV. Their capture cross-sections are determined by the frequency and temperature dependent positions of the corresponding resonances [10]. The voltage dependent interface resonances are less pronounced on the logarithmic scale, but remain still visible at lower frequencies and bias voltages.

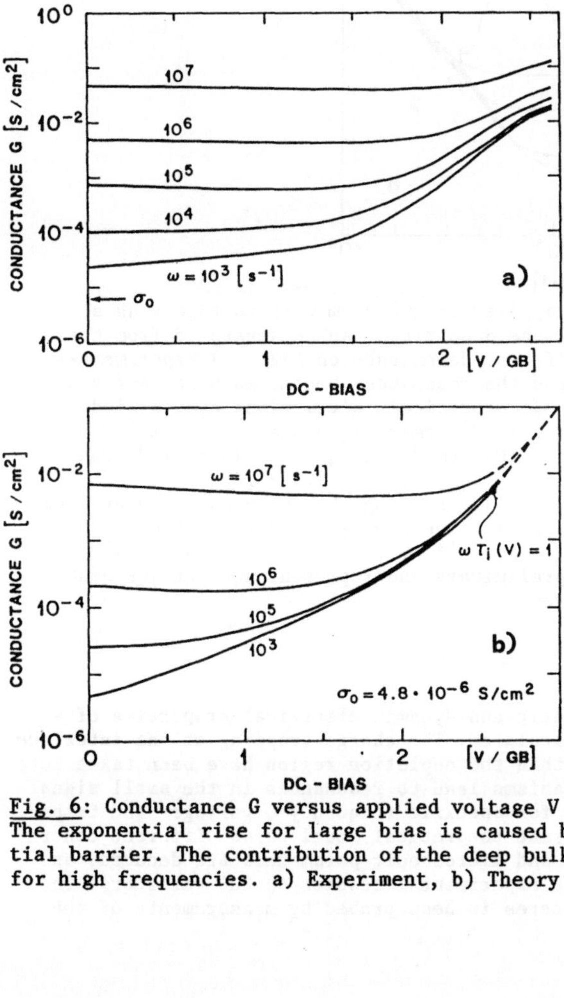

Fig. 6: Conductance G versus applied voltage V for various frequencies ω. The exponential rise for large bias is caused by the decay of the potential barrier. The contribution of the deep bulk traps becomes dominant for high frequencies. a) Experiment, b) Theory

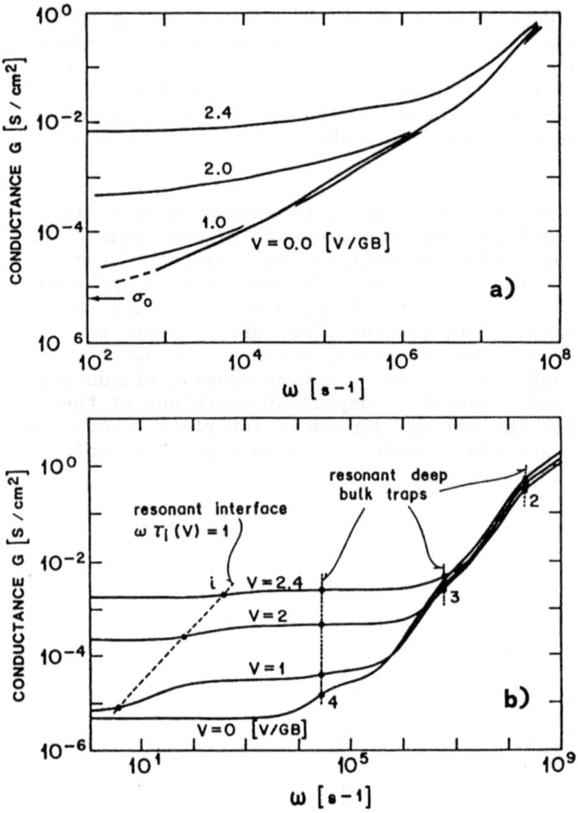

Fig 7: Conductance G versus applied frequency ω with voltage V as a parameter. The resonances of the deep traps can be separated from the interface effects by the different dependence on bias. a) Experimental results. At lower temperatures the resonances become much sharper and shift to lower frequencies. b) Theoretical calculations are carried out using the densities $N_0 = 2 \cdot 10^{17}$ cm^{-3}, $N_1 = 4.8 \cdot 10^{16}$ cm^{-3}, $N_2 = 4.2 \cdot 10^{16}$ cm^{-3}, $N_3 = 8 \cdot 10^{15}$ cm^{-3} and $N_4 = 5 \cdot 10^{15}$ cm^{-3}. The relaxation times are : $\tau_1 = 5.4 \cdot 10^{-10}$s, $\tau_2 = 5.4 \cdot 10^{-9}$s, $\tau_3 = 1.6 \cdot 10^{-7}$s, and $\tau_4 = 3.7 \cdot 10^{-5}$s using the energies ε_V mentioned in the text. The capture cross-sections are $c_1 = 3.8 \cdot 10^{-15}$cm^2, $c_2 = 4.4 \cdot 10^{-14}$ cm^2, $c_3 = 5.8 \cdot 10^{-14}$ cm^2, and $c_4 = 5 \cdot 10^{-14}$ cm^2. These parameter values are preliminary and represent not yet the best fit to the experimental curves

6. Summary and Conclusions

In the present paper the static and dynamic electrical properties of a grain boundary have been calculated. The charge trapping at the interface and by the defect states within the depletion region have been taken into account. Both trapping mechanisms lead to resonances in the small signal capacitance and conductance for specific frequency-, voltage- and temperature-ranges. With the present theoretical model these different contributions can be disentangled and microscopic parameters are deduced, as we have illustrated for some experimental results on ZnO-varistors. The dynamics of the interface states is best probed by measurements of the capacitance C(ω, V, T), where the direct contributions from the deep volume traps are only minor. The small signal conductance G(ω, V, T) on the other hand is strongly affected by the presence of deep bulk traps and provides a very sensitive method for their characterization.

The formalism presented in this paper can easily be extended to p-type materials and to include acceptor as well as donor states within the depletion region. The full theoretical treatment for several bulk trapping states of either donor- or acceptor-type will be given elsewhere.

The comparison of our calculations to the experimental data on polycrystalline ZnO reproduces all of the essential features seen in the admittance over a wide frequency- and voltage range and is a strong support for the presented model of the grain boundary. At high voltage, minority carriers are created by impact ionization. They strongly affect the electrical properties of the junction. The capacitance, for instance, shows a turn-over followed by a rapid decrease to negative values for low frequencies. This effect is not yet taken into account in our calculation. However, the turn-over and decrease of the capacitance at high voltage can also be obtained by a filling of the interface states. Thus care must be taken when experimental results are interpreted. The nature of the decrease is different for the experimental and the theoretical curve, since the latter approaches continuously the high frequency value C_{HF}.

The purpose of the above comparison between experiment and theory was to illustrate the complex electrical response of a grain boundary. We found that small changes in the microscopic parameters can significantly affect the results. The values of the parameters so far give the correct order of magnitude and further improvement is still possible.

Acknowledgment

The authors wish to thank J. Bernasconi for many helpful discussions and J. Waldvogel for his technical assistance.

References

1. See review articles in present proceedings.
2. C.H.Seager and G.E.Pike, Appl.Phys.Lett. 37, 747 (1980) and G.E.Pike, Grain Boundaries in Semiconductors, edited by H.J.Leamy et.al. (North Holland, Amsterdam, 1982), p.369, and Phys.Rev.B 30, 795 (1984).
3. J.Werner and H.Strunk, J.de Physique C1, 89 (1982), and Thesis, University of Stuttgart 1983, and references therein.
4. W.G.Oldham and S.S.Naik, Solid-State Electronics 15, 1085 (1972).
5. Y.Zohta, Solid-State Electronics 16, 1029 (1973).
6. D.L.Losee, J.Appl.Phys. 46, 2204 (1975).
7. C.H.Seager and G.E.Pike, Appl.Phys.Lett. 40, 471 (1982).
8. H.Morel, J.P.Chante, and J.P.Colinge, Proceedings MRS-Meeting 1984, J.de Physique, in press.
9. J.Werner, these proceedings and to be published in Phys.Rev.Lett..
10. F.Greuter, J.Bernasconi and R.S.Perkins, Ceramics Bulletin 63, 480(1984); F.Greuter and G.Blatter, to be published.
11. Any prefactors in (12) have been neglected for the present first comparison.
12. H.R.Philipp and L.M.Levinson, J. of Appl. Phys. 47, 1112(1976).

Beam Induced Current Characterization in Polycrystalline Semiconductors

C. Donolato
CNR-Istituto LAMEL, Via Castagnoli 1, I-40126 Bologna, Italy

1 Introduction

A finely focused light or electron beam can act as a localized source of excess carriers in semiconductors. If the semiconductor contains an internal field, for instance due to a p-n junction, the injected electrons and holes will be driven into opposite directions and produce a current in an external circuit. This signal carries considerable information on the electrical properties of the material and can be used, for instance, to determine carrier lifetime. By scanning the beam over the sample surface and monitoring the induced current, an image of the sample is obtained which can reveal the presence of crystal defects.

Extensive reviews on the electron beam induced current (EBIC) method are available /1-3/. These lectures will focus the attention on the application of EBIC and the related light beam (LBIC) technique to the assessment of polycrystalline materials. After an outline of the basic theoretical schemes of the two techniques, the use of EBIC/LBIC for the quantitative characterization of grain boundary and intragrain recombination in polycrystalline silicon for solar cells is described.

2 The Beam Induced Current Method

Figure 1 shows schematically the usual arrangement for EBIC observations in the scanning electron microscope (SEM), when charge collection occurs by means of a Schottky diode. The electron beam of the SEM penetrates the thin Schottky metallization and generates electron-hole pairs in the semiconductor. These pairs are separated and collected by the built-in field and produce a current that can be used, after amplification, to modulate the intensity of the SEM monitor. Since the sample and the display are scanned synchronously, the EBIC image gives a map of the current collected at each scanned point. At defects, the collected current is generally lower, because of the locally enhanced carrier recombination; therefore defects show up as dark features in the display.

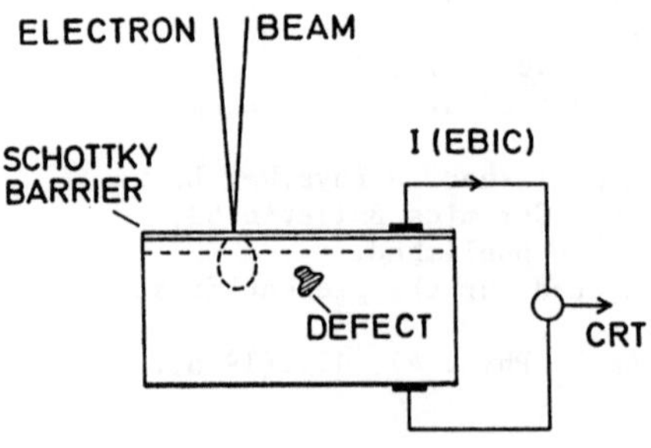

Fig.1 Schematic illustration of the electron beam induced current method

Similar images are obtained by probing the sample with a focused light beam. The raster pattern is provided either by deflection of the beam by servo-controlled mirrors /4/ or by displacing the sample under a stationary beam /5/. Comparable resolution ($\sim 1\ \mu m$) has been obtained in EBIC and LBIC images of dislocations /5/.

To interpret quantitatively induced current images of defects, a model for the contrast mechanism involved is required. The basic processes to be described are:

- the generation of carriers by the electron/light beam
- the transport and collection of beam-generated carriers
- their recombination at a defect.

Under some simplifying assumptions, an analytical discussion of these processes is feasible, as will be shown in the next Sections.

2.1 Carrier Generation

Electron Beam

The SEM beam is composed of 5÷40 kV electrons that can be focused to ~ 100 Å. These electrons penetrate the sample and undergo both elastic and inelastic scattering events; as a result they spread laterally and simultaneously lose energy by ionizing the sample atoms (Fig.2). In crystalline semiconductors this ionization process involves the raising of electrons from the valence to the conduction band, leaving holes in the valence band. The energy ε required to form an electron-hole pair is of the order of 3 times the band gap energy; the experimental value for silicon is $\varepsilon = 3.6$ eV. Hence a single electron with energy E generates E/ε electron-hole pairs; this shows that pair generation by kilovolt electrons is a very efficient process, having a gain $\sim 10^3$.

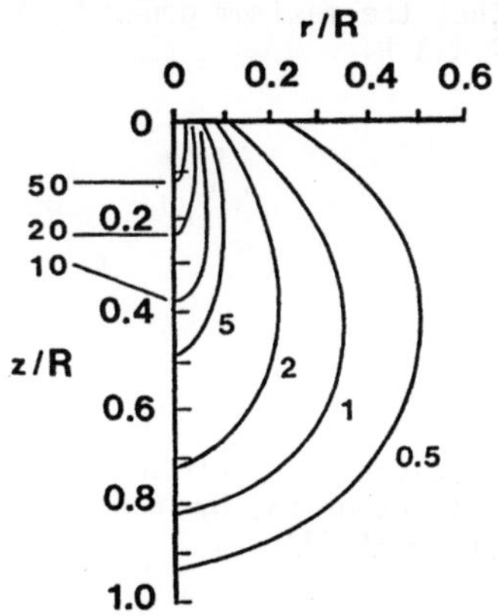

Fig.2 Normalized distribution of the generation due to a focused electron beam, displayed as contours of equal ionization (after /6/). The distance scales are normalized to the electron range R

A relevant parameter characterizing electron excitation is the depth of penetration of primary electrons, or electron range R, which also corresponds to the maximum depth of pair generation. In silicon, the following empirical relation has been found to hold for $5 < E < 25$ keV /6/

$$R = 0.0171\ E^{1.75} \qquad R \text{ in } \mu m,\ E \text{ in keV} \qquad (1)$$

Table 1 gives R for selected values of E according to (1). Similar rela-

Table 1 Values of the electron range in silicon for different beam energies, as computed from (1)

E(keV)	2.5	5	10	15	20	30	40
R(μm)	0.08	0.3	1	2	3.2	6.6	10.9

tionships have been given for other semiconductors /8/ and show that at a given beam energy the value of R is roughly inversely proportional to the semiconductor density. The knowledge of the approximate depth up to which electron-hole pairs are generated is useful to interprete quantitatively EBIC micrographs, especially in the case of planar structures.

Figure 2 shows that the electron beam generation is not uniform over the generation volume, but is highest closest to the surface and the beam axis. If we limit our attention to the depth distribution of the generation by integrating the distribution g(x,y,z) of Fig.2 over planes parallel to the surface at a depth z, we obtain the depth-dose function g(z). According to this definition, g(z)dz gives the number of pairs generated per second between z and z+dz. It has been shown /6/ that g(z) can be written as a universal function of the normalized depth z/R

$$g(z) = \frac{g_0}{R}\,\Lambda(z/R) \tag{2}$$

where $g_0(s^{-1})$ is the total generation rate. According to Everhart and Hoff /7/, $\Lambda(\zeta)$ with $\zeta = z/R$ is given by

$$\Lambda(\zeta) = 0.6 + 6.21\zeta - 12.4\zeta^2 + 5.69\zeta^3 \qquad 0 \le \zeta \le 1.1 \tag{3}$$

and is represented in Fig.3. This figure shows that the maximum generation occurs at z=0.3 R; the generation is zero for z > 1.1 R.

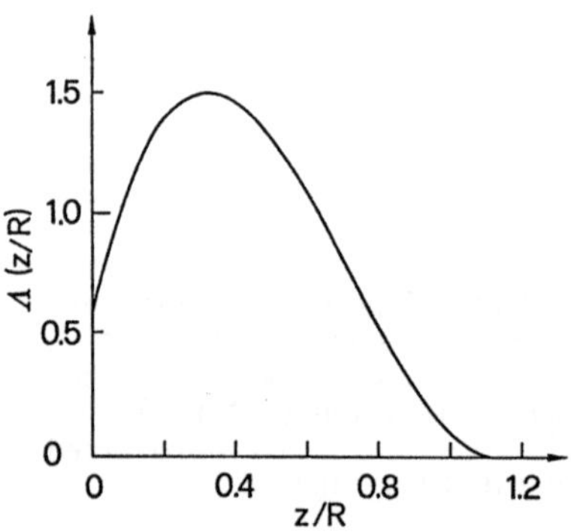

Fig.3 The universal electron depth-dose function (3) by EVERHART and HOFF /7/

The knowledge of g(z) is the starting point for the solution of the one-dimensional carrier transport problem in planar structures, when EBIC is used to evaluate bulk or surface properties of a device /9/. However, to evaluate quantitatively EBIC images of defects, the three-dimensional nature of the generation must be considered /10/. Although analytical approximations to the distribution of Fig.2 can be given /11/, it is often convenient to use a simplified, but still three-dimensional, generation scheme. In silicon a very useful scheme is that of the uniform generation sphere, where the generation is supposed to occur uniformly over a sphere of diameter R tangent to the surface /12/. Generations at a point or along a line with the distribution (3) can be assumed, to simplify the solution of the carrier transport problem. However, the generation scheme should be chosen keeping in mind the basic features of the EBIC experiment to be described.

Light Beam

Carrier generation by light has rather different properties in comparison to electrons. First, visible photon energies are of the order of a few eV, i.e. a factor $10^3 \div 10^4$ less than usual electron energies in an SEM. Thus a photon (with over-bandgap energy) generates a single electron hole pair, whereas an electron in the SEM typically generates thousands. In addition, it is not possible to define for photons a maximum penetration depth in the semiconductor, since light absorption follows an exponential law. Accordingly, the generation rate can be written as

$$g(z) = \phi_o \, \alpha \, \exp(-\alpha z) \qquad (4)$$

where ϕ_o is the number of photons incident /sec-cm^2 and α is the light absorption coefficient. It may be interesting to observe that g(z) of (4) is a universal function of αz, with α^{-1} playing the role of the electron beam range R in (2). The parameter α is a function of the photon energy (or wavelength λ). Table 2 gives the values of α^{-1} in Si for selected values of λ in the range $0.4 < \lambda < 1\ \mu m$. LBIC experiments are most frequently interpreted assuming generation along a line /13,14/, but the actual finite beam cross section and divergence has also been taken into account /16/.

Table 2 Values of the reciprocal light absorption coefficient in silicon for different wavelengths /15/

λ (μm)	0.4	0.5	0.6	0.7	0.8	0.9	1.0
α^{-1}(μm)	0.2	1.0	2.2	5.0	10.5	40	100

2.2 The Photovoltaic Effect

Electron or light excitation in a homogeneous semiconductor only produces an excess of free carriers. To give rise to a potential difference (or a current), the generated electrons and holes must be driven in opposite directions. In the photovoltaic effect (or in the related electron-voltaic effect) carrier separation arises through the action of the built-in field of a p-n (or Schottky) junction (Fig.4). By connecting an external circuit across the junction a photovoltage V or, if the load resistance is low, a photocurrent I (the beam-induced current) can be detected. This current arises both from the carriers generated in the depletion layer and from the minority carriers generated in the neutral n,p regions and diffusing into the space-charge region.

The one-dimensional modelling of charge collection in p-n junctions is well established, especially in connection with the development of solar cells /17/. However, the corresponding three-dimensional analysis, which is required to describe the EBIC imaging of defects, appears to be too complicated, so that the actual charge-collection process must be idealized further.

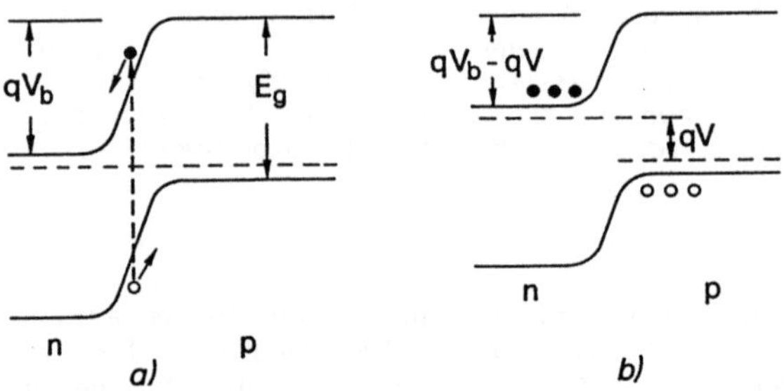

Fig.4 Generation of a photovoltage in an irradiated p-n junction. V_b is the built-in potential and V is the beam-generated voltage

In a Schottky diode (or a shallow p-n junction) with deep enough irradiation, most of the current arises from the base region, provided that bulk lifetime is not too small. The depletion region and emitter give a small contribution to the total current and do not require a detailed description. Thus the device can be assumed to be a homogeneous semiconductor and the collection of excess minority carriers from the base can be described as a condition of infinite surface recombination velocity. This approximation leads to a pure diffusion problem for minority carriers; the collected current is then the minority diffusion current to the surface. Carrier transport under these simplifications will be first discussed in one dimension, mainly for illustration purposes, although the usefulness of the above scheme will be more apparent in the three-dimensional case to be discussed in Sect.2.4.

2.3 One-Dimensional Charge Collection

The excess minority carrier density p(z) produced by an external generation g(z) obeys the steady-state diffusion equation

$$D \frac{d^2p}{dz^2} - \frac{1}{\tau} p = - g(z) \tag{5}$$

where D and τ are the diffusion coefficient and lifetime of minority carriers, respectively. The assumption of infinite surface recombination velocity at the surface yields the boundary condition

$$p(0) = 0 \tag{6}$$

The collected current is

$$I_\circ = D \left. \frac{dp}{dz} \right|_{z=0} \tag{7}$$

Equation (5) with (6) can be solved by standard methods; substitution of the solution into (7) leads to the final result

$$I_\circ = \int_0^\infty g(z) \exp(-z/L) \, dz \tag{8}$$

where $L=(D\tau)^{\frac{1}{2}}$ is the minority carrier diffusion length. Equation (8) can be interpreted by saying that the carriers generated at a depth z have a probability

$$\varphi(z) = \exp(-z/L) \quad (9)$$

of being collected by the junction.

As an application, it will be shown how (8) leads to a method of evaluating L from induced current measurements, both with electrons and light irradiation. The one-dimensional treatment given above implies uniform irradiation over the sample surface; however, it can be shown that (8) also holds in the case of localized irradiation by a narrow beam /18/. Therefore (8) can be used both for the evaluation of average and local diffusion lenghts, for instance inside single grains of polycrystalline materials.

Electron Beam

The generation function for electrons is given in (2) and (3); substituting these expressions into (8) we get

$$I_o = g_o \int_0^\infty \Lambda(\zeta) \exp(-\zeta R/L)\, d\zeta \;=\; g_o\, \eta(R/L) \quad (10)$$

The function $\eta(R/L)$ is called charge collection efficiency of the device and represents the fraction of the injected current g_o that is actually collected by the junction. Figure 5 shows η versus E for different values of L in silicon; use has been made of the relation (1) between R and E.

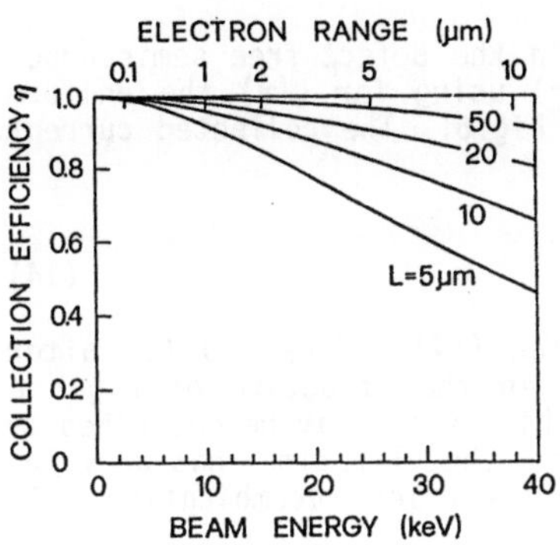

Fig.5 Charge collection efficiency at an idealized shallow p-n junction in silicon, as computed from (10)

The value of η can be determined experimentally by measuring I_o and evaluating g_o from the relation

$$g_o = I_b\, E(1-f)/\varepsilon \quad (11)$$

where I_b, E are the (measurable) beam current and energy and f is the backscattering coefficient of the sample. By comparing experimental values of η at different E with the curves of Fig.5 the value of L can be obtained. However, it appears that better accuracy is obtained for small values of L, i.e. for L comparable with the maximum R available /19/; Table 1 shows that at 40 keV, a typical maximum energy in an SEM, $R \simeq 11\ \mu m$. Diffusion lengths much longer than this value are determined more conveniently by a light beam, which allows higher penetration depths (see Table 2).

Light Beam

In the case of excitation by light, using (4) in (8), we obtain easily

$$I_\circ = \phi_\circ/\left[1 + (\alpha L)^{-1}\right] = \phi_\circ \eta(\alpha L) \qquad (12)$$

This expression is much simpler than the comparable (10), and the value of L can be obtained in this case by a graphical procedure, since $1/\eta$ is now a linear function of α^{-1}. Usually α is varied by selecting different light wavelengths and, at each wavelength, $\phi_\circ$ is adjusted (and measured) to give a constant value of I_0. If $\phi_\circ$ is plotted against α^{-1} at constant I_0, L may be obtained as the negative intercept on the α^{-1} axis. The constance of I_0 can be also established by measuring the correspondent open circuit voltage: this is the well-known surface photovoltage method /20/. Since α^{-1} can be of the order of $\sim$100 μm (see Table 2), diffusion lengths up to this order of magnitude can be evaluated. A detailed description of the method, of its implementation and inherent limitations is given in /20/.

2.4 Three-Dimensional Carrier Diffusion at a Defect

The analogous of (5), (6) in three dimensions is

$$\begin{aligned} &D\,\nabla^2 p_\circ(\underline{r}) - \frac{1}{\tau}\, p_\circ(\underline{r}) = -\, g(\underline{r}) \\ &p_\circ(\underline{r})\big|_{z=0} = 0 \end{aligned} \qquad (13)$$

where the index $\circ$ indicates that (13) holds in the defect-free semiconductor. An example of the distribution of $p_0(\underline{r})$ using for $g(\underline{r})$ the uniform generation sphere approximation is shown in Fig.6. The collected current is given by

$$I_\circ = \int_V g(\underline{r})\, \varphi(z)\, dV \qquad (14)$$

where V is the half space $z \geq 0$; the expression (14) reduces to (8) after integration over x,y. Let us now introduce in the structure of Fig.1 a defect enhancing the carrier recombination. This effect may be described by representing the defect as a region F where the lifetime has a value $\tau' \ll \tau$ (Fig.7) /18/. Inside F the minority carrier recombination rate will be

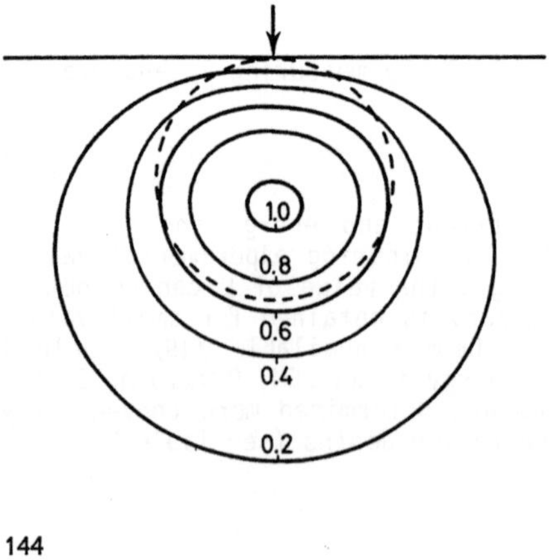

Fig.6 Distribution of the excess minority carrier density $p_0(\underline{r})$ produced by uniform generation within the dashed sphere, assuming L = ∞ (after /18/)

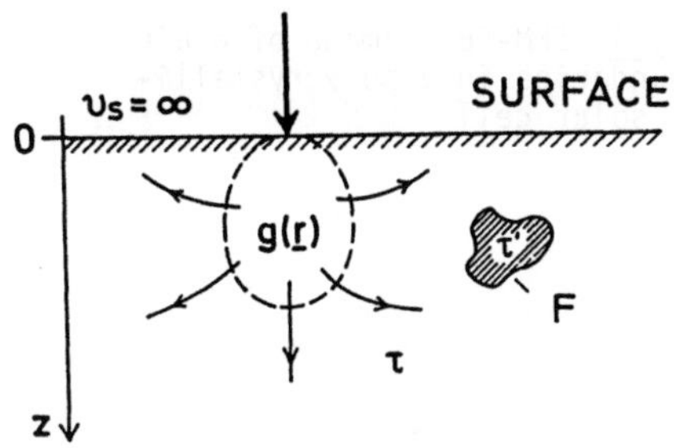

Fig.7 Schematic representation of the model used for calculating the EBIC contrast of a defect

$$- \frac{1}{\tau'} p(\underline{r}) \qquad (15)$$

where $p(\underline{r})$ is the minority carrier distribution in presence of the defect. Since $p(\underline{r})$ cannot be calculated easily, we may take as a first approximation in (15) $p(\underline{r}) \simeq p_o(\underline{r})$. Thus the influence of F can be represented by adding to the original source $g(\underline{r})$ over V a sink term $-p_o(\underline{r})/\tau'$ over F; this term is known once (13) has been solved. By applying (14) to the resulting net generation, we can express the collected current in presence of F as

$$I = \int_V g(\underline{r})\, \varphi(z)\, dV - \frac{1}{\tau'} \int_F p_o(\underline{r})\, \varphi(z)\, dV = I_o - I^* \qquad (16)$$

In (16) I^* represents the EBIC signal produced by F and is dependent on the the position of the beam relative to the defect, while I_o corresponds to the constant background current. It is useful to introduce the image contrast $i^*=I^*/I_o$ which gives an objective measure of the current reduction at a defect, independently of beam intensity or signal amplification. Figure 8 illustrates schematically the relations between I_o, I^* and i^*. Equation (16) applies to a defect of arbitrary shape and has been used, for instance, to simulate by computer the EBIC images of dislocations and stacking faults /21/. In the next Section, the expression (16) will be specified to the case of a grain boundary.

3 EBIC/LBIC Contrast of Grain Boundaries

The following discussion deals with large-grain polycrystalline semiconductors, as those used for solar cell fabrication, because in these materials induced current images show well separated, linear features of

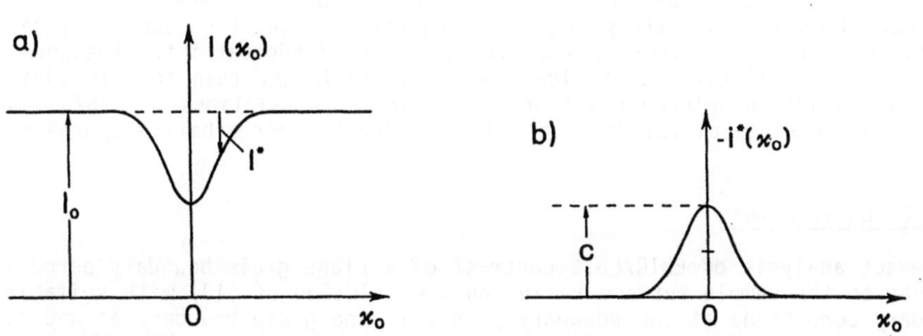

Fig.8 Typical beam induced current profile at a defect (a) and related contrast profile (b)

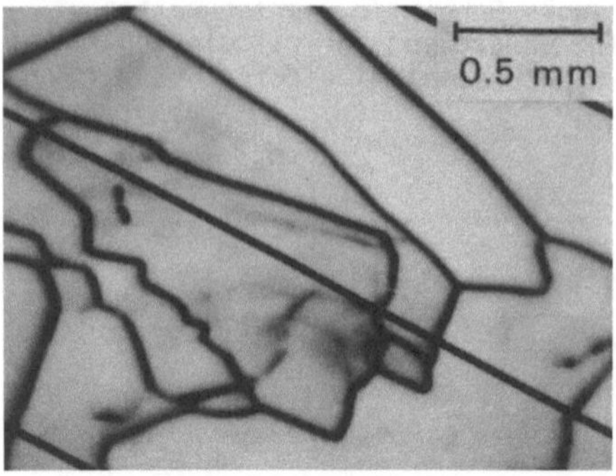

Fig.9 SEM-EBIC image of grain boundaries in a polycrystalline solar cell

constant intensity (Fig.9), which indicate a simple grain boundary configuration. In fact, for instance, in the Wacker "Silso" grains are of the order of some millimeters and have a columnar structure, so that grain boundaries have often plane portions perpendicular to the slice surface. The simple geometry will make a quantitative description of the image much easier.

3.1 First Order Treatment

In the framework of the defect scheme of Sect.2.4, a grain boundary can be described as a (plane) layer of thickness h much smaller than the extension of the generation regions. Hence the variation of p_o across h can be neglected and the integral over F in (16) can be replaced with a surface integral over the plane of the boundary. Accordingly, the image contrast can be written as

$$i^* = v_s \frac{1}{I_\circ} \int_\Sigma p_\circ(\underline{r}_\sigma)\, \varphi(z_\sigma)\, d\sigma \qquad \text{with} \tag{17}$$

$$v_s = h/\tau' \tag{18}$$

The parameter v_s gives a measure of the recombination strength of the grain boundary; the physical dimensions of v_s [cm s^{-1}]suggest the identification of this parameter with the interface recombination velocity. For a straight boundary, (17) gives the first order contrast profile. If the boundary is perpendicular to the surface, it is possible to treat the diffusion problem exactly, as will be shown in the next Section. The advantage of an exact treatment lies in its ability of describing strong (i.e. high contrast) defects, for which a first-order analysis may prove inadequate: for instance, for large v_s, (17) gives unphysical values of i* larger than one. In addition, a comparison between exact and first order calculations is useful to establish the range of validity of the more flexible perturbative approach.

3.2 Exact Treatment

The exact analysis of EBIC/LBIC contrast of a plane grain boundary perpendicular to the sample surface relies on the solution of (13) with suitable boundary conditions at the boundary plane. If the grain boundary at x=0 is described as an interface with recombination velocity v_s (Fig.10), the boundary conditions for $p(\underline{r})$ are

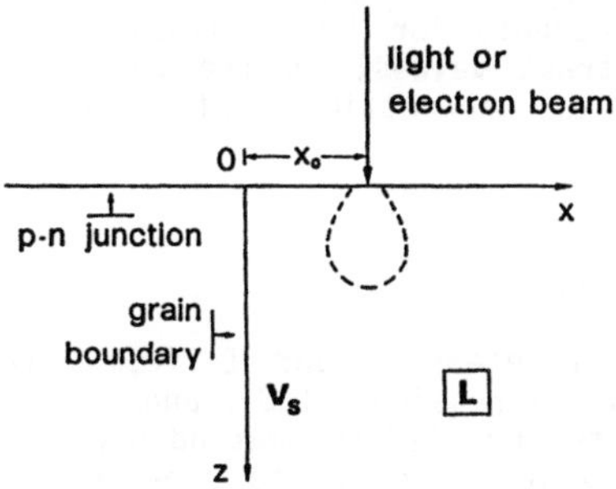

Fig.10 Schematic representation of a grain boundary as an interface with recombination velocity v_s

$$p|_{x=0^+} = p|_{x=0^-} \tag{19}$$

$$D \left.\frac{\partial p}{\partial x}\right|_{x=0^+} - D \left.\frac{\partial p}{\partial x}\right|_{x=0^-} = v_s \, p|_{x=0}$$

Because of the translational symmetry along y, the diffusion problem can be reduced to two dimensions. The related solution is described in /22/ and leads to the following expression for the contrast profile of the grain boundary

$$i^*(x_o) = \frac{2s}{\pi I_o} \int_0^\infty dk \int_0^\infty dz \int_{-\infty}^\infty F(x,z,k)\, h(x-x_o,z)\, dx \tag{20}$$

where

$$F(x,k,z) = \frac{k \sin(kz) \exp(-\mu|x|)}{\mu^2(2\mu+s)} \qquad \mu = (k^2+L^{-2})^{\frac{1}{2}} \tag{21}$$

$s=v_s/D$ and $h(x,z)$ is the projection of $g(\underline{r})$ onto the x,z plane. Equations (20) and (21) show that i^* depends upon the relevant recombination parameters in a complicated way; in particular the dependence of i^* on v_s is no longer linear as in (17).

A comparison between exact and first-order model is given in Fig.11 which shows the dependence of the grain boundary contrast $c=i^*(0)$ on the

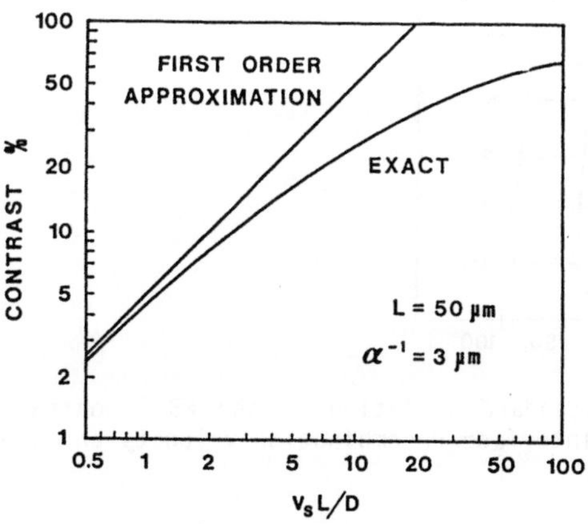

Fig.11 Comparison between first order and exact calculation of the LBIC contrast of a grain boundary

normalized recombinaton velocity $v_s L/D$ for light beam excitation. It is apparent that the linear approximation is adequate for low contrast, but becomes progressively worse for higher contrast values; in the case of Fig.11, for c=30% the linear model yields an underestimate of s by a factor $\simeq 2$.

3.3 Evaluation of s and L from a Contrast Profile

To determine the relevant recombination parameters s and L from an experimental profile using the theoretical expressions (20) and (17) several methods can be used. The more accurate, but tedious method uses a two-parameter (s,L) fit of the whole theoretical profile to experiment /14,16/. However, this requires the knowledge of the beam size, which influences the shape of the profile, especially in the LBIC case /16/. The assumption of negligible beam diameter, while simplifying the expression of $i^*(x_o)$, does not allow a good fitting near the maximum of the profile /14/.

A more rapid method relies on the evaluation of two properties of the measured peak, possibly weakly dependent on the spot size, and uses the theory to relate them to s, L. It has been shown /22/ that suitable properties are the profile area A and variance σ^2

$$A = \int_{-\infty}^{+\infty} i^*(x_o)\, dx_o \qquad \sigma^2 = \frac{1}{A}\int_{-\infty}^{+\infty} x_o^2\, i^*(x_o)\, dx_o \tag{22}$$

which are rather insensitive to the lateral distribution of the generation, provided that the lateral extension of the generation volume is small in comparison to L. The explicit expression of A and σ are obtained by substituting (20) into (21), and have been used to draw the diagram of Fig.12, which applies to electron excitation; a similar diagram holds for LBIC /22/. Figure 12 relates the area and variance normalized to the electron range A/R, σ/R to L/R and sL. In a given EBIC experiment, R is known as a function of the SEM beam energy, A and σ are evaluated from the experimental contrast profile. The couple of values (A/R, σ/R) defines a

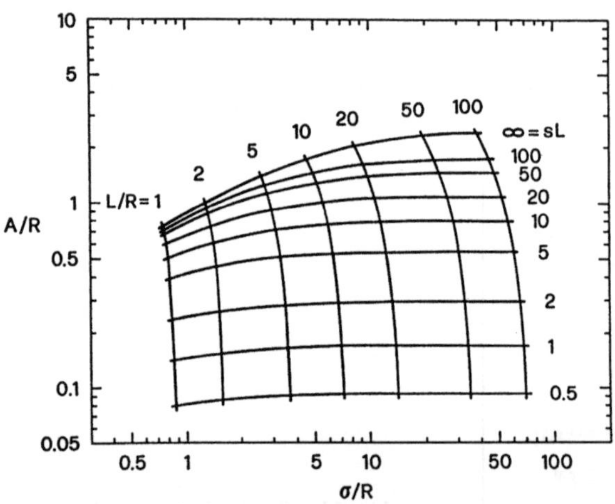

Fig.12 Diagram relating area and standard deviation of the EBIC contrast profile of a grain boundary to the interface recombination velocity and minority carrier diffusion length /22/

point in the rectangular coordinate system of Fig.12; the curvilinear coordinates of this point (L/R,sL) are sufficient to determine both L and s, since R is known. A practical example of application of this method is given in the next Section.

3.4 A Worked Example - Practical Considerations

Figure 13 shows the magnified EBIC image of a grain boundary and the associated line scan in a polycrystalline solar cell. The cell was fabricated on p-type Wacker "Silso", 5 ohm cm resistivity, by phosphorus diffusion and had a junction depth $\simeq 0.3\ \mu m$.

The quantitative evaluation of such a profile is made much easier by the use of a computer connected to the SEM for digital acquisition and processing of the data. It is also possible to store the diagram of Fig.12 in the same computer, so that the complete procedure described in the previous Section can be performed automatically, to yield as an output the values of s and L of a selected grain boundary /23/. In the case of Fig.13 this acquisition occurred through an expanded X-ray analysis system (Tracor-Northern TN-2000), which already includes a computer and profile processing and display facilities.

The profile of Fig.13 is slightly asymmetric, most probably because of the different value of L in the adjacent grains. However, this difference does not produce a visible change in the background current I_o in the two grains, since I_o is rather insensitive to variations of L for larger values of L (see Fig.5). For stronger peak asymmetry, a correlation between the steeper side of the profile and the grain with lower I_o can be observed /16/.

Although the model of Sect.3 assumes a unique value of L on either side of the grain boundary, it can be applied to a good approximation to each half of an asymmetrical profile. The results of the evaluation of the peak of Fig.13 are summarized in Table 3. It is interesting to observe that the values of v_s obtained by analyzing the two halves of the profile are nearly equal, although the values of L are rather different: this indicates that the model yields a consistent evaluation of v_s, which is a unique property of the interface.

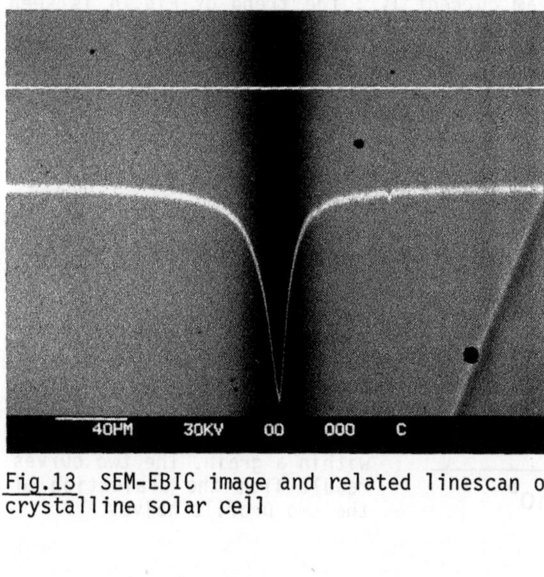

Fig.13 SEM-EBIC image and related linescan of a grain boundary in a polycrystalline solar cell

Table 3 Quantitative evaluation of the EBIC profile of Fig.13. The value D = 25 cm^2s^{-1} has been assumed

Profile side	A(μm)	σ(μm)	A/R	σ/R	L/R	sL	L(μm)	v_s(cm s^{-1})
left	2.0	21.1	0.30	3.19	4.6	2.2	30.4	1.8 10^4
right	1.4	16.4	0.21	2.48	3.4	1.4	22.4	1.6 10^4

In order to measure a contrast profile it is essential to establish the zero EBIC level (not shown in Fig.13), which is usually not relevant for imaging purposes. The evaluation of A and σ also requires the specification of the average base line; this is easier if the background is regular and noise-free. In particular the evaluation of σ is sensitive to the choice of the background and irregularities in the profile tails, because of the factor x_0^2 in its definition (22). The contribution of the profile tails to A and σ is best evaluated by using a fit to the experimental data /23/. It may happen that the profile quality is insufficient to obtain a reliable value of σ , yet good enough to evaluate A. In this case L should be determined independently, for instance by the method of Sect.2.3, and A used to determine s /24/.

3.5 Influence of the Injection Level

For the purpose of calculating the EBIC contrast profile, a grain boundary has been represented schematically as an interface with fixed recombination velocity v_s , independent of the external excitation. This holds for sufficiently weak excitation, but theoretical arguments suggest that v_s should decrease with the injection level at higher excitations /25,26/. Figure 14 gives experimental evidence for this circumstance. The value of v_s of a grain boundary has been determined by the method of Sect.3.3 for different values of the SEM beam current I_b ; the trend of Fig.14 is seen to be consistent with theoretical expectations. A similar effect has been observed in LBIC experiments /24,26/, though to a lesser degree, probably because of the lower injection usually obtained by light in comparison to electrons. Therefore, the true value of v_s is only obtained under low

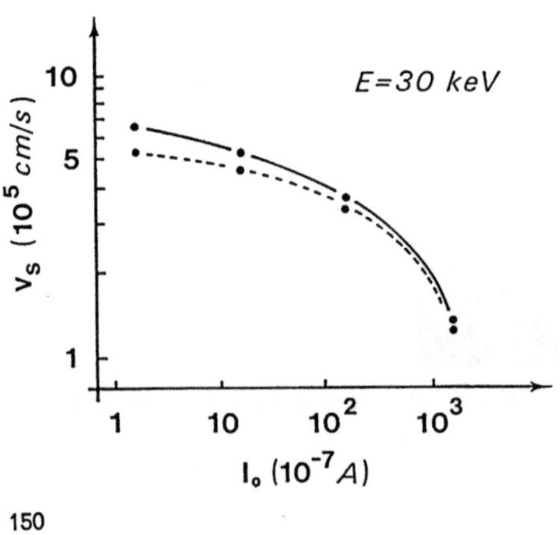

Fig.14 Dependence of the recombination velocity of a grain boundary on the injection level /27/. The injection level has been monitored through the EBIC signal I_o within a grain. The two curves result from the evaluation of the two profile halves

injection conditions, where unfortunately the signal to noise ratio is less favourable. For comparison purposes, a possibility is to operate with the same beam current, since v_s is not strongly dependent on I_b (in Fig.14 an increase of I_o by a factor $\sim 10^3$ gives a reduction of v_s by a factor of 4, /23/). However, the same excitation may give different injection conditions at grain boundaries with different recombination strengths /25,26/.

The effect shown in Fig.14 is a consequence of a real electrical structure of a grain boundary, which is not included in the purely phenomenological description of Sect.3.1, 3.2. It is known that grain boundaries are in general electrically charged by carriers trapped at the interface. This causes space-charge regions to form on both sides of the grain boundary plane (Fig.15). The electric field in these regions enhances the collection of minority carriers toward the grain boundary and consequently their recombination. Hence the parameter v_s introduced before should be regarded as an effective recombination velocity, which includes the effect of the two neighbouring depletion regions.

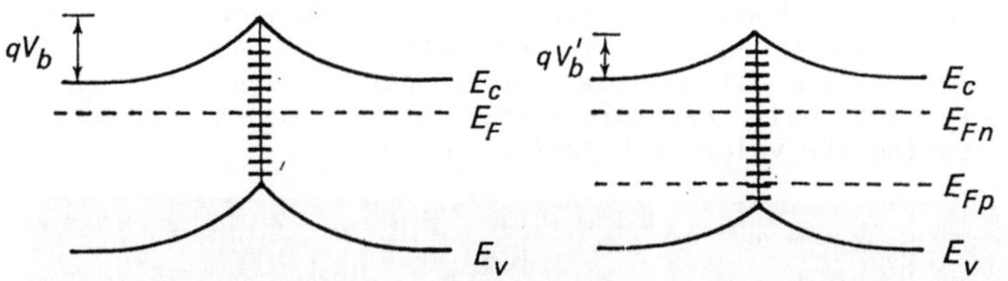

Fig.15 Energy band diagram in the neighborhood of a grain boundary in a polycrystalline semiconductor. (a) without external excitation (equilibrium); (b) under external excitation

It has been shown /25,28/ that

$$v_s = v_s(0) \exp(qV_b/kT) \tag{23}$$

where $v_s(0)$ is the recombination velocity at the boundary plane and V_b is the grain boundary potential barrier height, and

$$v_s(0) = N_i \, \sigma \, v_{th} \tag{24}$$

where $N_i(cm^{-2})$ is the density of the interface recombination centers, σ their cross-sections and v_{th} is the carrier thermal velocity. Equation (23) shows that the presence of a built-in potential (and field) enhances carrier recombination at a grain boundary. For low excitation, V_b remains essentially unchanged; however, for higher excitation, the minority carriers reaching the grain boundary plane neutralize part of the fixed charge and reduce the value of V_b. This produces a decrease of the effective recombination velocity v_s (see (23)), which is the parameter actually being measured in EBIC/LBIC experiments.

The above discussion indicates that a more realistic model of a grain boundary should include ab initio the dependence of v_s on the injection level, since the conditions at the grain boundary can change from low to high injection according to the position of the beam. This may affect the shape of the profile and consequently the values of A and σ ; however, the diffusion problem would involve non-linear boundary conditions /29/ and become rather difficult.

4 Application to Passivation Studies

The enhanced recombination of generated carriers at grain boundaries is known to decrease the efficiency of polycrystalline solar cells. Since the introduction of these defects cannot be avoided during the fabrication of the material, attempts have been made to neutralize their deleterious influence afterwards, by employing some passivation process. An effective passivation method uses annealing of the polycrystalline silicon in a hydrogen plasma. For instance, in EFG (Edge-Defined Film-Fed Growth) silicon ribbon, a treatment at $\sim 300°C$ for a few minutes in a Kaufman ion source was shown to give a substantial efficiency increase /30,31/.

Beam induced current methods are useful for monitoring passivation processes. Figure 16 gives the EBIC images of an EFG sample before and after passivation: it is seen that all boundary-related contrast virtually disappears, which indicates that passivation extends at least up to the electron beam range $R(30\ keV) \simeq 6.5\ \mu m$. This sampled depth is significant for the actual cell operation, since about 60% of the incident sunlight photons would be absorbed within this depth /3/. Before complete passivation, when some grain boundary contrast is still observable, it is possible to use the method of Sect.3.3 to evaluate the influence of the H-treatment both on v_s and L. It has been shown that passivation reduces both grain boundary and bulk recombination, by decreasing v_s while simultaneously increasing the value of L /23/.

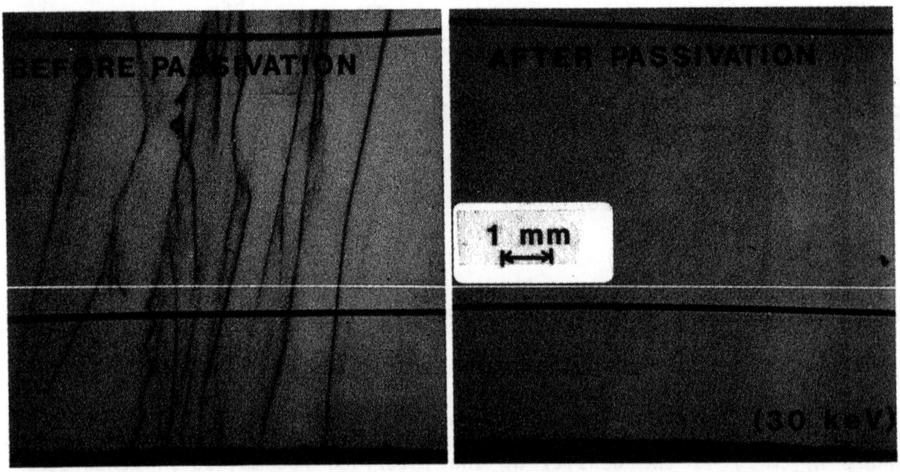

Fig.16 SEM-EBIC image of an EFG silicon ribbon solar cell before and after hydrogen passivation (courtesy of J.I. Hanoka)

Recently, EBIC has been employed to determine directly the depth of passivation /32/. The method consists of producing hydrogen penetration through the cleaved edge of a polycrystalline sample and observing the EBIC contrast in the usual plane view. Figure 17 shows, for instance, an effective passivation of the grain boundary up to a depth of about 20 μm. It has also been shown that the passivation depth of the grain boundary is dependent on the associated recombination velocity, being higher for lower values of v_s. By measuring the passivation depth (i.e. the hydrogen diffusion depth) as a function of a subsequent annealing time, the hydrogen diffusivity along grain boundaries can also be estimated: values between $10^{-8} \div 10^{-9}\ cm^2/s$ at 400°C have been obtained /32/.

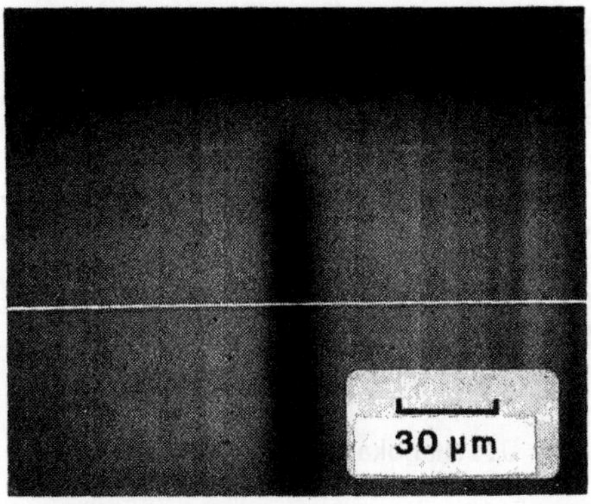

Fig.17 EBIC micrograph of recombination along a grain boundary after hydrogen passivation through the sample cross section. The horizontal black-gray limit corresponds to the sample edge (courtesy of J.I.Hanoka)

Acknowledgements

The author wishes to thank J.I.Hanoka of the Mobil Solar Energy Corporation for providing the micrographs of Figs.16 and 17.

References

1 D.B.Holt: in *Quantitative Scanning Electron Microscopy*, ed by D.B.Holt, M.D.Muir, P.R.Grant, and I.M.Boswarva (Academic, London 1974) Chap.8
2 H.J.Leamy: J.Appl.Phys.53,R51(1982)
3 J.I.Hanoka and R.O.Bell: Ann.Rev.Mater.Sci.11,353(1981)
4 J.R.Szedon, T.A.Temofonte and T.W.O'Keeffe: Solar Cells 1,251(1979/80)
5 T.Wilson, W.R.Osicki, J.N.Gannaway and G.R.Booker: J.Mater.Sci.14, 961(1979)
6 A.Cohn and G.Caledonia: J.Appl.Phys.41,3767(1970)
7 T.E.Everhart and P.H.Hoff: J.Appl.Phys.42,5837(1971)
8 G.Oelgart and H.Scholz: Phys.Stat.Sol.(a)75,547(1983)
9 G.E.Possin and C.G.Kirkpatrick: in *Scanning Electron Microscopy 1979*, ed. by O.Johari (SEM Inc., AMF O'Hare, Illinois 1979) p.245
10 C.Donolato: *ibidem*, p.257
11 C.Donolato: Phys.Stat.Sol.(a)65,649(1981)
12 J.F.Bresse: in *Scanning Electron Microscopy 1972*, ed. by O.Johari (IIT Res.Inst., Chicago 1972) p.105
13 J.D.Zook: Appl.Phys.Lett.37,223(1980)
14 C.H.Seager: J.Appl.Phys.53,5968(1982)
15 S.M.Sze: *Physics of Semiconductor Devices* (Wiley, New York 1969)
16 J.Marek: J.Appl.Phys.55,318(1984)
17 H.J.Hovel: *Solar Cells* (Academic, New York 1975)
18 C.Donolato: Optik 52,19(1978/79)
19 C.J.Wu and D.B.Wittry: J.Appl.Phys.49,2827(1978)
20 A.M.Goodman, L.A.Goodman, and H.F.Gossenberger: RCA Review 44,326(1983)
21 C.Donolato and H.Klann: J.Appl.Phys.51,1624(1980)

22 C.Donolato: J.Appl.Phys.54,1314(1983)
23 C.Donolato and R.O.Bell: Rev.Sci.Instr.54,1005(1983)
24 J.Oualid, H.Amzil, J.P.Crest, J.Dugas, G.Mathian, M.Zehaf and S.Martinuzzi: J.Physique 43,C1-313(1982)
25 C.H.Seager: J.Appl.Phys.52,3960(1981)
26 J.Oualid, C.M.Singal, J.Dugas, J.P.Crest, and H.Amzil: J.Appl.Phys.55, 1195(1984)
27 C.Donolato and R.O.Bell: unpublished results
28 H.C.Card and E.S.Yang: IEEE Trans.Electron Devices ED-24,397(1977)
29 J.G.Fossum and F.A.Lindholm: IEEE Trans.Electron Devices ED-27, 692(1980)
30 C.H.Seager, D.J.Sharp, J.K.G.Panitz and J.I.Hanoka: J.Physique 43, C1-103(1982)
31 J.I.Hanoka, C.H.Seager, D.J.Sharp and J.K.G.Panitz: Appl.Phys.Lett.42, 618(1983)
32 C.Dubé, J.I.Hanoka and D.B.Sandstrom: Appl.Phys.Lett.44,425(1984)

Part III

Properties and Applications of Polycrystalline Silicon

Optical Properties of Polycrystalline Silicon Films

Günther Harbeke

Laboratories RCA Ltd., Badenerstraße 569, CH-8048 Zürich, Switzerland

1. Introduction

Polycrystalline silicon layers have many important applications in microelectronics, such as gates, interconnects, and load resistors. The morphology and surface topography of these layers can strongly influence device performance especially in thin-film layered structures with fine device geometries. Structural properties, as influenced by deposition and processing parameters, can have an appreciable effect on the electrical properties of the layers for a given level of dopant [1]. Surface asperities and "bumps" on the initial silicon layer surface may lead to a lowering of the dielectric field strength and enhanced tunneling currents in interfacing dielectrics [2, 3].

Low pressure chemical vapour deposition (LPCVD) is the standard preparation method for polycrystalline films in integrated-circuit manufacturing. For many years, films have usually been prepared directly in the polycrystalline state at a temperature of about 620 - 650°C. They have characteristically rough surfaces which may not only degrade the dielectric properties of silicon dioxides grown on the surface but may introduce problems in photolithographic processing.

Recently, several publications have appeared on the properties of polycrystalline silicon films which were prepared by LPCVD in the amorphous phase at about 560 to 580°C and subsequently crystallized at higher temperatures [4-9]. It has been reported that these films are superior in several respects to films prepared in the polycrystalline phase. Some of the advantages cited are: much smoother surfaces, higher structural perfection, higher dielectric strength of thermal oxides on the polycrystalline silicon films and a decrease in oxide tunneling currents. Improved properties are claimed for undoped, diffusion doped, in-situ doped, and ion implanted films.

Among the various characterization methods the determination of the optical properties is extremely valuable in elucidating structure and morphology of the films. In this paper we will review the results of light scattering, optical absorption, reflectance, surface roughness, and ellipsometry measurements.

2. Light Scattering

2.1. Raman Scattering

The technique of inelastic light scattering from optical phonons, termed Raman scattering [10], has been applied by various groups to polycrystalline

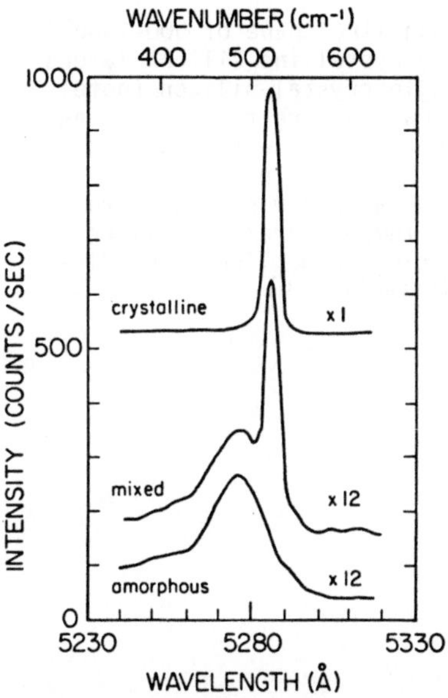

Fig. 1. (Bottom) Raman spectrum of amorphous LPCVD silicon layer, as deposited at 570°C. (Middle) Mixed amorphous-crystalline layer deposited at 590°-600°C. (Top) Crystalline layer, deposited at 570°C with subsequent annealing at 900°C. (From Ref. 5)

[4, 5, 7] and microcrystalline [11-14] silicon films. The degree of crystallization (amorphous, crystalline, or mixed) can immediately be determined from the Raman spectrum. This makes use of the fact that in the first order Raman scattering process in crystalline silicon only the zone center (q=o) optical phonon is Raman active. It is observed as a sharp line at 522 cm^{-1} as seen, e.g., in Fig. 1 (top) as obtained on an LPCVD film deposited at 570°C in the amorphous phase and subsequently annealed at 900°C. The crystallization of LPCVD amorphous silicon is a temperature activated process which takes less than 30 minutes for temperatures above 660°C. In the amorphous state the q-selection rule does not apply due to the loss of lattice periodicity. Scattering from all vibrational modes is thus permitted and the spectrum resembles the one-phonon density of states with a broad hump centered at 483 cm^{-1}, as seen, e.g., in Fig. 1 (bottom). Intermediate states between the two forms (i.e. partly crystalline, partly amorphous) are easily recognizable as a superposition as found, e.g., in Fig. 1 (center).

The results presented in Fig. 1 have been obtained by means of the standard Raman scattering technique, with 100 mW Argon-ion-laser light of wavelength 5145 Å focussed to a spot of 50 μm x 2 mm on the film surface. It was verified that these conditions are sufficient to avoid thermal crystallization by the beam. The scattered light is dispersed in a double grating monochromator and detected by photon counting. The light penetration depth is of the order of 1000 Å to more than 2000 Å depending on the degree of crystallinity (see section 4) so that there is no doubt that the volume is probed.

The conclusions concerning the degree of crystallization have been corroborated by X-ray diffraction and transmission electron microscopy. The average grain size of films deposited at 570°C and annealed at 900°C as in Fig. 1 (top) is about 800 Å whereas the average grain size of the as-grown mixed films as in Fig. 1 (center) is about 250 Å. These differences in

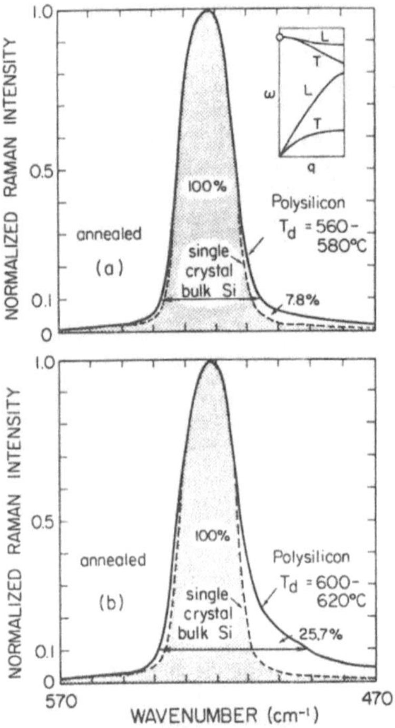

Fig. 2. Raman line shape of 900-1000°C annealed polycrystalline silicon layers and bulk single crystal silicon (note that all linewidths are to be taken as a relative measure since they are influenced by the averaging process). Top: Low-temperature/amorphous grown. Bottom: High-temperature/crystalline grown exhibiting large line shape distortions. (From Ref. 4)

the average grain size lead to marked differences in the Raman line shape depending on the deposition temperature. It is seen in Fig. 2 that films grown in the amorphous phase at 560 to 580°C and subsequently annealed/crystallized show a Raman line shape which is very close to that of bulk single crystal silicon. The phonon dispersion curves of crystalline silicon are given in the inset. The films grown directly in the polycrystalline state, on the other hand, show a highly distorted line shape with additional intensity on the high- and mainly on the low-wavenumber side. It appears that these films have a considerably higher fraction of material with a perturbed lattice and internal strain than the originally amorphous films [4, 5]. It was found that in-situ phosphorus doped films grown directly in the polycrystalline state show only half as much additional intensity in the low-frequency wing as the corresponding undoped films [7]. Phosphorus doping thus seems to have a beneficial effect on the internal strain which develops during the crystallization process. This finding is also reflected in the average grain size of about 800 Å for in-situ phosphorus doped growth-crystallized films as opposed to about 300 Å for undoped films; the corresponding numbers after annealing at 900 - 1000°C are 2000 Å and 380 Å, respectively.

Far larger line shape distortions and also frequency shifts have been observed for microcrystalline silicon which has become the commonly accepted term for grain sizes below 100 Å. Such films have been prepared by chemical transport in a dc hydrogen plasma [12, 13] or by plasma decomposition of SiF_4 and H_2 [14]. The observed Raman line shift to low frequencies and the line broadening both increase with decreasing crystallite size. RICHTER et al. [13] presented a model to account for these effects in terms of a relaxation of the q-vector selection rule for excitation of the Raman active optical phonons which are confined to the volume of the small crystallites. Peak frequency shifts observed in silicon which was heavily disordered by silicon implantation and subsequently laser annealed have similarly been interpreted by the finite-size effects of microcrystallites existing in the annealed region [11].

2.2. Elastic Scattering

The technique of elastic light scattering, i.e., observation of light scattered without energy loss, is responsive to all effects resulting in local or temporal variation of the refractive index. Such effects can be of intrinsic nature, such as temperature or entropy fluctuations, or of extrinsic origin, due to all kinds of structural imperfections of the bulk or surface, such as grain boundaries, stacking faults, dislocations, crystallographic twinning, mosaic misorientation, as well as more localized defects.

Elastic light scattering measurements were performed to determine the structural perfection of LPCVD silicon layers. In one method the same spectrometer arrangement, laser wavelength, power, and spot size were used as given in Sect.2.1. for Raman scattering except for an additional 10^6 to 10^8 attenuation filter inserted in front of the photomultiplier. Fig. 3 shows that the smallest intensity of elastically scattered light is found for the amorphous films grown at 560 to 580°C and the highest intensity for fully crystalline films grown at deposition temperatures $T_d \geqslant 620$°C. We must conclude that the amorphous films contain far less local or temporal variations of the refractive index than growth-crystallized polycrystalline films. Next we have the amazing result that this relation is maintained after annealing, i.e., crystallization of the amorphous films. It means that the newly formed boundaries between grains of an average size of 800 Å do not introduce a noticeable amount of additional scattering centers unless there is a fortuitous compensation against the relatively few, un-

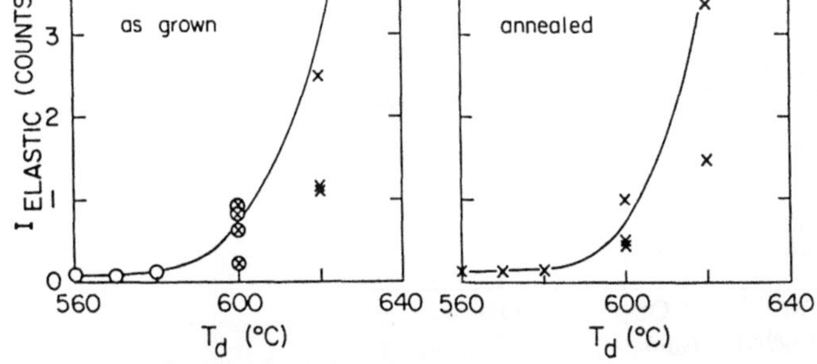

Fig. 3. Elastic peak intensity for as-grown (left) and 900-1000°C annealed (right) polysilicon layers vs. deposition temperature: O = amorphous, X = crystalline, ⊗ = mixed. (From Ref. 4)

identified scattering centers in the amorphous silicon. In any case it is justified to claim a higher structural perfection for the polycrystalline films which have been made via the amorphous phase. The gross features for in-situ phosphorus doped films are quite similar and the elastic intensity is again unchanged after annealing, as for undoped material. In a second method a laser scanner [15] is used in the quality control mode [16] where laser light falls onto the spinning wafer while being translated. The scattered elastic light originating from the bulk is used as a measure for the crystalline perfection.

3. Reflectance and Surface Roughness

Fig. 4 gives a comparison of the reflectance spectra of single crystal silicon, growth-crystallized polycrystalline silicon, and amorphous silicon from 250 to 600 nm. The two prominent maxima in the reflectance spectrum of single crystal silicon lie at 280 nm ($\hbar\omega$ = 4.4 eV) and 365 nm ($\hbar\omega$ = 3.4 eV). They are caused by optical interband transitions at the X point and along the Γ - L axis of the Brillouin zone, respectively [17]. These maxima can still be discerned in the curve taken from the growth-crystallized polycrystalline silicon film but at a much lower level. In particular, the reflectance difference between these two curves increases with decreasing wavelength which is a strong indication for reflection loss due to light scattering because the scattering intensity is typically increasing with the fourth power of the frequency. Since the penetration depth of light of wavelengths below 380 nm is shorter than 10 nm [18], one has to conclude that the reflection loss, at least in this range, in the polycrystalline film is largely due to light scattering from the rough film surface. The reflection curve of an LPCVD amorphous film does not display any structure in this range and is lower than the single crystal curve at the two maxima but higher in other spectral regions, e.g., at λ = 400 nm. It should also be noted that it is much higher than the curve of the

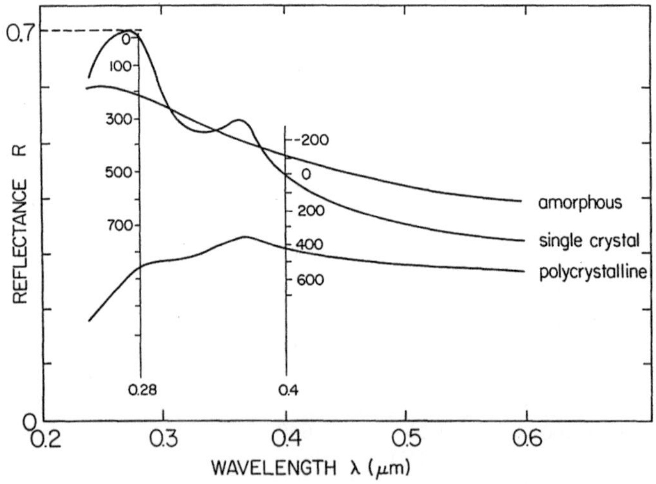

Fig. 4. Reflectance of single crystal silicon, a growth-crystallized polycrystalline LPCVD silicon film and an amorphous LPCVD silicon film versus wavelength. (From Ref. 18)

growth-crystallized polycrystalline silicon which is suggestive of a far lower surface roughness.

Since it has been shown by various authors [2, 3] that "bumps" and asperities on the initial polycrystalline silicon surface are determining factors for the interface to the silicon dioxide film grown on it and for the dielectric strength of the oxide, it is obvious that a quantitative characterization of the surface roughness is of particular importance for subsequent processing. There is a vast amount of literature on the measurement of surface roughness by mechanical, optical, and other techniques. Optical techniques are preferable because of higher sensitivity and accuracy combined with their contactless nature. The total intensity and angular dependence of light scattered from surfaces with microirregularities whose heights are much smaller than the wavelength of the incident light and whose lateral dimensions are comparable to or greater than the wavelength of the incident light can be explained by diffraction theory [19]. Another optical technique to measure the surface roughness makes use of the coupling of light to surface plasmons at a metal surface. Plasmons are the quanta of the longitudinal polarization field of the electron gas. The dispersion relation of surface plasmons, as found in a semi-infinite electron gas, is sketched in Fig. 5 for illustrative purposes. The upper branch starting at the frequency ω_p for $k_x = 0$ is that of the radiative (transverse) mode. In an infinite ideal electron gas there are no excitations at $k \rightarrow 0$ in the frequency range between $\omega = \omega_p$ and $\omega = 0$. The very existence of a sample surface gives rise to a polarization wave propagating along the surface and carrying an electric field that is partly longitudinal and partly transverse. It therefore couples to the electromagnetic field resulting in the lower dispersion branch of Fig. 5. On an ideally smooth surface this coupling is, however, only of platonic character and no coupling to an incident light beam is possible because for any frequency the surface plasmon has a higher momentum than the light and momentum conservation can thus not be achieved. There is then a stop band of anomalous dispersion for light between $\omega = 0$ and $\omega = \omega_p$ and the reflection coefficient is equal to 1 in this region, if damping effects and interband transitions are disregarded. This is actually the same situation as for optical phonons in the stop band region between the transverse and longitudinal phonon frequency. The situation is however different for a rough surface. Since a randomly rough surface can be considered to consist of a distribution of sinusoidal corrugations of various wavelengths,

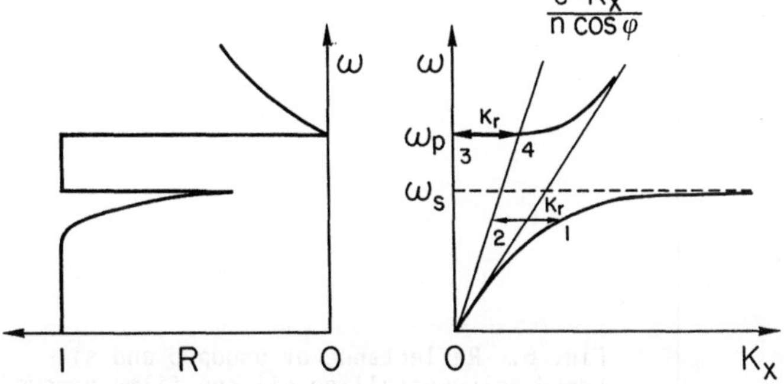

Fig. 5. Plasmon dispersion curves, coupling of plasmons with photons via surface roughness (right side) and corresponding reflectance curve (left side)

the Fourier transform of this roughness spectrum will consist of a distribution of surface momenta such as k_r in Fig. 5. In the process $1 \rightarrow 2$ k_r will provide the missing momentum to couple a surface plasmon to light incident under an angle φ. This coupling allows light to flow in and out at frequencies ω_s and below, i.e., within the band of anomalous dispersion. This leads to a reflection minimum in the spectral region of the surface plasmon frequency, as shown on the left hand side of Fig. 5 [20]. It can also be seen from the dispersion curves that the coupling of light to surface plasmons is strongest for roughness wavevectors larger than the wavevector of light, i.e., for vertical corrugation on a scale smaller than the wavelength of incident light. Hence, the surface plasmon technique covers mainly the lateral roughness dimensions which cannot be reached by light scattering but there is also some overlap. The lower cut-off is of the order of the inverse Fermi wavevector, k_F^{-1}, or a few Ångstroms [19].

Many authors [21] have used silver films as metal layers covering the surface under study because the surface plasmon frequency of silver corresponds to a very convenient wavelength region of λ = 350 nm. Equally important is the evidence that the evaporated silver film conforms to the contours of the original surface with an accuracy of a few Å and furthermore, it adds no structure of its own [22]. This claim has been supported by comparing the surface roughness of the evaporated silver layer with that of a silver replica of the covered surface which gave nearly identical results.

The reflectance of silver layers of 700 to 1000 Å thickness evaporated onto the surface of various as-grown and annealed LPCVD silicon films has been measured [4, 5, 7]. To illustrate the results, Fig. 6 gives the reflectance spectra of two silvered polycrystalline silicon films in the wavelength range from 0.2 to 0.5 µm. Common to both curves is the long-wavelength part and the lower part of the steep volume plasma edge of λ = 0.32 µm. The solid line represents an as-grown film deposited at T_d = 560°C. It shows only a small indention at the surface plasmon wavelength λ = 0.35 µm. The dashed line is from an undoped film grown at T_d = 620°C and annealed at T_a = 950°C. We observe a dramatic loss of reflection, ΔR, due to stronger surface plasmon excitation which, in turn, is caused by much higher surface roughness in comparison to the low-T_d film.

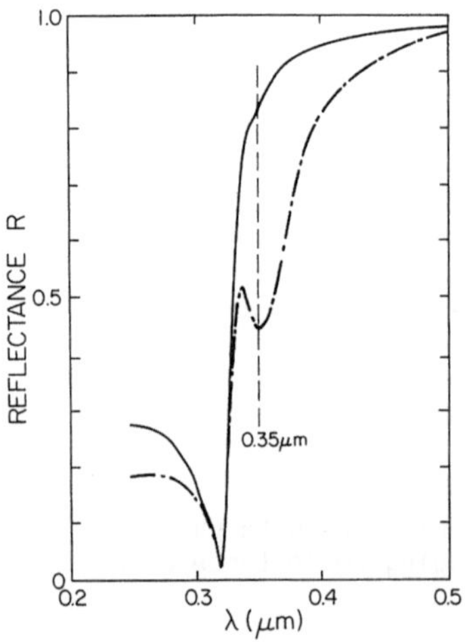

Fig. 6. Reflectance of undoped and silvered polycrystalline silicon films versus wavelength. The solid line is for T_d = 560°C; the dashed line is for T_d = 620°C, annealed at 950°C. (From Ref. 7)

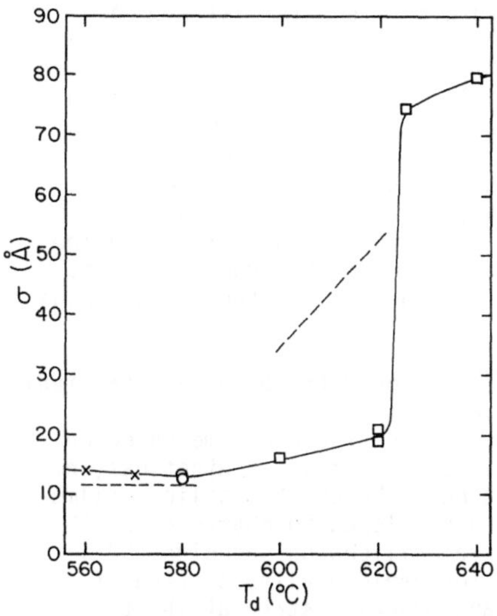

Fig. 7. Root-mean-square surface roughness σ of in-situ phosphorus doped polysilicon films (as grown, annealed at 800-1000°C) versus deposition temperature; X = amorphous, ○ = mixed, □ = crystalline. The dashed curve illustrates data obtained on undoped films. (From Ref. 7)

The root-mean-square surface roughness σ_{rms}, in the following denoted as σ, is defined by $\sigma^2 = \langle Z^2 \rangle$ where Z is the surface height relative to the mean surface level Z = 0. The σ values of our films were obtained by using the calibration curve between ΔR and σ established by CUNNINGHAM and BRAUNDMEIER [23] who used the interferometric method of observation of fringes of equal chromatic order (FECO) to determine σ.

Fig. 7 shows the σ values of in-situ phosphorus doped films (points and solid line) [7] compared to undoped films [5] (dashed line) versus deposition temperature T_d. There is no difference between as-grown and annealed films outside of a small local variation of 1 to 2 Å. We see that σ values less than 15 Å can be reliably achieved only at deposition temperatures $T_d \leqslant 580°C$, both for undoped and doped films. These films are amorphous or mixed as deposited, but the important point for critical device applications is that the smooth surface is preserved in the annealing process. A sharp transition to very rough films occurs at T_d = 625°C or T_d = 600 to 620°C for doped and undoped films, respectively. There seems to exist a correlation between excessive surface roughness and the occurrence of a strong texture. Undoped films had a very rough surface for T_d = 620°C where a strong <110> texture and columnar growth [5] were observed. In-situ phosphorus doped films grown at T_d = 640°C again have a high roughness and a strong texture, this time of the <311> type [7]. For both types of films the surface roughness is not changed by the annealing process. This is surprising, particularly for the supersmooth, initially amorphous films that crystallize during annealing. It is even more surprising for the phosphorus doped films, since the grain growth is strongly enhanced by the presence of phosphorus. The average grain size for $T_d \leqslant 570°C$ after annealing is 2000 to 3000 Å, which is a factor of 3 to 4 higher than for undoped films. There is even an appreciable number of grains of a size around 5000 Å, i.e., of the order of the film thickness, and the surface roughness is still no more than σ = 15 Å. Since the peak-to-valley roughness is only a few times the value of σ, it amounts to just about 1% of the largest grain sizes. This surprising result can only be understood by crystallization phenomena near the surface such that the nucleation and growth of the crystallites are accommodated within the initially smooth contour [7].

The photon-surface-plasmon technique yields one number characterizing the vertical roughness; it does not give numerical information on the lateral structure. For this reason it is necessary to complement it by other techniques such as scanning electron microscopy [7] and cross-sectional transmission electron microscopy [8]. Such measurements not only confirmed the results of the optical technique but also revealed the characteristic lateral roughness dimensions to be of the order of 200 to 2000 Å. As discussed above, this is just the range where the surface plasmon technique is well suited.

It has also been shown [18] that the reflection difference between single crystal silicon and polycrystalline silicon at λ = 280 nm (see Fig. 4) is closely correlated to the root-mean-square surface roughness measured on the same films. This correlation forms the basis for a fast, contactless, non-destructive and quantitative characterization technique for silicon films by U.V. reflection. It had originally been developed to characterize the near-surface crystallinity of heteroepitaxial silicon on sapphire (SOS) films [24] but was found to be applicable for rapid quality testing of LPCVD silicon films both in the as-grown and annealed state at the two wavelengths indicated by the reflectance difference scales in Fig. 4.

The benefit of obtaining extremely smooth film surfaces by LPCVD of amorphous films and maintaining this smoothness after crystallization to polycrystalline films can be exploited to improve the insulating properties of SiO_2 films thermally grown on such silicon films [9]. The histograms of destructive breakdown field E_{bn} shown in Fig. 8 represent examples of the improvement that can be attained. For diffusion doped samples there is a factor of two increase in E_{bn} over polycrystalline silicon deposited in the crystalline phase, see Fig. 8(a) and (b). Correspondingly, the Fowler-Nordheim tunneling current through the oxide at a constant field of $3 \cdot 10^6$ V/cm was reduced by more than two orders of magnitude for oxides

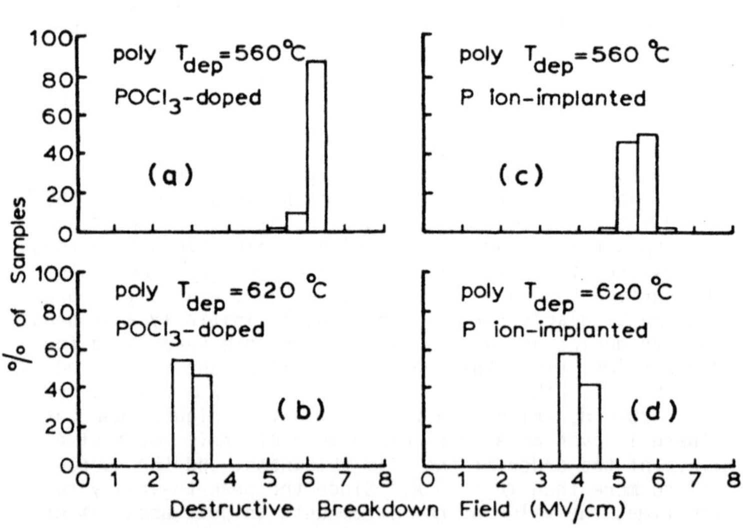

Fig. 8. Destructive breakdown field histograms for samples oxidized in dry O_2 at 1000°C.(From Ref. 9)

grown on initially amorphous films. A similar improvement of the insulating properties of SiO_2 grown on amorphous-deposited films was found for ion implanted samples, as is evident from Fig. 8(c) and (d). Obviously, the reduced texture of the silicon surface leads to a reduced roughness at the interface between polycrystalline silicon and its oxide and thus to less electric field enhancement at the asperities than for strongly textured surfaces of grown-crystallized films. This growth-roughness relation gives also the knowledge base for handling those situations where not the smoothest possible but a well-defined surface texture has to be reliably manufactured.

4. Optical Absorption

In 1974 KÜHL et al. [25] have determined the absorption coefficient of epitaxial, polycrystalline, and amorphous silicon films by evaluating the interference extrema in the reflectance spectra or by Kramers-Kronig analysis. They reported that the absorption coefficient of ambient-pressure CVD polycrystalline silicon deposited at 680°C is markedly higher than that of epitaxial silicon grown on sapphire at 1050°C at photon energies from 1 to 4 eV. The absorption of evaporated amorphous silicon was in turn found to be higher than that of the polycrystalline films. It was argued that these results can be understood by considering the degree of crystal disorder of the films. It seems indeed that the CVD polycrystalline films are heavily disordered because they do not show the critical-point maxima as we see them in LPCVD polycrystalline films in reflectance (see Fig. 4) or in the dielectric function spectra as obtained by ellipsometry (see Sect. 5).

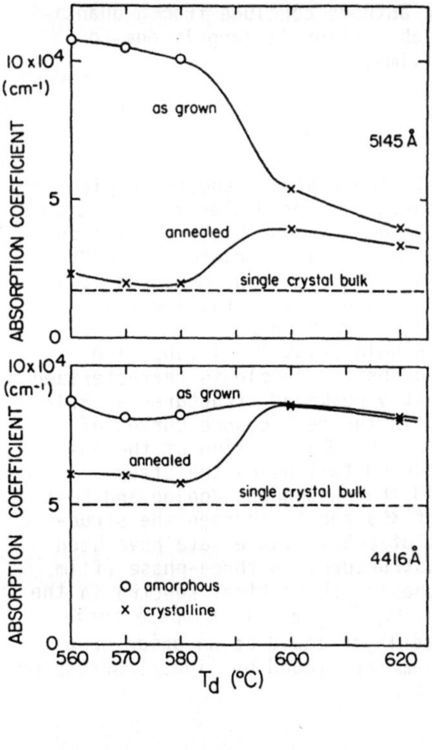

Fig. 9. Absorption coefficient of polycrystalline silicon vs. deposition temperature, as grown and annealed (900°-1000°C), at 5145 Å (top) and at 4416 Å (bottom). (From Ref. 5)

More detailed measurements of the optical absorption coefficient of LPCVD silicon films grown in the amorphous or polycrystalline state have been reported in Ref. [5]. To this end the silicon layers have been deposited on quartz glass substrates concurrently with layers grown on silicon/silicon dioxide substrates for other optical measurements. The absorption coefficient was then determined from transmission and reflection measurements at 5145 Å and 4416 Å. Fig. 9 shows the results for undoped films for $560^{o}C \leqslant T_d \leqslant 620^{o}C$ in comparison to those for single crystal silicon. As is well known [26], the absorption coefficient of amorphous silicon is considerably higher than that of single crystal and polycrystalline silicon at these wavelengths. After annealing, however, the absorption of the previously amorphous films is close to that of bulk silicon while it is markedly higher for the growth-crystallized polycrystalline silicon films. The additional effective absorption for the latter ones must be due to a sizeable amount of internal scattering from strained regions and grain boundaries in the fine-grained films. Similar results have been reported for in-situ phosphorus doped films [7]. From the absolute values of the absorption coefficient in Fig. 9 the effective penetration depth can be deduced. Values of more than 2000 Å at λ = 5145 Å and more than 1000 Å at 4416 Å ascertain that for the light scattering data presented in Sect. 2 volume effects dominate over surface effects.

The effective absorption coefficient of hydrogenated microcrystalline silicon, prepared by chemical transport in a dc hydrogen discharge, has been found to be anomalously high for photon energies up to 2 eV [27]. Under certain preparation conditions it can even be higher than that of amorphous silicon. Annealing at temperatures up to 800°C resulted in a further, albeit small increase of absorption. The intensity of forward scattered light, measured as a function of angle for various wavelengths, was found to correlate with the enhanced absorption. Also the surface roughness parallels the absorption but the authors conclude from a quantitative estimate that the enhanced optical absorption is largely due to diffuse scattering and absorption in the films.

5. Ellipsometry

ASPNES and STUDNA [28] have described a rotating-analyzer spectroscopic ellipsometer for obtaining the complex reflectance and dielectric function spectra $\epsilon = \epsilon_1 + \epsilon_2$ in the energy range from 1.5 to 6.0 eV. This technique has also been applied to characterizing the microstructure of LPCVD amorphous and polycrystalline silicon films. Fig. 10 shows measured ϵ_2 spectra displayed from 2.75 to 6.0 eV for undoped polycrystalline films deposited at 650°C and for heat-treated and phosphorus doped films [29]. Compared to the intrinsic ϵ_2 response of single crystal silicon, also shown in Fig. 10, the spectrum for the as-deposited sample is characterized by a marked decrease in amplitude with greatly weakened structures at 3.4 and 4.2 eV. This behaviour is the same as in the reflectance curves of similarly prepared samples shown in Fig. 4. The ϵ_2 function of the as-deposited film in Fig. 10 also has a pronounced tail below the steep threshold of c-Si. The principal effects of the phosphorus doping and heat treatment are to increase the amplitude of ϵ_2 and to sharpen the structures but the curve is still far from that of c-Si. These data have been interpreted in terms of the material microstructure. A three-phase (film, overlayer, ambient) model was used to synthesize theoretical spectra in the Bruggeman effective-medium approximation (EMA) [30] and to compare these with the measured spectra. The assumed overlayer could be an oxide or a "layer" of microscopic roughness. The parameters found by linear regression

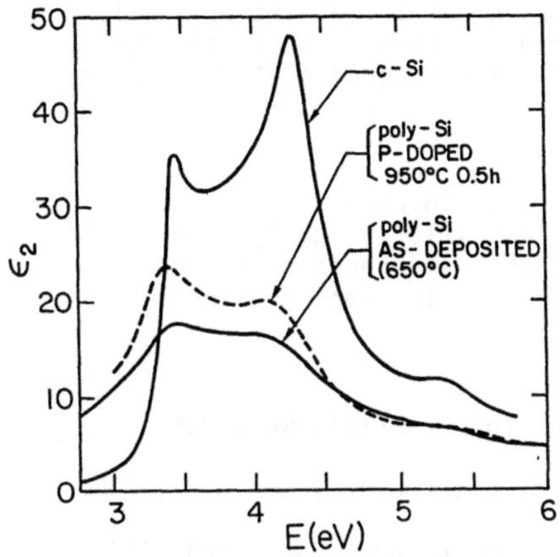

Fig. 10. The imaginary part ϵ_2 of the dielectric function vs. energy for as-deposited and phosphorus doped polycrystalline silicon and single crystal silicon (from Ref. 29)

analysis for the as-deposited sample were volume fractions of 0.25 for voids, 0.14 for c-Si, and 0.61 for a-Si for the film, and a thickness of d = 12 Å for the overlayer. Upon heat treatment and phosphorus doping the c-Si fraction increased to 0.57 and the void fraction decreased to 0.08 leaving 0.35 for a-Si. At the same time the overlayer thickness increased to d = 38 Å and was unambiguously modeled as microscopic roughness rather than SiO_2 with 50% film material and 50% voids. The derived volume fractions should be viewed as average components of the microstructure and not reflect a particular disposition. For example, the calculated excess volume could be distributed as voids, vacancies, or an excess volume per atom over that of maximally dense material. Laser annealing of the same type of as-deposited films led to a crystalline volume fraction of 0.88 and a relative density of 96% [31]. In another investigation specific emphasis was put on the determination of surface roughness of polycrystalline films deposited at 740 or 870°C [32]. Since the electron diffraction patterns indicated a random orientation of grains smaller than 100 Å (microcrystalline!), all layers were assumed to consist of EMA mixtures of the measured dielectric function of a-Si, obtained from an amorphous sample grown at 575°C, and voids. The best-fit parameters for samples grown at 740°C are a volume fraction f_v = 0.19 for voids in the film and a 53 Å thick overlayer with f_v = 0.44. Considerably higher roughness was found in the best fit for samples grown at 870°C, namely a first layer of d = 213 Å, f_v = 0.49 and a second layer on top of d = 180 Å, f_v = 0.76. Although a description in terms of a stratified structure seems somewhat arbitrary, the optical measurement demonstrated the existence of density gradients in the surface region of a microscopically rough material. The best-fit results for the rough surface were also used to show that the density profile is characteristic of hemispherical rather than pyramidal irregularities. This hemispherical profile indicates three-dimensional nucleation growth at 870°C and 350 Torr pressure and is markedly different from the columnar growth structure observed by cross-sectional TEM for LPCVD films grown at 620°C [8].

More recently, spectroscopic ellipsometry has been applied to silicon films which were saturation doped with phosphorus either by in-situ doping or from the PBr_3 gas phase [33]. A film grown in the amorphous phase at 591°C and heat treated at 1000°C in the gas-phase doping process was found to be polycrystalline with grain sizes ranging from 500 to 2000 Å, in agreement with previous observations for heat-treated, in-situ phosphorus doped films [7]. In this case the modeling has to be done with the dielectric function of c-Si and, in fact, the films turned out to consist primarily of material that is optically indistinguishable, except for carrier concentration effects, from lightly doped c-Si. The roughness of these films is exceedingly small; the best-fit parameters give a "roughness layer" thickness of $d = 6.5 \pm 1$ Å with $f_v = 0.6$. The microstructural parameters of an as-grown LPCVD amorphous film doped in situ with phosphorus and deposited at 571°C include a somewhat larger surface layer thickness of $d = 19.5$ Å and $f_v = 0.5$ [33]. These roughness values may differ slightly from run to run but it should be noted that they are in the same range as measured by the surface plasmon technique on films prepared by the same growth technique, as shown in Fig. 7.

1. K.C. Saraswat, Symposium Proceedings of the Materials Research Society, Vol. 5, p. 261 (1982).
2. E.A. Irene, E. Tierney and D.W. Wong, J. Electrochem. Soc. 127, 705 (1980).
3. R.B. Marcus, T.T. Sheng and P. Lin, J. Electrochem. Soc. 129, 1282 (1982).
4. G. Harbeke, L. Krausbauer, E.F. Steigmeier, A.E. Widmer, H.F. Kappert and G. Neugebauer, Appl. Phys. Lett. 42, 249 (1983).
5. G. Harbeke, L. Krausbauer, E.F. Steigmeier, A.E. Widmer, H.F. Kappert, and G. Neugebauer, J. Electrochem. Soc. 131, 675 (1984).
6. M. Sternheim, E. Kinsbron, J. Alspector and P.A. Heimann, J. Electrochem. Soc. 130, 1735 (1983).
7. G. Harbeke, L. Krausbauer, E.F. Steigmeier, A.E. Widmer, H.F. Kappert and G. Neugebauer, RCA Review 44, 287 (1983).
8. M.T. Duffy, J.T. McGinn, J.M. Shaw, R.T. Smith, R.A. Soltis and G. Harbeke, RCA Review 44, 313 (1983).
9. L. Faraone, in: Insulating Films on Semiconductors, J.F. Verweij and D.R. Wolters, Eds., North-Holland, Amsterdam, 1983, p. 252.
10. See, e.g., Light Scattering in Solids, M. Cardona and G. Güntherodt, Eds., Vol. 1, 1983; Vol. 2, 1982, Vol. 3, 1982, Vol. 4, 1984, Springer, Berlin.
11. J.F. Morhange, G. Kanellis and M. Balkanski, Solid State Commun. 31, 805 (1979).
12. Z. Iqbal and S. Veprek, J. Phys. C 15, 377 (1982).
13. H. Richter, Z.P. Wang and L. Ley, Solid State Commun. 39, 625 (1981).
14. R. Tsu, M. Izu, S.R. Ovshinsky and F.H. Pollak, Solid State Commun. 36, 817 (1980).
15. E.F. Steigmeier and H. Auderset, RCA Review 44, 5 (1983).
16. E.F. Steigmeier and H. Auderset, J. Electrochem. Soc. 131, 1693 (1984).
17. See, e.g., D.L. Greenaway and G. Harbeke, Optical Properties and Band Structure of Semiconductors, Pergamon Press, Oxford, 1968
18. G. Harbeke, E. Meier, J.R. Sandercock, M. Tgetgel, M.T. Duffy and R.A. Soltis, RCA Review 44, 19 (1983).
19. J.M. Elson and J.M. Bennett, Opt. Engineering 18, 116 (1979).
20. R.H. Ritchie, Surf. Sci. 34, 1 (1973).
21. See, e.g. P.J. Zanzucchi and M. Duffy, Appl. Opt. 17, 3477 (1978).
22. D.L. Decker, J.M. Bennett, M.J. Soileau, J.O. Porteus and H.E. Bennett, Opt. Engineering 17, 160 (1977).
23. L.J. Cunningham and A.J. Braundmeier, Jr., Phys. Rev. B14, 479 (1976).
24. M.T. Duffy, J.F. Corboy, G.W. Cullen, R.T. Smith, R.A. Soltis, G. Harbeke, J.R. Sandercock and M. Blumenfeld, J. Cryst. Growth 58, 10 (1982).
25. Ch. Kühl, H. Schlötterer and F. Schwidefsky, J. Electrochem. Soc. 121, 1496 (1974).
26. D.E. Aspnes, A.A. Studna and E. Kinsbron, Phys. Rev. B29, 768 (1984).

27. Z. Iqbal, F.A. Sarott and S. Veprek, J. Phys. C 16, 2005 (1983).
28. D.E. Aspnes and A.A. Studna, Appl. Opt. 14, 220 (1975); Rev. Sci. Instrum. 49, 291 (1978).
29. B.G. Bagley, D.E. Aspnes, A.C. Adams and C.J. Mogab, Appl. Phys. Lett. 38, 56 (1981).
30. D.A.G. Bruggeman, Ann.Phys. (Leipzig) 24, 636 (1935).
31. B.G. Bagley, D.E. Aspnes, G.K. Celler and A.C. Adams, in: Laser and Electron-beam Interaction with Solids, B.R. Appleton and C.K. Celler, Eds., Elsevier, 1982.
32. D.E. Aspnes, J.B. Theeten and F. Hottier, Phys. Rev. B20, 3292 (1979).
33. D.E. Aspnes, A.A. Studna and E. Kinsbron, Phys. Rev. B29, 768 (1984).

Polycrystalline Silicon in Integrated Circuits

H.C. de Graaff

Philips Research Laboratories, NL-5600 JA Eindhoven, The Netherlands

1. Introduction

The use of polycrystalline silicon for integrated circuits makes it possible to produce multilayer or three-dimensional structures. Another advantage is the large range of resistivities (more than 7 orders of magnitude). Polysilicon also has drawbacks: the grain boundaries cause dopant segregation and the grain boundary states may act as trapping centres, recombination centres and scattering centres [1] .

The large recombination has hitherto prevented the realization of a complete bipolar transistor in polysilicon.The increased scattering lowers the mobility, which hampers the operation of thin-film field effect transistors, as does the trapping of mobile carriers. Laser-beam annealing, however, can minimize these drawbacks.

In the following sections we will treat the application of polycrystalline silicon in circuit elements for integrated circuits, like resistors, diodes and thin-film transistors. Other applications refer to polysilicon (tunnel) emitters for vertical bipolar transistors, and polysilicon as diffusion source for realizing small dimensions. This review ends with a few examples of miscellaneous use.

2. Polysilicon integrated circuit elements

2.1. Resistors in polysilicon films

Resistors can be made in thin polysilicon films, usually deposited by low-pressure CVD on a SiO_2 layer, e.g. 0.5 µm thick, on a monocrystalline silicon substrate. This substrate may already contain several integrated circuit elements which contact the polysilicon film via windows in the SiO_2 layer.

A great deal has already been published on the theory of the electrical conduction mechanism in polysilicon films. The dominant effect appears to be thermionic emission over the potential barriers at the grain boundaries [2-5] . These potential barriers arise from the trapping of mobile carriers. Other mechanisms are sometimes added, like thermionic field emission and tunnelling [6] , quantum-mechanical reflection [7], inclusion of diffusion [8] . It has also been proposed to consider the grain boundary as a thin piece of amorphous silicon, with conduction based on an extended-state mobility [9] . In order to keep our treatment as simple as possible we shall disregard all these refinements and take only the thermionic emission into account. Then we can write for the current density

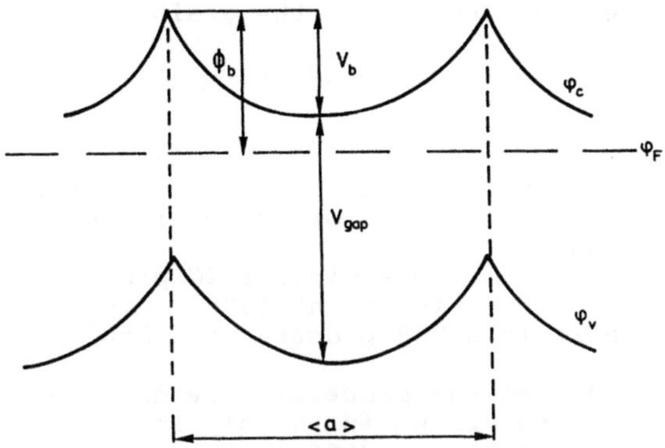

Fig. 1. Band diagram of a polysilicon grain: ϕ_b and V_b are the potential barrier heights with reference to the Fermi level and the conduction band edge respectively, $\langle a \rangle$ is the mean grain size.

$$J = 2A^*T^2 \exp(-q\phi_b/kT)\,\sinh\left(\frac{qV_a}{2kT}\cdot\frac{\langle a \rangle}{L}\right) \qquad (1)$$

A^* is Richardson's constant and ϕ_b is the potential barrier height; see fig. 1.
L is the sample length, $\langle a \rangle$ the mean grain size and V_a the total applied voltage. For small values of V_a eq. (1) can be linearized and we get for the resistivity

$$\rho = \frac{V_a}{LJ} = \frac{k}{q\langle a \rangle A^*T} \exp(q\phi_b/kT) \qquad (2)$$

Eq. (2) clearly shows the influences of the barrier height ϕ_b and the mean grain size $\langle a \rangle$. Small grains may be fully depleted, but larger grains are usually only partly depleted. In that case eq. (2) can be rewritten:

$$\rho = \frac{k}{q\langle a \rangle A^*T}\,\frac{N_c}{N_G} \exp(qV_b/kT) \qquad (3)$$

The bottom of the conduction band in the middle of the grain is here taken as the reference for the barrier height V_b. The effective density of states $N_c \propto T^{3/2}$ [10] . N_G is the electrically active dope concentration. The barrier height V_b is a function of temperature: [4,11]

$$V_b \approx V_{bo}(1-\gamma T)\ , \quad \gamma \approx 1.5 * 10^{-3}\ K^{-1} .$$

This makes

$$\rho/\sqrt{T} \approx \text{const.}\,\frac{1}{\langle a \rangle N_G} \exp(qV_{bo}/kT) \qquad (4)$$

Measuring the temperature dependence of ρ and using eq. (4) gives us a value for V_{bo}. This barrier height is a function of the dope concentration and the trap density at the grain boundaries.

2.1.1. Loss and segregation

The electrically active dope concentration N_G is not equal to the implanted impurity concentration: it depends on the kind of impurity (As, P, B) and the type of processing of the integrated circuit technology.
We have studied low-pressure CVD films in a bipolar IC process with high-temperature steps (emitter drive-in at 1000°C and base drive-in at 1160°C) and also in a MOS process with 950°C as the highest temperature step.
The difference in temperature during the processing leads to a difference in mean grain size: $\langle a \rangle \approx 0.2$ µm in the bipolar and $\langle a \rangle \approx 0.1$ µm in the MOS process. We also get different segregation of impurities at grain boundaries. Mandurah et al. [12] have derived that the ratio of N_G and $N_{chemical}$ is given by

$$N_G / N_{chem} = \frac{1}{1 + A \frac{Q_s}{N_{Si}} \exp(Q_0 / kT_a)} \tag{5}$$

Here Q_s/N_{Si} is the ratio of the number of sites in the grain boundaries and the total number of sites: $Q_s/N_{Si} \approx 1.6 * 10^{-7} / \langle a \rangle$ Q_o is the heat of segregation and is $6.95*10^{-20}$ Joule/atom, A is a constant ($A \approx 1$) and T_a is the anneal temperature. Following eq. (5) we may expect less segregation at grain boundaries in the bipolar process than in the MOS process. Part of the implanted dose may also be lost during the processing, either by direct evaporation or by segregation in oxide layers that are etched away in following steps. So we get for the impurities:

$$N_{implant} = \underbrace{N_G + N_{segr}}_{= N_{chem}} + N_{lost}.$$

The chemical impurity concentration is known from SIMS measurements. For our processes we then arrive at the following results:

- approximately 50% of the implanted boron is lost in the bipolar and somewhat less in the MOS process.
- 30-40% of the phosphorus in the bipolar and
- 10-15% of the phosphorus and arsenic in the MOS process is lost.

Moreover, not all the electrically active dope atoms give rise to mobile carriers, because some of the carriers are trapped in the grain boundary states:

$$N_G = n + n_t.$$

The mobile carrier concentration n is obtained from Hall measurements. With n known we can calculate n_t and N_G. If we assume, for the sake of simplicity, a rod-like structure of the grains with a square cross-section (see fig. 2), we can derive:

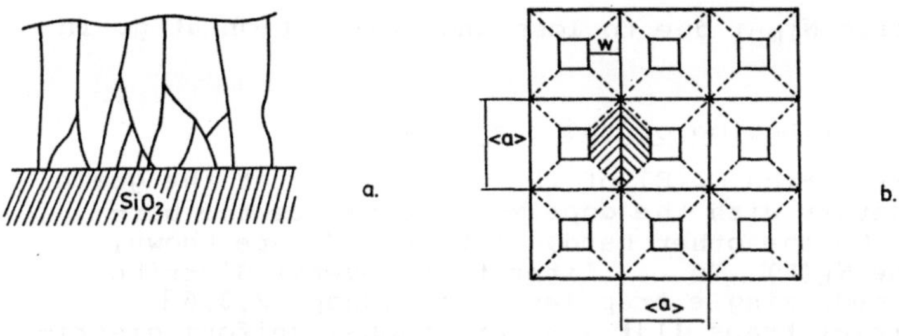

Fig. 2. a. Rod-like structure of polycrystalline silicon on SiO_2.
b. Assumed square cross-sections with (hatched) depletion regions.

$$\frac{lbh}{\langle a\rangle}\int_{\varphi_0}^{\varphi_F} N_t(E)\,dE = 2\,W\left(\langle a\rangle - W\right)\frac{lbh}{\langle a\rangle^2}\,N_G \qquad (6)$$

Here l, b and h are the dimensions of the polysilicon body, $N_t(E)$ is the energy distribution of the trap levels at the grain boundaries and W is the thickness of the depletion layers at both sides of the grain boundary.
If the grains are not fully depleted, we can write

$$W=\sqrt{\frac{2\varepsilon V_b}{q N_G}} \qquad (7)$$

Eq. (6) simply states that the space charge in the depletion regions of the grains is compensated by the trapped charge in the grain boundaries. We have assumed furthermore that traps are filled below the Fermi level φ_F and empty above it. The trapped carrier density n_t is then

$$n_t = \frac{2}{\langle a\rangle}\int_{\varphi_0}^{\varphi_F} N_t(E)\,dE \qquad (8)$$

Bearing in mind that $N_G=n+n_t$, we can combine eqs. (6), (7) and (8) and get:

$$n = N_G - \frac{4}{\langle a\rangle}\sqrt{\frac{2\varepsilon V_b}{q}}\sqrt{N_G} + \frac{8}{\langle a\rangle^2}\frac{\varepsilon V_b}{q} \qquad (9)$$

Eq. (9) is a quadratic equation in $\sqrt{N_G}$ and can be solved, because we know n from Hall measurements and V_b from temperature measurements (see eq. 4). Once N_G is known, we also know the ratio N_G/N_{chem} and thus have an idea of segregation at grain boundaries.
We arrive at the following conclusions for our processes:

- boron shows hardly any segregation in both processes;
- phosphorus shows minor segregation in the bipolar process;
- phosphorus and arsenic have strong segregation in the MOS process.

Summarizing, we can say that the electrically active dope concentration N_G may be considerably less than the implanted dope concentration N_{impl} due to loss and segregation at grain boundaries.

2.1.2. Energy distribution of the trap levels

The energy distribution $N_t(E)$ of the trap levels at the grain boundaries, together with the dope N_G, determines the barrier height V_b. On the other hand, if V_b and N_G are known, we can determine $N_t(E)$. In literature several distributions are proposed: single trap level at midgap [2,3,6], discrete multilevel traps [13], a continuous, uniform distribution [3,5], an exponential distribution [5] or a U-shape distribution [11,14,15].

Following Pike and Seager [5] we will use a simplified deconvolution method, starting with eq. (6). The constant φ_o refers to the initial trap charges in the "unjoined grains" situation. The Fermi level for fully depleted grains is

$$\varphi_F = V_{gap} - \Phi_b \tag{10a}$$

and for partly depleted grains

$$\varphi_F = V_{gap} - \frac{kT}{q} \ln \frac{N_c}{N_G} - V_b \tag{10b}$$

When N_G and V_b are known from measurements, W can be calculated from eq. (7). Then the right-hand side and the upper integral boundary of eq. (6) are known. Taking several samples with different dope concentrations, we vary N_G and therefore also φ_F, V_b and W. In this way we obtain an approximate function for $N_t(E)$.

In fig. 3 we have plotted the thermal activation voltage V_{act} as a function of N_G as obtained from temperature measurements of the resistivity. There are two distinct curves: one for the bipolar and one for the MOS process. The consequence is that we have also two distinct trap level distributions, see fig. 4. Both show a U-shape, but the MOS process has an

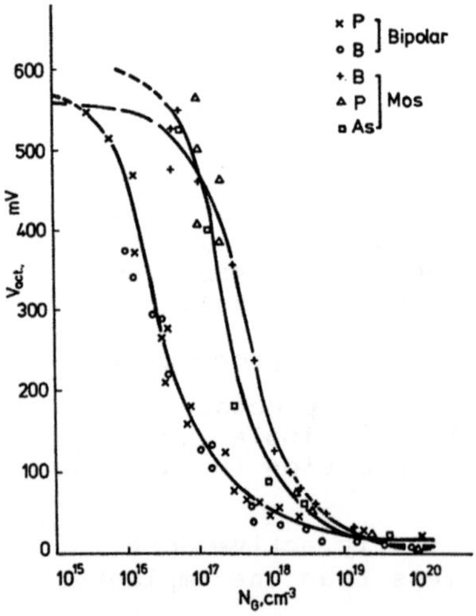

Fig. 3. Thermal activation voltage as a function of the electrically active dope concentration.

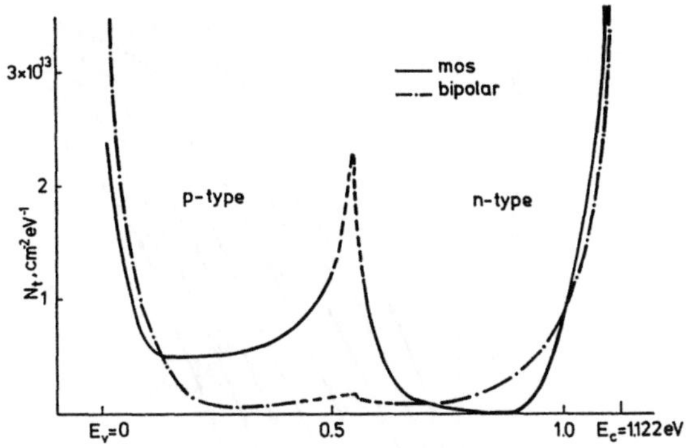

Fig. 4. Grain boundary trap density distribution for different processes.

additional peak at midgap! The resistivity, measured as a function of $N_{implant}$ is shown in fig. 5. Now the distinction between the bipolar and the MOS-process is less clear.

2.1.3. Polysilicon resistors in integrated circuits

From fig. 5 it is obvious that we can achieve very high resistance values with polysilicon, but that the accuracy is less than in monocrystalline material. Laser annealing [16] improves the accuracy but limits the resistivity range. Eq. (1) shows that polysilicon resistors become non-linear at high voltages. For good linearity the applied voltage V_a must be kept below $2\frac{kT}{q}\frac{L}{\langle a\rangle}$, so small grains are required. The temperature coefficient $\frac{1}{R}\frac{dR}{dT}$ in the linear regime follows from eq. (4):

$$\frac{1}{R}\frac{dR}{dT} = \frac{1}{T}\left(1/2 - \frac{qV_{bo}}{kT}\right) \qquad (11)$$

It is often negative and large; if $V_{bo} \approx 200$ mV, $\frac{1}{R}\frac{dR}{dT} \approx -2.5\%\ K^{-1}$ at room temperature against e.g. $+\ 0.2\%\ K^{-1}$ for monocrystalline silicon. At the highest doping levels V_{bo} becomes small (~ 10 mV) and $\frac{1}{R}\frac{dR}{dT}$ approaches the monocrystalline values.

At high frequencies we get a distributed RC network, because the silicon film is deposited on an oxide layer above a monocrystalline substrate.

A peculiar effect of high-ohmic polysilicon ($> 10^3\ \Omega$ cm) is that its resistivity can be lowered by several orders of magnitude, irreversibly, by a current pulse of high density ($\sim 10^4$ A/cm^2). This may be useful in PROM circuits [17] .

2.2. Polysilicon diodes

The I-V characteristics of polysilicon diodes are largely influenced by the presence of grain boundary states. The difference with monocrystalline diodes is that they have a non-ideality factor m 1.5 to 2 instead of m=1 and that the reverse leakage current is much higher. One can make vertical diodes in thick polysilicon films [18,19] , but more interesting for IC applications are the lateral diodes in thin polysilicon films (thickness 0.5 to 2.5 µm) [11,20,21] . We

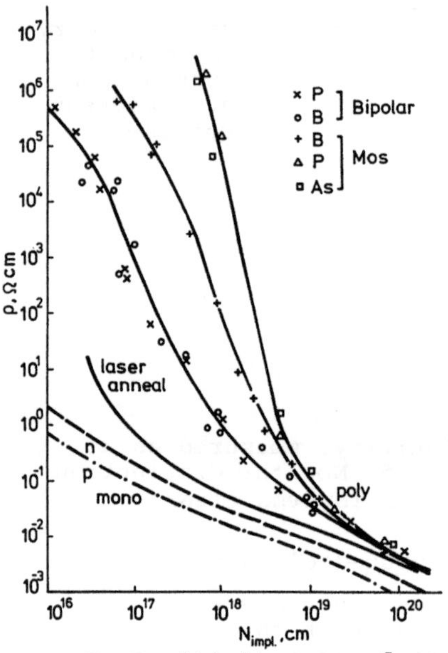

Fig. 5. Resitivity as a function of N_{impl}. The "laser anneal" curve refers to B-implanted polysilicon [16].

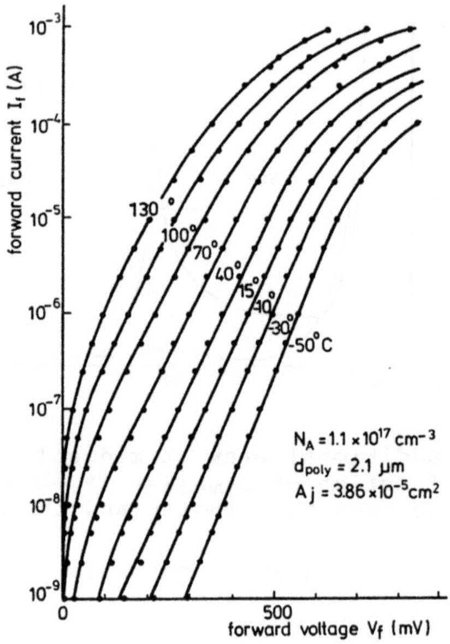

Fig. 6. Forward characteristics of the lateral diode, at different temperatures.

have studied lateral $n^+p^-p^+$ diodes made in the bipolar process (see section 2.1).
Fig. 6 shows the forward characteristics.
The forward current can be written as

$$I_f = I_{fo} \exp\left(qV_f/mkT\right) \tag{12}$$

with $I_{fo} = \frac{1}{2}\, q n_i\, N_t\, qV_f\, v_{th} \sqrt{\sigma_n \sigma_p}\; A$

$$\approx \frac{1}{2}\, q n_i \frac{W_{depl.}}{\tau_o}\, A$$

N_t is the trap density of the grain boundary states, in cm^{-2} eV^{-1} and $\sqrt{\sigma_n \sigma_p}$ is their mean capture cross-section, A is the junction area and W_{depl} the depletion layer thickness of the forward-biased junction.
In our samples $N_t\, v_{th} \sqrt{\sigma_n\, \sigma_p}\, qV_f \sim 10^3$, so the effective lifetime $\tau_o \sim 10^{-9}$ to 10^{-11} s, which is rather short in comparison with that of monocrystalline diodes. The non-ideality factor $m \approx 2$, which makes it clear that recombination dominates over diffusion. The temperature dependence of I_{fo} in eq. (12) is governed by n_i and the thermal velocity v_{th}, so

$$I_{fo} \propto T^2 \exp\left(-qV_{gap}/2kT\right) \tag{13}$$

The reverse current is strongly voltage dependent, and is ascribed to field-enhanced emission (Poole-Frenkel effect) from grain boundary traps. It can then be written as

$$I_r = I_{ro} \exp\left(q\beta\sqrt{E_{max}}/kT\right) \tag{14}$$

By plotting I_r in fig. 7 as a function of $q\sqrt{E_{max}}/kT$, where E_{max} is the maximum electric field occurring near the metallurgical junction, we find $\beta=(3.5-6)\times10^{-4}(\mathrm{Vcm})^{1/2}$. The Pool-Frenkel $\beta_{PF}=\sqrt{q/\pi\varepsilon}=2.26*10^{-4}$ (Vcm) and that is smaller than we have found. By including thermally assisted tunnelling through the potential barrier, the β-values increase to the ones we found experimentally [11] . The tunnelling component is only weakly temperature dependent; if it is neglected we have

$$I_{ro} \propto T^3 \exp\left(-qV_{gap}/2kT\right) \tag{15}$$

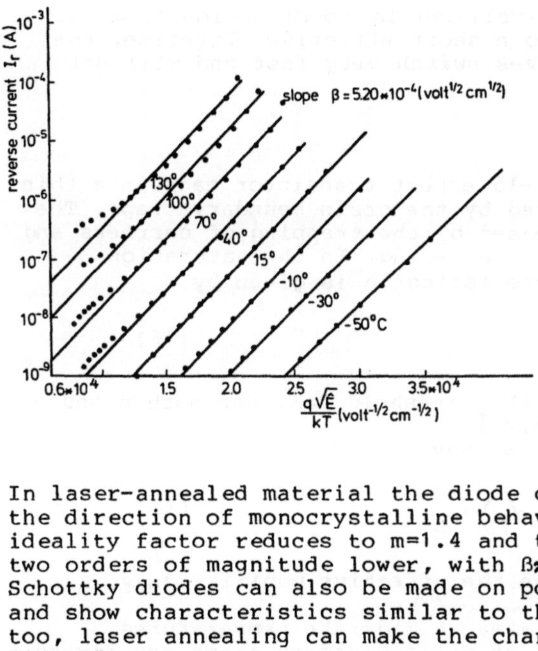

Fig. 7. Reverse characteristics, plotted against $q\sqrt{\hat{E}}/kT$: $\hat{E}$ is the maximum electric field.

In laser-annealed material the diode characteristics move in the direction of monocrystalline behaviour[22].The non-ideality factor reduces to m=1.4 and the reverse current is two orders of magnitude lower, with $\beta\approx2.8*10^{-4}(\mathrm{Vcm})^{1/2}$. Schottky diodes can also be made on polysilicon films [23,24] and show characteristics similar to those of pn diodes. Here too, laser annealing can make the characteristics more "ideal".

Polysilicon diodes can easily be incorporated in an integrated circuit, especially when polysilicon is also used for the interconnections. As an example we mention here their use in an I^2L circuit [25] , where polysilicon diodes are used as series diodes in the outputs, (see fig. 8). Such a series diode lowers the voltage swing at the input of the next stage, thus making the circuit faster (at the expense, however, of a smaller noise margin). The diodes have a horizontal pn

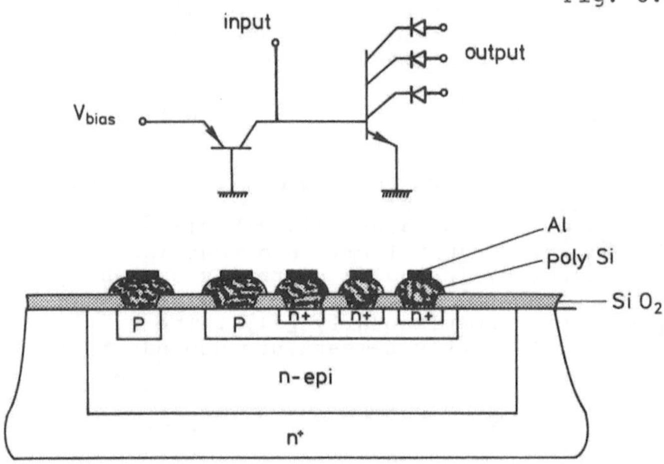

Fig. 8. Schematic diagram and cross-section of an I^2L gate with polysilicon diodes.

junction, formed in the polysilicon by up-diffusion from the n^+ collector region. Due to a short effective lifetime, the polysilicon diodes themselves switch very fast and will not hamper the gate delay.

2.3. Thin-Film Transistors

The operation of a field effect transistor made in a thin polysilicon film is hampered by the grain boundary traps. The threshold voltage is increased by the trapping of carriers and the mobility is lowered by scattering. In the saturation region of the I_D-V_D characteristics I_D is given by

$$I_D = \text{const.}\ (V_G - V_T)^2 \tag{16}$$

which enables us to find the threshold V_T. For both n and p channels $|V_T| \sim 5$ volts. [26,27].
At very low drain voltages we have

$$I_D = \mu_{eff} \frac{W}{L} C_{ox} (V_G - V_T) V_D \tag{17}$$

from which we can determine the effective mobility μ_{eff}. A typical value here is $\mu_{eff} \sim 1$ cm^2/Vs.
Annealing in a hydrogen plasma at moderate temperatures (300°C) reduces the number of grain boundary traps and lowers V_T (~ 2 volts) and increases μ_{eff} (10 cm^2/vs) [27]. With laser recrystallization the situation improves still further ($\mu_{eff} \sim 300$ cm^2/Vs) and approaches the monocrystalline situation [28]. In such laser-annealed material the number of grain boundaries between source and drain becomes important. As a rule of thumb we may write :

$$\mu_{eff} \approx \frac{\mu_{single\ cr.}}{1 + L/\langle a \rangle} \tag{18}$$

The mobility depends not only on the annealing treatment and the channel length, but also on the gate voltage V_G. At higher gate voltages we have shallower channels and therefore more surface scattering. So μ_{eff} decreases. On the other hand, higher gate voltages lower the potential barriers of the grain boudaries at the channel surface and that makes μ_{eff} higher. Mercier et al. [26] have found experimentally:

$$\mu_{eff} = \text{const.}\ V_G^{-4.4} \exp\left(- \frac{2.08}{kT\, V_G^{1.19}}\right) \qquad (19)$$

According to eq. (19), μ_{eff} (V_G) shows a maximum (see fig. 9).

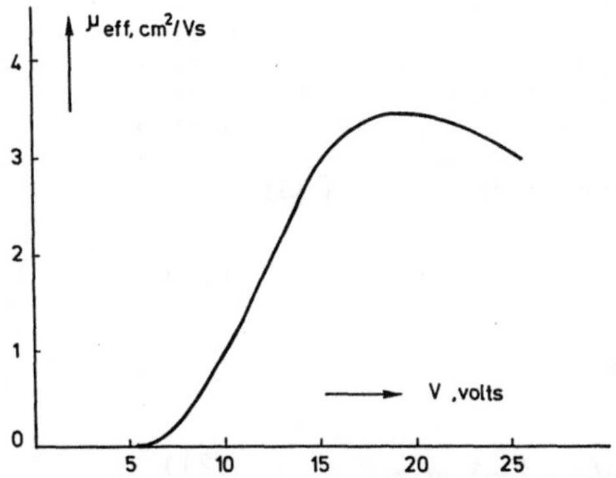

Fig. 9. Field-effect mobility (μ_{eff}) as a function of gate voltage (V_G) of a polysilicon thin-film transistor.

Possible structures for an n-channel MOS transistors are given in [26,27] .
We can have a metal (e.g. Al) gate electrode or a polysilicon gate, heavily doped with phosphorus ($N_D \approx 10^{20}$ cm^{-3}). In practice the latter shows a better long-term stability with regard to V_T .

With regard to high packing densities in VLSI circuits, the structure in fig. 10 is very interesting. Here we have a three-dimensional complementary CMOS structure [29] . The p-channel device is made in the monocrystalline substrate, the

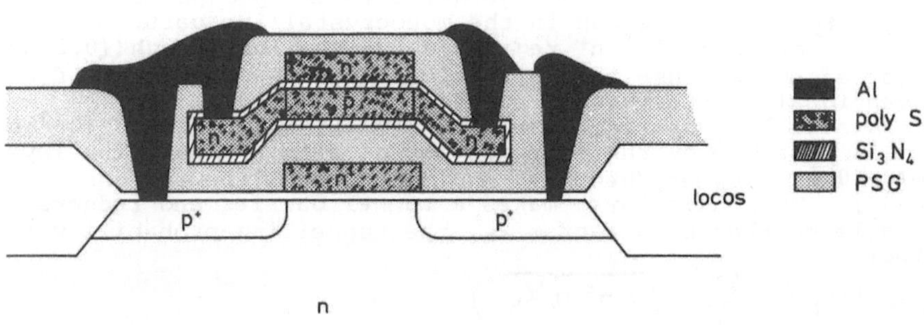

Fig.10. Three-dimensional complementary MOST gate with the n-channel device in polysilicon [29].

n-channel device in a laser-recrystallized polysilicon film right on top. Both devices have polysilicon gate electrodes, their W/L ratio is 18/2 μm.

3. Other applications

In the previous section we treated circuit elements consisting entirely of polysilicon. Here we will discuss the role of polysilicon as an auxiliary tool. Examples are polysilicon emitters in bipolar transistors, polysilicon as diffusion source and diffusion mask and some miscellaneous things.

3.1. Polysilicon emitters

Polycrystalline silicon can be used as the material for the emitter in a bipolar technology. In the DOPOS technique [30] a doped polysilicon layer is used, in the POLYSIL technique an undoped layer is deposited first and later on doped by ion implantation [31], followed by a drive-in step. The main feature of the polysilicon emitter is its higher efficiency [31-32]. In an npn transistor holes are injected into the emitter and their current density is [33]

$$J_p = \frac{q n_{io}^2}{G_e} \tag{20}$$

with
$$G_e = \int_0^{W_e} \frac{N_D(x)}{D_p(x)} \frac{n_{io}^2}{n_i^2(x)} dx$$
$$= \int_0^{W_e} \frac{N_D}{D_p} exp(- q \Delta V_{gap}/kT) dx \tag{21}$$

G_e is the so-called Gummel number for the emitter: if the emitter region is thin enough, most of the recombination will take place at the aluminium contact and the emitter is said to be "transparent". Eq. (21) will apply in that case. For monocrystalline emitters typical values of G_e are 2-5 * 10^{13} s cm^{-4} For heavily doped ($N_D = 10^{20}$ cm^{-3}) polycrystalline emitters we have seen in section 2.1 that the mobility is lower, (fig. 5), so $D_{poly} \sim (1/2$ à $1/3)$ D_{mono}. This makes the typical value of G_e for polysilicon emitters $G_e \approx 8*10^{13}$ s cm^{-4}.

In fig. 11 we have sketched the cross-section of a polysilicon emitter; during the drive-in the phosphorus or arsenic forms an n-region in the monocrystalline part. If the monocrystalline n^+ region is not shallow enough ($<$0.2 μm) G_e decreases, because this situation then resembles that of a conventional emitter with a polysilicon contact.

A still further increase of G_e to values of (2-10)*10^{14} s cm^{-4} is possible by the introduction of a thin insulating layer ($\delta \sim$20-40 Å) between the poly- and monocrystalline regions [34-35]. This thin layer makes a tunnel barrier and reduces the hole current by a factor P_h, the tunnelling probability for holes:

$$P_h = exp\left(- \frac{4\pi\delta}{h} \sqrt{2m^* q \chi_h}\right)$$
$$\approx exp\left(- \delta \sqrt{\chi_h}\right) \tag{22}$$

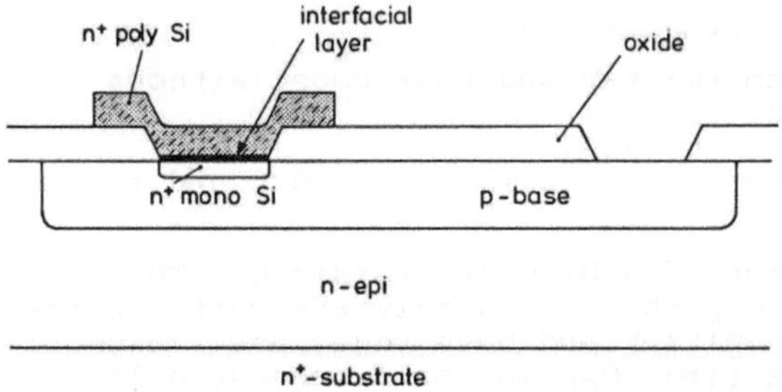

Fig. 11. Bipolar transistor structure with SIS tunnel emitter.

where δ is expressed in Angströms and χ_h in volts.

On the other hand, the required electron current is also hampered, though less, and this causes an extra voltage drop, which must remain within bounds, e.g. $\Delta V_n < 25$ mV.

The improved emitter efficiency may be used to make a device with a high current gain ($h_{FE_{max}} = G_e/G_b$) or to increase base doping, thus increasing

$$G_b = \int_0^{W_b} \frac{N_A(x)}{D_n(x)} \frac{n_{io}^2}{n_i^2(x)} \, dx \tag{23}$$

while $h_{FE_{max}}$ remains constant.
The latter choice enables us to decrease the base resistance and to improve the quasi-saturation behaviour. This is illustrated in fig. 12, where two different (I_c, V_{ce}) characteristics are shown. Both devices had the same structure: emitter 80x100 µm, $H_{FE_{max}}$ 200. For the conventional, monocrystalline emitter $G_e = 5*10^{13}$ s cm^{-4} and $Gb = 2.5*10^{11}$ s cm^{-4}, for the polysilicon tunnel emitter $G_e = 8*10^{14}$ s cm^{-4} and $G_b = 4*10^{12}$ s cm^{-4}.

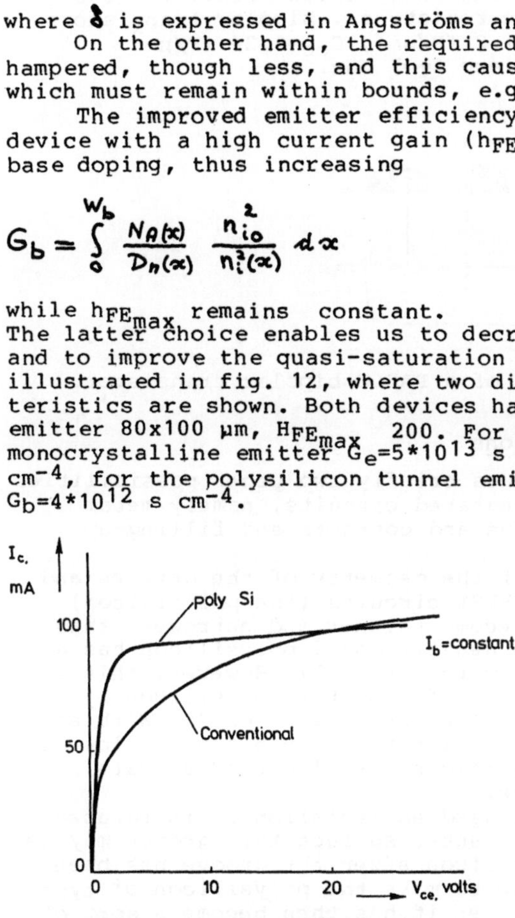

Fig. 12. I_c-V_{ce} characteristics of a conventional and a SIS tunnel (polysilicon) emitter transistor.

3.2. The PABLO process [36]

PABLO (Perfect Alignment By Lateral Oxidation) is a technological process that uses two polysilicon layers. The first layer is used as a diffusion source [37] and for contacting emitter, base and collector. The second layer is an auxiliary layer, which is partly oxidized for masking purposes and removed later on.

The typical features of the process are:

- submicron distance between emitter and base boost without using submicron lithography;
- emitter and base doped from polysilicon;
- easy combination with polysilicon resistors diodes and n^+ poly interconnections.

Fig. 13 gives a cross-section of a bipolar transistor. The capacitances are small: in C_{eb} there is hardly any contribution from the periphery because emitter and base edges are separated from each other by an oxide ring. C_{cb} is small because a large part of the base boost (p^+) is in the polysilicon layer and separated from the collector region by thick oxide. The base boost itself reduces the base resistance. Bipolar transistors have been realized with C_{oe}=1.2 fF/μm^2, C_{oc}=0.24 fF/μm^2, current gain h_{FE}=200 and maximum f_T=7 GHz.

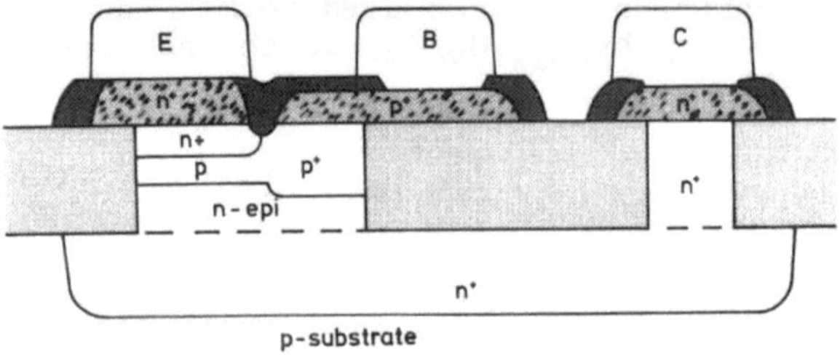

Fig.13. Cross-sectional view of a PABLO bipolar transistor.

3.3. Miscellaneous applications

In this section we will discuss briefly two other possibilities for using polysilicon in integrated circuits, namely metal silicides for interconnections and contacts and filling-up U-grooves.

With the down-scaling of the geometry of the devices and the high packing density in VLSI circuits (the polysilicon) interconnections must also become thinner and narrower, thus raising the resistance. Even heavily doped polysilicon has a resistivity not below 10^{-3} Ωcm (see fig. 5). However, this can be reduced considerably by silicidizing the polysilicon by means of an appropriate metal like Pt, Ti, Wo or Mo. With an extra temperature anneal the resistivity can become as low as $2.1*10^{-5}$ Ωcm [38] . Metal silicides are also used as gate electrodes in MOS transistors.

U-grooves are sometimes used as isolation in integrated circuits. In order to get a flatter surface this groove may be filled-up with undoped polysilicon after the groove has been oxidized. It is also possible to make the polysilicon p^+ type and connect it to the substrate; it has then become a sort of guard ring.

4. Conclusion

The main feature of polycrystalline silicon in integrated circuits is the possibility of introducing a second and even a third layer, in which active and passive components can be realized, and which are isolated from the monocrystalline substrate and from each other. Resistors can be made in a very wide range of resistivities, but they are not very precise and are largely influenced by the process technology. More precision requires laser-beam recrystallization.

Active devices like MOS transistors are only compatible with their monocrystalline counterparts if made in recrystallized material. Complete bipolar transistors are not yet made with great success in polysilicon material, but as a diffusion source, or as an intermediate step in the processing, polysilicon already plays an essential role.

References

1. D.J. Bartelink, Symp. Proc. Materials Research Soc., vol. 5, 249 (1982).
2. J.Y.W. Seto, J.A.P. 46, 5247 (1975).
3. G. Baccarani, B. Ricco and G. Spadini, J.A.P. 49, 5565 (1978)
4. C.H. Seager and T.G. Castner, J.A.P. 49, 3879 (1978).
5. G.E. Pike and C.H. Seager, J.A.P. 50, 3415 (1979).
6. N.C.C. Lu, L. Gerzberg, C.Y. Lu and J.D. Meindl, IEEE Trans. ED-30, 137 (1983).
7. C.M.M. Wu and E.S. Yang, IEEE Trans. ED-29, 1598 (1982).
8. C.Y. Wu and W.D. Ken, Solid-State Electr. 26, 675 (1983).
9. D.M. Kim, A.N. Khondker, R.R. Shah and D.L. Crosthwait, IEEE Electr. Dev. Ltrs, EDL-3, 141 (1982).
10. W. Shockley, Electrons and Holes in Semiconductors, Van Nostrand, New York 1963,
11. H.C. de Graaff and M. Huybers, Solid-St.Electr. 25, 67 (1982).
12. M.M. Mandurah, K.C. Saraswat, C.R. Helms and Th. J. Kamins, J.A.P. 51, 5755 (1981).
13. A. Broniatowski, J. de Phys. 43, C1-63 (1982).
14. S. Hirae, M. Hirose and Y. Osaka, J.A.P. 51, 1043 (1980).
15. J. Werner, W. Jantsch, K.H. Froehner and H.J. Queisser, Proc. Mat. Res. Soc. Ann. Meeting, Boston 99 (1981).
16. M.E. Roulet and M. Duboit, Appl. Pys. Lett. 37, 737 (1980).
17. M. Tanimoto, J. Murota, Y. Ohmori and N. Ieda, IEEE Trans. ED-27, 517 (1980).
18. H.K. Charles, Ch. Feldman, F.G. Satkiewicz, Techn. Digest IEDM, paper 4-4, 71, Washington (1976).
19. T.L. Chu, S.S. Chu, G.A. van der Leeden, C.J. Lin and J.R. Boyd, Solid-St. Electr. 21, 781 (1978),
20. J. Manoliu and T.J. Kamins, Solid-St. Electr. 15, 1103 (1972).
21. N. Dutoit and F. Sollberger, J. Electrochem. Soc. 125, 1648 (1978).
22. G.E.J. Eggermont and J.G. de Groot, IEEE Electr. Dev. Ltrs, EDL-3, 156 (1982).
23. C.M. Wu, and E.S. Yang, Appl. Phys. Lett. 37, 945 (1980).
24. H. Schaber and D. Cutter, J.A.P. 53, 8827 (1982).
25. H. Mikoshiba, IEEE J. Solid-St. Circ. SC-13, 483 (1978).
26. A. Mercier, M. El Koosy, A. Le Glaunec and E. Le Tiran, J. de Phys. 43, C1-369 (1982).
27. S.D.S. Malhi, R.R. Shah, H. Schichijo, R.F. Pinizotto, C.E. Chen, P.K. Chatterjee and H.W. Lam, Electr. Letters 19, 993 (1983).
28. R.C. Frye and K.K. Ng, Symp. Proc. Mat. Res. Soc., 5, 275 (1982).

29. S. Kawamura, N. Sasaki, T. Iwai, R. Mukai, M. Nakano and M. Takagi, Techn. Digest IEDM, paper 14.5, 364 Washington (1983).
30. M. Takagi, K. Nakayama, C. Tevada and H. Kamioko, J. Jap. Soc. Appl. Phys. (suppl.) 42, 101 (1972).
31. H. Murrman and A. Glasl, 6th Int. Congr. Microelectr., Munich (1974.
32. T.H. Ning and R.D. Isaac, IEEE Trans. Electr. Dev. ED-27, 2051 (1980).
33. H.C. de Graaff, J.W. Slotboom and A. Schmitz, Solid St. Electr. 20, 515 (1977).
34. H.C. de Graaff and J.G. de Groot, IEEE Trans, Electr. Dev. ED-26, 1771 (1979).
35. A.A. Eltoukhy and D.J. Roulston, IEEE Trans. Electr. Dev. ED-29, 1862 ((1982).
36. H.G.R. Maas and J.A. Appels, submitted to IEEE Electr. Dev. Letrs.
37. W.J.M.J. Josquin, P.R. Boudewijn and Y. Tamminga, Appl. Phys. Lett. 43, 960 (1983).
38. K.L. Wang, T.C. Holloway, R.F. Pinizzotto, Z.P. Sobczak, W.R. Hunter and A.F. Tasch, Jr., IEDM Techn. Digest, paper 3.3, 58, Washington (1981).

Part IV

Applications of Polycrystalline Semiconductor Compounds

Electroluminescence in Polycrystalline Semiconductors

R. Mach

Zentralinstitut für Elektronenphysik, AdW der DDR, Hausvogteiplatz 5-7
DDR-1086 Berlin, German Democratic Republic

1. Introduction

Research and development in electroluminescence (EL) of polycrystalline semiconductors is almost as old as semiconductor research as a whole. The aim from the very beginning was to obtain EL from large areas. As early as 1936 *Destriau* [1] started the run for "cold-light" emitting devices based on ZnS powders as the luminescent polycrystalline (pc) material. In the fifties and early sixties several thousand publications on "Destriau luminescence" were published [2,3]. Relying on the convincing optical and luminescent properties of the II-VI compounds [4], which had been so successfully exploited from then on in CRT screen phosphors, the ambitious aim was to replace the inefficient incandescent lamp by Destriau panels. But that period ended with the general belief that the lifetime problem of these powder devices based on copper activated ZnS embedded in a dielectric binder was insurmountable, and little effort was devoted to the study of EL in II-VI compounds during the last 15 years. The big success of the light emitting diode (LED) and the semiconductor laser based on monocrystalline III-V material was one reason more for using III-V compounds. These devices rely on the efficient injection of minority carriers via some kind of p-n junction and the following radiative recombination of e-h pairs. One prerequisite is expensive, highly perfect, single crystal chips or even higher quality epitaxial layers, due to the stringent requirements which must be placed on the minority carrier lifetime in order to obtain the necessary luminescence efficiency [5].

It is of course impossible to produce on such an operating principle large area electroluminescent devices (production costs would be tremendous), even for information display, not to mention lighting applications. Since the middle of the 70s, on the other hand, there is a big demand for large area active displays to replace the heavy and inconvenient cathode ray tubes (CRT).

In principle, the II-VI compounds are attractive candidates for use in pc form. Their low surface recombination velocity and their high optical transition probabilities due to the direct gap would favour them by far relative to the II-Vs. However, it has been well known for a long time that no p-n junctions could be formed in the wide gap materials capable of visible emission. And the minority carrier injection yield for all possible metal-semiconductor contacts and heterojunctions had been proven to be much too low, even with monocrystals [4]. The prospects seem to be even worse if one thinks about the application of pc thin films deposited on cheap glassy substrates at necessarily comparatively low temperatures, where the minority carrier lifetime and luminescence efficiency are expected to be influenced by the imperfection of the thin films obtainable.

Indeed, a lot had to be learnt in order to live with such thin pc layers for large area luminescent applications. But in 1983 mass production of ac driven thin film electroluminescent displays (ac TFD) was started. Research activities are known from the literature to go on at various laboratories [6-17].

The luminescent active pc material in all commercial devices is ZnS:Mn leading to a yellow emission (590 nm). The best performance figures given for these devices are 10^4fL at 5 kHz, 10^4h lifetime with only 30% degradation in brightness [18] and up to 8 lm/Watt efficiency [19]. On the market are already 240×320 pixel matrices of about 9×12 cm^2 active area addressed "a line at a time" with solid state drive electronics coping with television signals and a usual microcomputer interface. From the brightness point of view even more lines are possible; the limitations do not come from the physics but from the problem of production yield for such large area multilayer thin film devices.

The external efficiency values reached are quite high for information displays, but of course still too low to raise the question of lighting applications.

As will be outlined in Sect.3 the two keys to this solution of the old problem were (i) direct impact excitation by hot electrons of luminescent inner shell transitions which are *not* affected by factors which affect the minority carrier lifetime (which is inherently low in pc materials), and (ii) a design of the structures used which is optimum for the generation of hot carriers.

The breakdown stabilization problem of these high field devices has been solved by using a "charge control" and not by "current limiting". Concerning the impact excitation of inner shell transitions, in 1968 [20] work was started on electron impact excited *Lu*minescence of *Mo*lecular *Cen*ters (Lumocen) —e.g., TbF_3 in ZnS —but was considered in 1972 to be "inherently ineffective" [21]. At the same time a development was carried out [22] on ZnS:Mn.

There are of course some sidewalks, as for instance the dc powder displays and the dc thin film device now under discussion, and we shall in Sect.2 briefly compare these different types of electroluminescent devices, but after that (Sect.3), we concentrate on the present day understanding and problems of ac TFD, as these devices are the best understood and technologically closest to a simple concept. In Sect.4 we shall elaborate on the influence of technology, especially on the growth method of the active ZnS layer, i.e., the influence of polycrystallinity.

2. Comparison of Different Solid State EL Devices

All solid state ELDs are high field ones. As the principal aim was to make large area devices, all are made from pc II-VIs. The two processes used for excitation of luminescence are direct impact excitation of luminescence centers with inner shell radiative transitions [d-d transitions in Mn^{2+} and f-f transitions in rare earths $(RE)^{3+}$] by hot carriers and/or minority carrier injection and/or production in high field regions and their subsequent radiative recombination mainly in regions of smaller fields via usual dopants.

All four possible EL displays have been under discussion: ac powder, dc powder, ac thin film (ac TFD), and dc thin film displays, but only dc powder and ac TFD have been successful enough to be presently commercially available.

In powder displays the luminescent active layer consists of a suitably doped ZnS powder (about 1 μm grain size) suspended in a dielectric binder, whereas the luminescent active ZnS layers of thin film structures are produced by different deposition techniques (e-beam, thermal evaporation, low temperature CVD, MOCVD, sputtering). In Fig.1a,b the basic design of powder

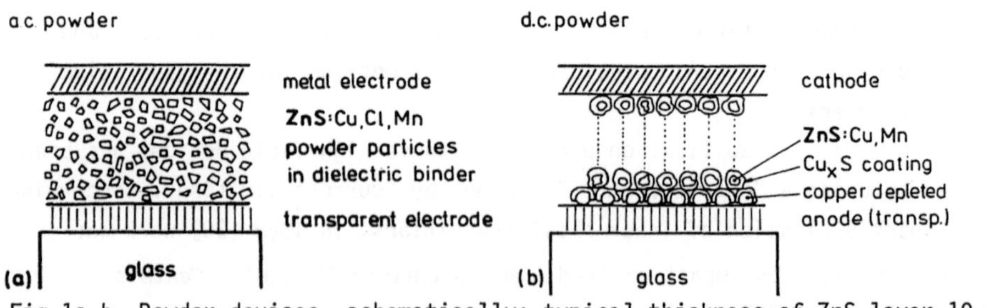

Fig.1a,b. Powder devices, schematically; typical thickness of ZnS layer 10 μm

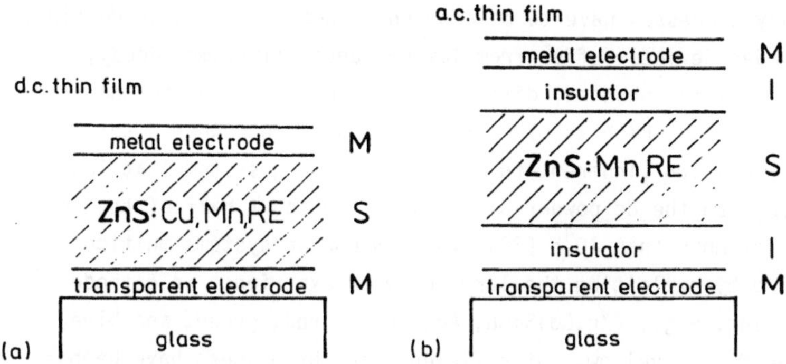

Fig.2a,b. Thin film devices, schematically; typical ZnS layer thickness d = 0.6 μm

displays is shown, whereas Fig.2a,b shows the same for thin film devices. The principal difference between a powder layer and a thin film is its usual appearance — the thin film is completely transparent and optically homogeneous, the powder layer opaque and diffusely reflecting. Besides the optical homogeneity, all other properties too are by far more homogeneous in the microcrystalline, seemingly one phase thin film.

A comprehensive compilation of facts about the different display types has been given by *Theis* [14], the ac TFD has been reviewed by *Mach* and *Mueller* [23-25], *Howard* [26], and *Alt* [27], whereas recent developments in powder displays have been outlined by *Vecht* [28]. The ac driven powder display is just the EL structure of Destriau reviewed some years ago by *Fischer* [29] and relies (as do the dc driven powder devices of Vecht) on a heavy Cu doping of the ZnS layer, which, in the forming process, leads to an inhomogeneous distribution of one or more of the numerous Cu_xS phases which are conducting (or semiconducting) and concentrates, in the case of the ac powder devices, the field in needle-like inhomogeneities, so that devices with a very inhomogeneously distributed field result. At high field spots hot carriers and/or minority carriers or electron-hole pairs are produced, resulting in direct impact excitation of luminescent centers and radiative recombination of the generated e-h pairs in lower field parts as well.

In just the high field regions almost no luminescence is generated as the activators are field-ionized (field quenching). A step in the direction of a laterally homogeneous field distribution is the dc powder device, where a thin Cu_xS-free layer (see Fig.1b), which is obtained by an initial forming process, consumes by voltage division the major part of the applied one. The rest of the (ZnS + Cu_xS) layer acts as a distributed series resistance stabilizing the electrical breakdown of the "active" high-resistance part.

Again both primary processes have to be assumed to act for light production. Both types of powder devices suffer from luminescence "lifetime" decay, mainly due to the (field enhanced) diffusion of Cu but also due to the large vulnerable surface area, which is of course inevitable in powder displays. Only for pulsed operation (inherent to matrix addressing systems) at low brightness (30 fL) can the dc powder displays retain about half of the initial brightness for more than 10^4h [28]. In both powder systems multicoloring has been tried by exploiting the large body of experience in cathode ray phosphors, using, e.g., (Zn,Cd)S:Cu, Ag, Br for red, green, and blue, respectively, and Mn for yellow, but efficiency and brightness have been found highest for Mn doping (10^3Cd/m^2, 0.5-1 lm/Watt). Only very recently *Vecht* published a result of 400 fL for green emission for CaS:Ce dc powder devices [28] (Fig.3). These experimental results clearly demonstrate the advantages of inner shell luminescence transitions (Mn,Ce) in comparison to free-bound or bound-bound (e.g., donor-acceptor) radiative recombination in high field devices, where the latter is strongly diminished by field quenching.

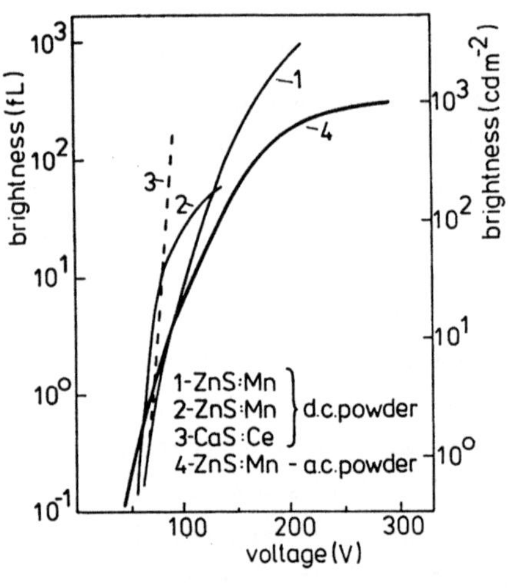

Fig.3. Comparison of brightness-voltage characteristics of dc powder cells – *Curve 1* [30], *Curves 2,3* [28] – and ac. powder structures – *Curve 4* [2]

An inherent common feature of powder devices is the good optical outcoupling of light, but the price to be paid is the poor contrast in bright ambients due to the diffuse reflection and scattering of light by the powder particles.

Decisive for the multiplexibility (number of addressable lines) is the steepness of the brightness (B)-voltage (V) characteristic because of the cross-talk so common to LCDs. The nonlinearity of ac powder devices is not high enough to give a good discrimination ratio, whereas dc devices are well multiplexable due to the more homogeneous field distribution and the resulting comparatively steep rise of B(V) (Fig.3). At the moment, a range of dc powder devices are commercially available, as, for instance, a 256 character matrix display [31].

Dc thin film structures are the simplest imaginable — a ZnS layer, e.g., doped with Mn — between two electrodes (Fig.2a). They promise, at first sight, a lot of advantages in comparison to powder devices: a high uniform electric field at low voltages (no need for Cu doping), high contrast, and a smaller vulnerable surface than powder devices.

Using a single crystal model structure, even the direct impact excitation of Mn^{2+} in ZnSe could be proved by the fact that, at a reverse voltage below the gap of the host, the Mn^{2+} luminescence could be excited [32] (see Fig.12). Of course, these devices have to be stabilized against electrical breakdown by a series resistance. And this resistance has to be distributed laterally, not just one for a large area, as such a homogeneity cannot be achieved. So in fact, even in the laboratory, such structures can only be made stable for some hours by going back to dc powder like structures, using Cu drift forming, with all its control and stability problems [33,34] or by using some other kind of distributed current limiter, as for instance a ZnSe layer [35]. The driving force behind these activities is the potential use of cheap low voltage drive circuitry. As luminescent dopants, Mn [36] and RE [33] have been used (Fig.4). The brightness can be as high as 1000 fL at least for Mn, Er, and Tb doping (inner shell luminescence transitions) but the highest efficiency published for dc structures is only 0.15 lm/Watt. It

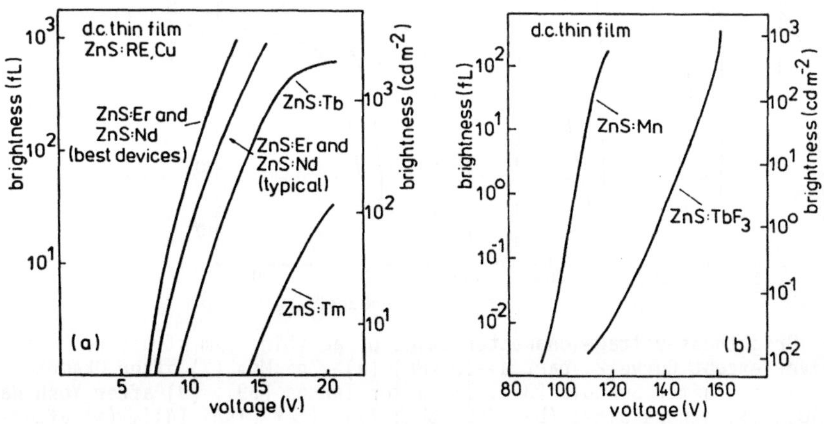

Fig.4a,b. Comparison of B(V) characteristics of different types of dc thin film devices (**a**) after Bryant and Krier [33]; (**b**) after Ohnishi et al. [36]

has to be stressed that these dc TFD structures are rather less developed than the ones discussed above. Nobody claims at present to have reached reliability.

By far the most convincing and even commercial success has been reached by sandwiching the luminescent active thin film (S) between two insulating layers (I) as schematically illustrated in Fig.2b, forming an MISIM-ac TFD. In this case, breakdown is controlled by the charge transferred through the semiconductor to the insulator capacitances, not by any resistance. If the insulator layers stand the applied voltage, no failures can occur. No forming is necessary, Cu can be completely avoided, and the electric field situation is a completely homogeneous one. Besides the basic advantages already mentioned in comparison to powder displays, brightness and efficiency of these devices are high (ZnS:Mn 10^4 fL at 5 kHz, 8 lm/Watt), the steep B(V) characteristic is extremely (Fig.5) favorable for multiplexing (more than 1/1000 possible), and the lifetimes claimed for a commercial (!) large area product are 10^4h with only 30% [18] degradation for operation in the saturated (high brightness) part of the B(V) characteristic. The main (and commercial) success with the figures of merit mentioned has been reached with ZnS:Mn, using the direct impact excitation of Mn^{2+} by hot electrons, for EL. The success with other colors is rather limited, although some encouraging results have been reported by different groups using RE, especially

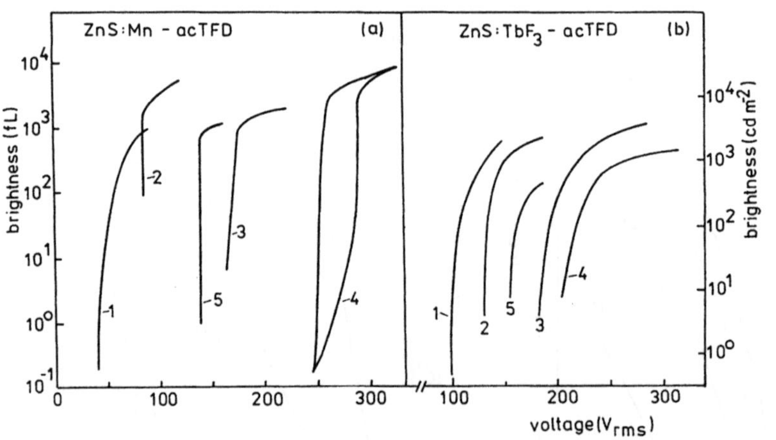

Fig.5a,b. Brightness-voltage characteristics of ac thin film structures under 5 kHz drive (except Curve 2, part a – 10 kHz) (**a**) – ZnS:Mn; (*1*) after Okamoto et al. [37], (*2*) after Suntola [38], (*3*) after Tannas [39], (*4*) after Yoshida et al. [40], (*5*) own results; (**b**) – $ZnS:TbF_3$; (*1*) after Oishi [41], (*2*) after Hale et al. [42], (*3*) after Miura et al. [43], (*4*) after Okamoto et al. [44], (*5*) after Kobayashi et al. [45]

TbF_3 [41-44]. The highest brightness is 1000 fL at 5 kHz for the ZnS:TbF_3 green emission [43] (Fig.5b). The efficiency for these bright green devices (curve 3) is 0.8 lm/Watt. However, one should note that nothing has been written about maintenance in the RE-doped structures.

Ac TFD can be made fully transparent giving extremely high contrast, which can even be enhanced by using a black dielectric [22]. There is on the other hand a disadvantage inherent to thin films — the bad optical out-coupling of the generated light. The major fraction of the generated light is lost by internal reflection due to the high refractive index of ZnS.

Ac TFDs are very liable for pinholes (weak points) in each of the layers and they may lead to catastrophic breakdown. In this connection, the importance of high quality insulator layers cannot be overemphasized. Since random defects leading to electrical breakdown of the insulators are unavoidable, the whole thin film stack (nowadays up to 10 layers) must form a system in which pinhole burn-out is self-limiting rather than propagating. The problem is attacked by composite insulators and careful matching of layer properties. These problems determine more or less the production yield and the costs of the device.

Summarizing this comparison, it appears fair to state that the ac TFD is the most promising solid state display.

In the following we will try to understand the physical reasons for the success so far reached in these polycrystalline MISIM structures, but also discuss the physical limitations of the material combinations used up to now. We shall concentrate on ZnS:Mn as the luminescent active layer.

Finally, it should be mentioned that ac TFD based on ZnS:Mn as active layer are unique in showing hysteresis in the B(V) characteristic if certain preparational conditions are met (Fig.6). The benefits of this characteristic (long known and used in ac plasma displays) for the driving of large

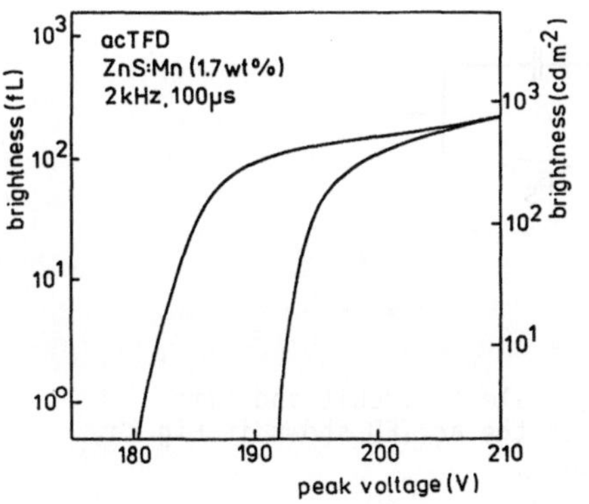

Fig.6. Dependence of brightness on voltage of a hysteretic sample of ZnS:Mn [49]

area displays have long since been recognized [46], and since then a lot of technological and scientific effort has been put into these special structures [47-49] but up to now, to our knowledge, nobody has solved the maintenance problem of the memory behavior. Therefore we shall not discuss this feature in the following.

3. A Simple Picture of ac TFD

3.1. Phenomenology

The ac TFD is a stack of thin films, as schematically illustrated in Fig.2b. Sandwiched between two insulator layers I, light is emitted from suitable luminescence centers within the semiconductor layer S. Judging from the published experimental results, the most suitable material for the S-layer is ZnS doped with Mn. The physical reasons for this will become clear later on by looking at the basic processes in these structures. A large variety of insulators is applicable; the materials point of view has been discussed in detail in [23]. Voltage can be applied via two contacts, one of which at least should be transparent. The applied voltage has to be an alternating (ac) voltage, as the insulators make the device almost capacitive. If a voltage is applied, nothing happens below the threshold U_{th}, but starting at this voltage, light is emitted in the form of a pulse in each half period of the ac voltage. Above the threshold, the device is no longer purely capacitive, and a dissipative current component starts to increase. The equivalent circuit, together with the time dependences of applied voltage (a), light output (b), and current (c), are shown in Fig.7 schematically. The threshold behavior of the brightness dependence on voltage is evident from Fig.5. The details depend on the wave form of the driving voltage. A typical time behavior of a ZnS:Mn structure during *one* applied 100 µs pulse is shown in Fig.8.

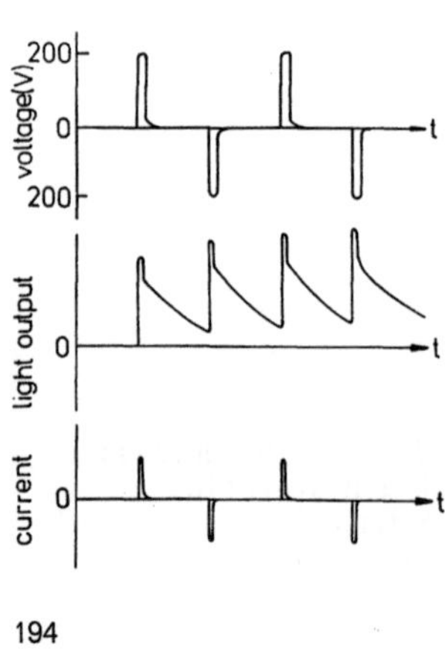

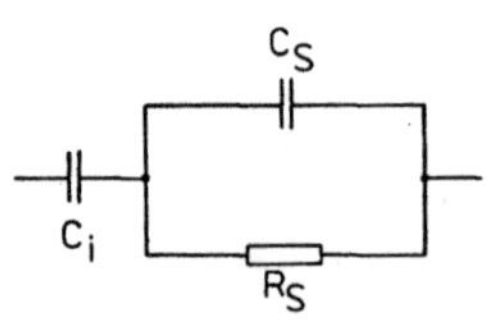

Fig.7. Equivalent circuit and time behavior of the ac TFD shown in Fig.2b, schematically

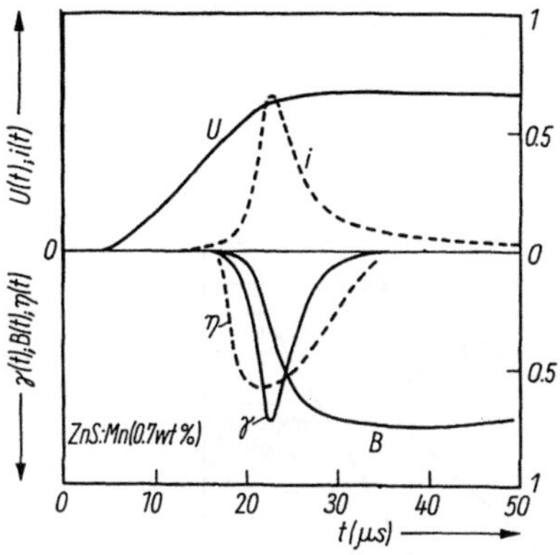

Fig.8. Time dependence during one 100 μs excitation pulse of applied voltage U, current i, brightness B, excitation function $\gamma(t) = dB/dt + B/\tau$, and excitation yield $\eta(t) = \gamma(t)/i(t)$, with $\gamma(t)$ giving the number of Mn^{2+} centers excited per second, after Mach et al. [50,51]

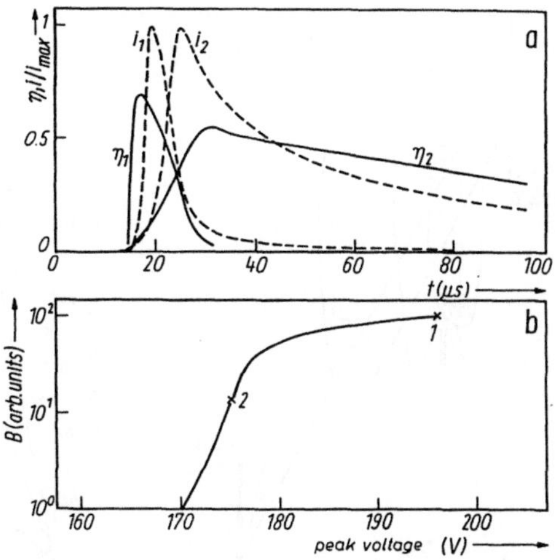

Fig.9. (a) Time behavior of excitation yield η and current i at two different drive levels indicated in (b), ZnS:Mn, 200 Hz; after Mach and Mueller [25]

The detailed time behavior depends of course on the voltage applied. Figure 9 illustrates this for the current and yield function. More details and the physical interpretation of these time resolved measurements can be found in [50,51]. The most interesting feature for the user of the device is, of course, the (time averaged) B(V) characteristic; Fig.5 shows a summary of data from the literature, which suggests an upper limit for the achievable brightness at about 10^4fL for ZnS:Mn. To understand this we look firstly at the basic processes occurring in ac TFD.

3.2 Basic Processes

According to the present understanding of the physical mechanisms in ac TFD, which has been outlined in detail in [23-25], the simplest description of the mechanism of ac TFDs has to include the six basic processes schematically depicted in Fig.10:

1) tunnel emission of electrons from the interface states of the cathode IS interface,
2) ballistic acceleration of the electrons within the S-layer,
3) impact excitation of the $3d^5$ electron of Mn^{2+} ions (on a Zn^{2+} lattice site) by the highly accelerated electrons,
4) radiative (or non-radiative) de-excitation of the excited $(Mn^{2+})^*$ centers,

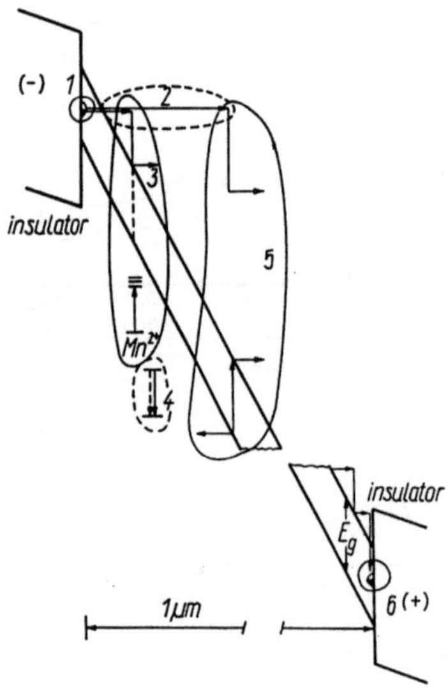

Fig.10. Simple model of the physical processes in MISIM structures under applied voltages above the threshold of the B(V) characteristic of Fig.5. The basic processes are indicated and numbered — see text; after Mueller [24]

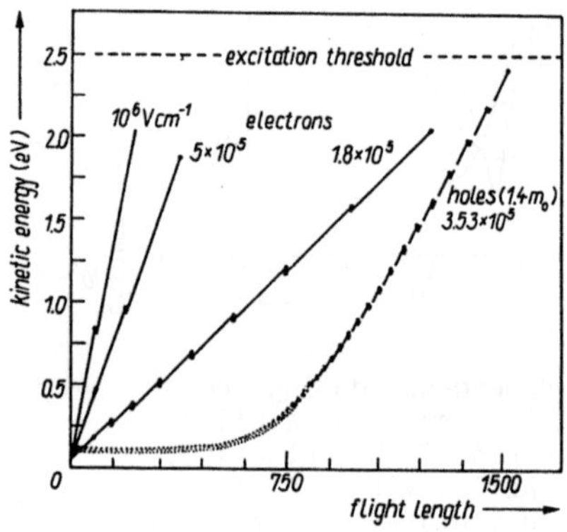

Fig. 11. Computer plot of carrier trajectories for ZnS in the electric fields indicated. LO forward scattering and a parabolic energy dependence in a ZnS model have been used; after Mueller [24]

5) carrier multiplication by valence electron impact ionization leading to an accumulation of positive space charge trapped in the bulk,
6) carrier capture by interface states at the anode (IS) interface.

Two of these processes exhibit a threshold behavior upon applied voltage — at a threshold field strength E_t tunnel emission of electrons from the IS interface starts, and above the threshold field strength E_a a ballistic, i.e., loss-free, acceleration of charge carriers sets in. It is the latter point which makes ZnS extremely suitable for application in ac TFD structures. The dominant energy-loss mechanism of the drifting carriers is to LO phonons via Fröhlich coupling. The characteristic feature of this process is the decreasing loss rate with increasing electron energy, leading to an acceleration which is already almost loss free for 10^6V/cm. That means, in the end, as outlined in [24], that the 2.5 eV necessary for impact excitation of Mn^{2+} can be got from a 10^6V/cm field within a 25 nm flight length (Fig. 11). After having gained 2.5 eV from the field, the electron is able to impact-excite an Mn^{2+} center by raising the outermost electron to its first excited state. Experimental evidence for this impact excitation of Mn^{2+} has been given in [32] by using a metal-ZnSe:Mn diode driven in the reverse direc-

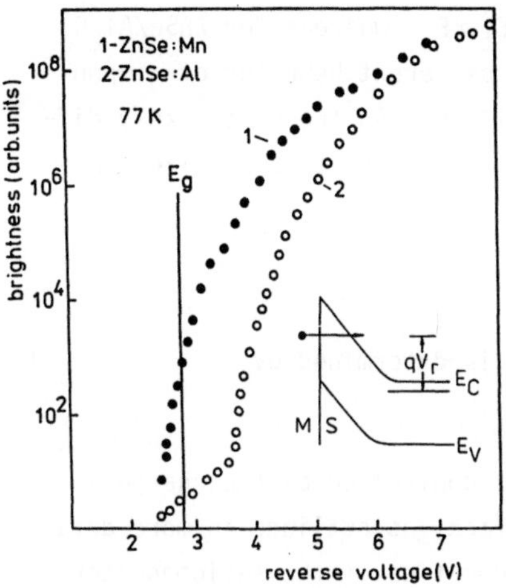

Fig.12. Brightness versus applied (reverse) voltage for an MS tunnel diode on ZnSe:Mn and ZnSe:Al, respectively, evidencing impact excitation of Mn^{2+}, after Mach et al. [32]

tion (Fig.12), the current of which was purely tunnel emission controlled. For Mn doping at voltages below the gap E_g of the host Mn^{2+}, emission could be excited, whereas for all other dopants (as for instance Al, I, Cu) the brightness starts to increase steeply at voltages somewhat higher than the gap.

The electrons which do not meet an Mn^{2+} after an acceleration length l_a, will gain further energy and reach the anode, or are inelastically scattered by a valence electron, ionize this to the conduction band, thereby multiply, and generate a hole.

Mainly based on time averaged efficiency measurements, with an efficiency defined as

$$\eta = \frac{\text{photon flux}\left[\frac{\text{photons}}{\text{cm}^2\text{s}}\right]}{\text{transferred charge}\left[\frac{\text{electrons}}{\text{cm}^2\text{s}}\right]}$$

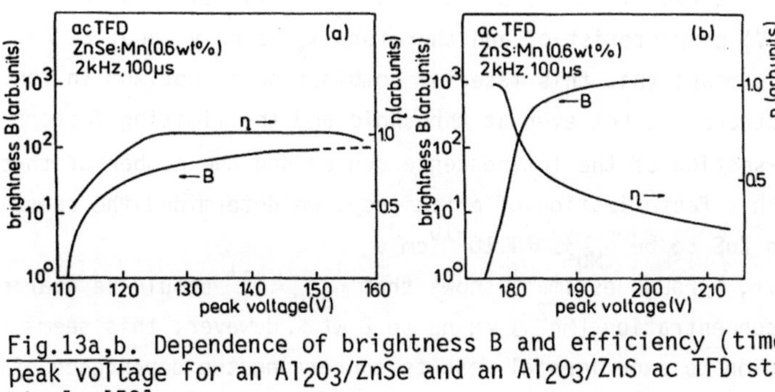

Fig.13a,b. Dependence of brightness B and efficiency (time averaged) η upon peak voltage for an Al_2O_3/ZnSe and an Al_2O_3/ZnS ac TFD structure; after Mach et al. [53]

we could show that for ZnS/Al_2O_3 structures $E_t > E_a$, whereas for $ZnSe/Al_2O_3$ structures $E_t < E_a$ [51,52], leading to a very different behavior of η upon voltage for both material combinations (Fig.13a,b). As the most likely difference between the two structures is in E_t, these results emphasize the importance of the IS interface.

3.3 Efficiency and Saturation

The maximum achievable brightness of ac TFDs is determined by

$$B_{max} = \eta Q_{max}$$

where the maximum transferable charge Q_{max} is controlled by the charge at breakdown of the insulator and the I/S interface, as outlined in more detail in [24,25]. η is mainly determined by the properties of the semiconductor layer and we will concentrate on these problems.

By a factorization into partial efficiencies,

$$\eta = \eta_{exc}\eta_{opt}\eta_{lum} \quad ,$$

characterizing the three independent fundamental processes involved — (i) impact excitation of centers η_{exc}, (ii) radiative decay of excited centers η_{lum}, and (iii) outcoupling of the emitted photons η_{opt} —, we obtained [25] some insight into the limiting physical mechanisms. The main results are the following.

$\underline{\eta_{exc}}$: Generally speaking, η_{exc} might be limited either by the impact length $l_i = (\sigma N)^{-1}$ (σ-impact cross-section of the luminescence centers, N concentration of these centers) or by the acceleration length l_a over which an electron attains the 2.5 eV necessary for the excitation of Mn^{2+}, and this length is determined by the electric field strength (at threshold of the B(V) characteristic) as illustrated by Fig.11. In the concentration range of practical interest in $ZnS:Mn/Al_2O_3$ structures η_{exc} is controlled by $\sigma_{Mn^{2+}}N$, due to the fact that in these structures $E_t > E_a$ is realized at the threshold of the B(V) characteristic, and therefore l_a is only about 25 nm. In other words, this means that this material combination is optimum in the sense that all electrons are hot even at threshold and the limiting factors are only the cross-section of the luminescence center and the number of these centers. Based on this factorization of efficiency, we determined the cross-section for Mn^{2+} in ZnS to be $\sigma_{Mn^{2+}} = 4.10^{-16} cm^2$.

Using this result, a rough estimate shows that $\eta_{exc} = l_i^{-1}$ should be guaranteed for a dopant concentration $[Mn^{2+}] = N$ up to 2 wt %. However, this seems to be in contradiction to experimental results concerning the dependence of

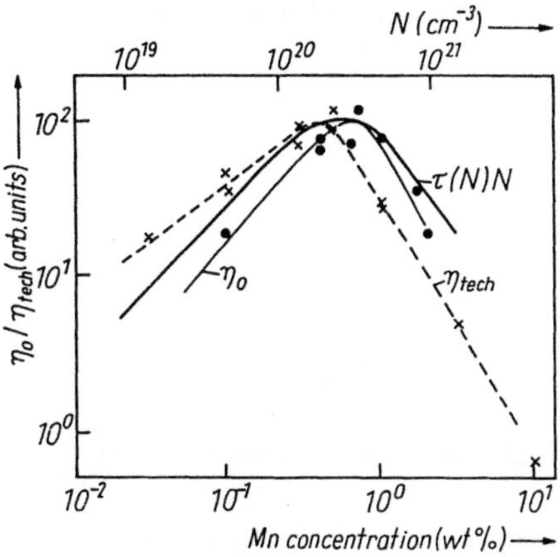

Fig.14. Efficiency in dependence on Mn concentration, η (×), after Sasakura et al. [51]; η_0 (•) own results, $\tau(N)N$ with τ from Fig.15; after Mach and Mueller [25]

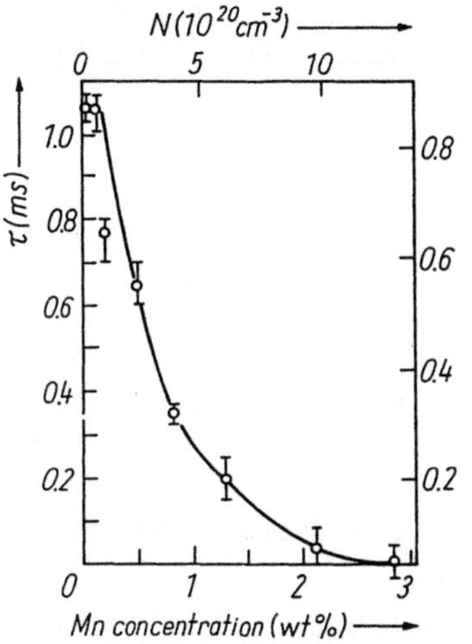

Fig.15. Decay time of cathodoluminescence of ZnS:Mn (measured at low excitation level) in dependence on Mn concentration N (τ-time for half decay $B/B_0 = 0.5$); $\tau_r = 1.77$ ms after [57], for cubic environments); afer Mach and Mueller [25]

η on the Mn concentration, as illustrated by the data given in Fig.14. For low Mn concentrations an increase of η with N is observed as expected, but already for $N \approx 0.6$ wt % the efficiency goes through a maximum. This can be shown to be due to influences on η_{lum}. As is well known [54,55] and shown in Fig.15, the luminescence decay time of Mn^{2+} in ZnS decreases strongly with increasing Mn concentration. This can be related to a dependence of luminescence yield $\eta_{lum}(N) = \tau(N)/\tau_r$ (where τ_r is the radiative decay time) on the center concentration N.

As outlined by *Goede* [56] the reason is a resonance energy transfer from the excited Mn^{2+} ions via unexcited ones to centers responsible for unwanted transitions. This transfer appears to be controlled solely by the Mn concentration and not much influenced by the concentration of killer centers, resulting in an almost technology-independent decrease of luminescence yield with the increase of N. In Fig.13 the product $N\tau(N)$ has been included and obviously it matches rather well the experimental data of different authors.

The luminescence yield in the system ZnS:Mn furthermore depends strongly on the concentration of excited centers N^*, and that means on the brightness level.

An example is given in Fig.16. This phenomenon has of course a severe influence on the maximum achievable brightness and is the main reason for the

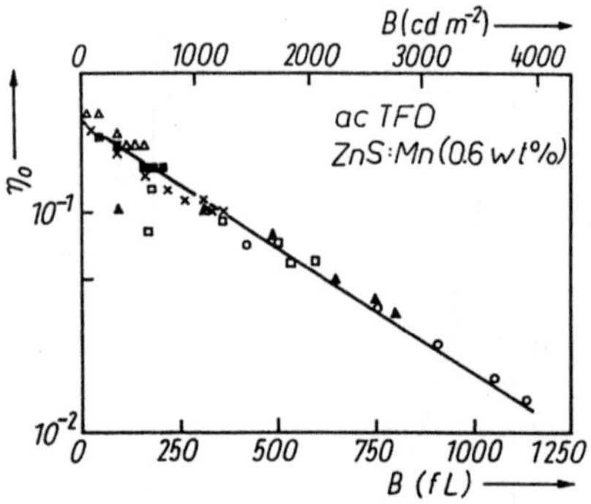

Fig.16. Efficiency η of an ac TFD in dependence on brightness B, frequency as parameter: (△) 500 Hz, (■) 1, (×) 2, (▲) 10, (○) 20 kHz

decrease of efficiency with increasing voltage, which was illustrated in Fig.13b.

Possible explanations are discussed in [25].

η_{opt}: As the emitting ZnS layer is of high refractive index, total reflection seriously influences the efficiency:

$$\eta'_{opt} = 1 - \sqrt{1 - n_s^{-2}} = \eta_{opt}/d$$

relates the flux incident on an inner surface of the ZnS layer to the flux outside the structure. This gives roughly $\eta'_{opt} = 0.1$ ($n_s = 2.4$).

Summarizing these results, we can state:

- ZnS is a material well suited for getting hot carriers at comparatively low electric fields of some 10^6V/cm.
- By a suitable choice of the I/S interface this can be fully exploited. If $E_t > E_a$ is guaranted at the threshold of the B(V) characteristic (and this condition can be met by different insulators) the efficiency is only limited by the number of available centers and not by hot carrier availability.
- Mn^{2+} on Zn site is a suitable luminescent center for several reasons: (i) the luminescent transition is an inner shell transition, well shielded against electric fields, (ii) Mn^{2+} is neutral, avoiding drift in the high electric fields, (iii) the impact cross-section is relatively large, (iv) the solubility is high.
- The main limitations arise from the influences on the luminescence efficiency: (i) concentration quenching is high in ZnS:Mn and (ii) at high excitation levels a new nonradiative decay channel is opened.
- The promising optical and luminescent properties of II-VIs cannot be used because of the field quenching of normal luminescent centers.

Having the results of factorization in mind, the main question left and to be answered in the scope of this paper is whether the partial efficiencies can be influenced by technology, or more precisely, by the microstructure of the polycrystalline semicondutor films.

Surprisingly enough, until recently, exceptionally few workers have concentrated on the material science of the thin films in electroluminescent displays, especially the semiconductor layer. Only very recently has the pioneer work of *Theis* been published [58,59].

4. The Influence of Technology

4.1. The Microstructure of Polycrystalline ZnS Films

From a technological point of view, the ac TFD is a large area multilayer thin film stack (Fig.2b) and all technological steps to produce this device have to be compatible with the demands on every single component of the system. The main problems are the interdiffusion and interface reactions between the different layers. Therefore, and in order to be compatible with a cheap soda glass as a substrate, the temperature of all technological steps should not exceed 400 to 500°C. For this low temperature deposition of ZnS different methods have been employed: e-beam and thermal evaporation, sputtering, atomic layer epitaxy (ALE) [23], and most recently CVD and also MOCVD [16].

The substrate (the transparent electrode or the insulator layer) is in any case polycrystalline or even amorphous. For the physical deposition methods the most important parameters are thought to be the substrate temperature T_s, the thermal annealing temperature T_{an}, and the layer thickness. The most often used T_s is about 150-200°C [23]. This temperature is a compromise between the decrease of deposition rate and the increase of grain size with increasing temperature. *Theis* et al. [58] investigated in detail the influence of T_s on the microstructure of e-beam deposited ZnS and compared this with ZnS layers deposited by the ALE technique. The microstructure was investigated by transmission electron microscopy (TEM) of thin cross-sectional specimens. According to these results, independent of the substrate (Y_2O_3, $BaTiO_3$, Al_2O_3, glass), ZnS layers deposited by e-beam evaporation exhibit a very fine grained substrate-near region, upon which, with increasing thickness, larger columnar grains begin to grow. The grain size is related to the substrate temperature T_s (Figs.17,18).

An increase of mean grain size results in an increase of surface roughness (Fig.18). The microstructure of ALE deposited films is, according to these TEM investigations, rather different. Figure 19 illustrates the differences between both technologies. Most striking are the absence of the fine grained

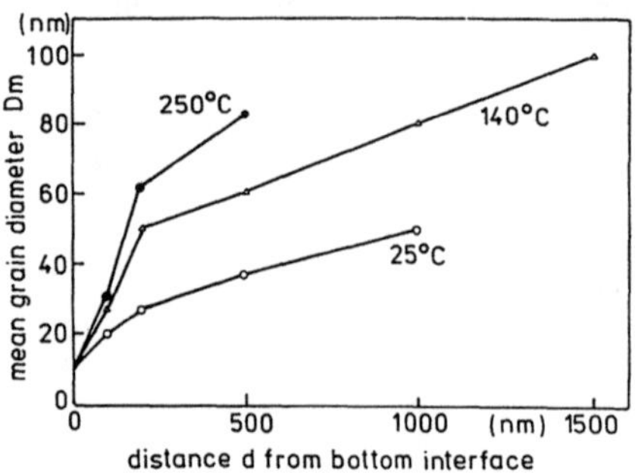

Fig.17. Mean grain diameter D_m parallel to the film surface versus distance d from bottom interface for electron beam deposited ZnS at various substrate temperatures; after Theis et al. [58]

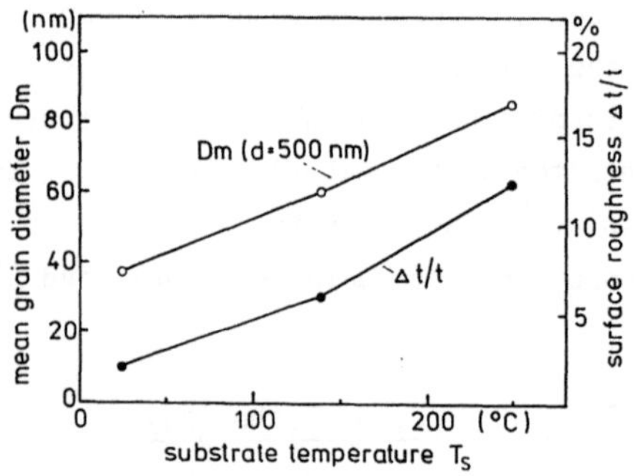

Fig.18. Mean grain diameter D_m parallel to the ZnS film surface at a distance from the bottom interface of d = 500 nm and surface roughness $\Delta t/t$ versus substrate temperature of electron beam deposited films; after Theis et al. [58]

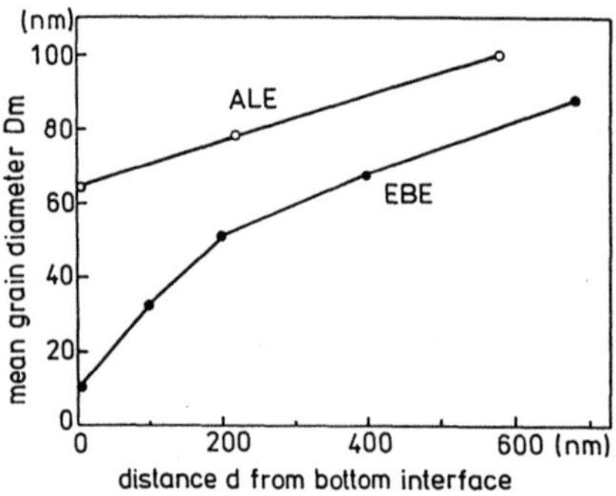

Fig.19. Comparison of electron beam deposited ZnS (EBE) and atomic layer epitaxy (ALE) ZnS; mean grain diameter parallel to film surface versus distance from bottom interface; after Theis et al. [58]

region in ALE samples and the comparatively large columnar grains. These results could be confirmed by X-ray diffraction measurements performed by *Tanninen* et al. [60]. According to *Theis* [61] the microstructure of e-beam layers remains almost unaffected up to annealing temperatures of 550°C; only at or above 850°C is strong grain growth in the bottom part of the film observed. In spite of this, the operational properties of ac TFDs can be influenced by an annealing procedure at about 400°C. The underlying reasons are not clear yet. There are only first attempts to improve the layer properties by laser treatment [62,67].

One other point of interest is the stoichiometry of the growing layer. As expected, the published results are quite technology-dependent. According to AES and SIMS profile analysis the S/Zn ratio of e-beam deposited samples ($T_S > 140$°C) is always smaller than unity, as compared to a single crystal standard [58].

Preisinger and *Pulker* [63] report the opposite for thermal evaporation at $T_S < 90^{\circ}C$. This unstoichiometry is said to be the reason for the humidity sensitivity of ZnS layers.

Up to now nothing is known about the homogeneity of the Mn distribution within the polycrystalline layers and the interface reactivity between the different thin film layer components of the system.

4.2. Correlation of Microstructure to the Performance of the Devices

According to the results presented in the last section, at least in e-beam and thermally deposited ZnS layers, a very inhomogeneous situation is encountered. Compared to the inhomogeneity of powder devices the scale is of course completely different. All the films described in the last paragraph show excellent optical interference patterns even at large angles and light scattering is extremely small. The fine grained region on the bottom has been discussed in connection with a "dead layer": the only convincing experiment, however, confirming the expected low luminescent yield from this part of the ZnS layer has been one by *Marello* et al. [66] using a probe layer technique in which only a small fraction of the S-layer thickness was doped with Mn and the position of the probe layer was changed stepwise (Fig.20). According to this result only the first deposited 200 nm near the substrate seem to be luminescence-dead, whereas this is not observed for the top layer. Unfortunately nothing is said about the drive conditions used which influence markedly the field distribution (operational inhomogeneities [25]) in this experiment.

Figure 21 compares the dependence of brightness on semiconductor thickness measured by different authors on samples prepared by e-beam (Sasakura,

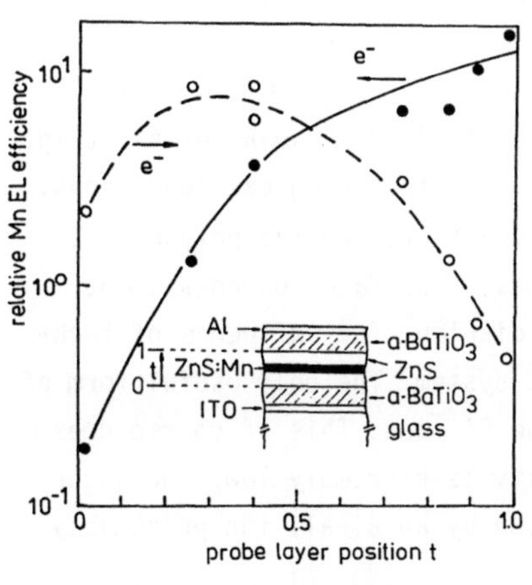

Fig.20. Relative EL intensities for ac TFD with Mn doped probe layers in undoped ZnS (cross section shown schematically in the inset). The EL emission for each half cycle of excitation has been analyzed separately and the direction of electron propagation is noted; after Marello et al. [66]

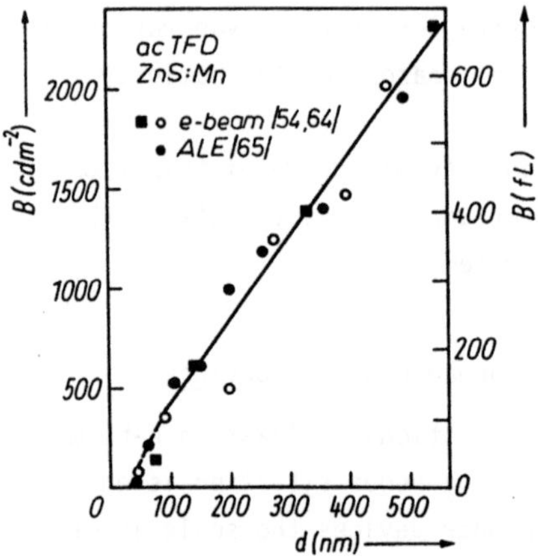

Fig. 21 The dependence of brightne on semiconductor thickness d for ac TFD structures with ZnS:Mn layers c posited by the different techniques indicated; after MACH and MUELLER [

Okamoto) and ALE. The agreement is remarkable. According to the simple picture of ac TFD developed in Sect.3 we would expect $\eta(d) \sim d$ as long as $d \gg l_i$ and it should be very small for $d \leq l_a = 250$ Å. If the experiments summarized in Fig.21 had been done under the meaningful condition of constant transferred charge, B(d) should depict this $\eta(d)$ behavior.

Really, the experiments hint that the dramatic decrease of brightness (or efficiency) with layer thickness for $d < 200$ Å might not mainly be connected with the luminescence efficiency but with the excitation efficiency of such thin layers.

Summarizing, we can conclude that in spite of the significant differences in microstructure of e-beam and ALE ZnS:Mn layers (as two extreme cases) the electrooptical properties (including the published data on luminescence efficiency) of ac TFD differ not so very much. The reason is that in this system the luminescence efficiency is not significantly influenced by the structural properties of the ZnS layer, but by interaction processes in the Mn system. This has been confirmed by an estimate of the maximum brightness level achievable based on the factorization of efficiency in this system [25] which gives about the values already published 1977 by *Inoguchi* et al. (Fig. 5). If we do not take into consideration very bad technological conditions, the partial efficiency, which could be mainly affected by the polycrystallinity, seems to be η_{opt} due to the influences of surface roughness, the extreme case in this respect being a powder display. The influence of technology is much more pronounced in the ZnS:RE system. The only useful form of REs for emission in the visible region is the RE^{3+} ion. This of course does not fit into the ZnS lattice and so solubility is extremely low. The incorporation is possible under charge compensation by an alkali ion presumably sitting on a nearest cation site. The great trick of the "Lumocen" period was to incorporate LnF_3 molecules presumed to be incorporated as neutral entities into the ZnS lattice. Even in this case, heavy stress and strain are likely to occur and limit the "solubility". There are strong indications that clustering and even a formation of a separate LnF_3 phase happens. Clustering has been suspected even in the gas phase after e-beam evaporation. In some recent publications [41,43] it is claimed that both effects can be avoided by sputter deposition of a ZnS/TbF_3 mixture (see Fig.6), but up to now nothing is known about the stability of these structures.

The main point to stress is that MnS is miscible with ZnS whereas that is not the case for RE or any RE compounds which have been used [45]. For this reason the color problem of ac TFD appears to be an open problem at the moment, but this is true not only for ac TFD but for all systems based on the direct impact excitation of luminescence centers by hot electrons.

Conclusions

Summarizing and neglecting the learning factor one might say that the ac TFD is an excellent example for coping with the special properties of pc material: it uses just the processes and the materials which are adopted to work well in structurally partly disordered systems.

The starting point in this sense was the need for a large area EL device. Unescapably the economics of the problem demanded the use of pc material on a cheap, i.e., glassy, substrate. Looking from a nowadays "learned" point of view, it is readily obvious what the considerations had to be: the only radiative deexcitation process which is almost completely untouched by its surrounding is the one in inner shells of the 3d, 4f, or 5f configurations. A look at the spectroscopic tables tells us that visible emission can be got only from Mn^{2+} or RE^{3+} and eventually Eu^{2+}. In all the possible centers, the visible transitions are quantum-mechanically forbidden, so it is highly probable that one needs high concentrations to get the necessary intensities out. High solubility of the dopants, therefore, should be one factor in the selection of the host. Excitation: these inner shell transitions can be excited by two processes effectively—by transfer from activators or by direct impact of carriers not forbidden by dipole selection rules. Activator transfer would mean having a high activator recombination traffic; that means high minority carrier concentrations would have to be supplied by injection, which seems to be impossible for pc materials. So direct impact excitation should be the solution and this demands high fields, which in turn quench activator processes. The next selection factor for the host is its suitability for carrier acceleration: a polar material is wanted, to take advantage of the Fröhlich coupling. However, it should have a small effective electron mass and in order not to increase that, no big polarons should form — the Fröhlich coupling constant α should be intermediate. Looking through the well-known materials, ZnS fulfills these requirements very well: it is miscible with MnS, $m^* = 0.3$, and $\alpha = 0.64$, and appears to be a good compromise.

Certainly it has been a fortuitous accident that ZnS was available and known to be a good phosphor if Mn doped; the feeling that it should be good in a polycrystalline form because of small surface recombination velocity and high oscillator strength did not hinder the development. And so the "unlearned" choice appears to be a rather optimal one. The problems which lie in the transfer between the dopant ions are just starting to be understood and concern the second iteration of the problem. At first glance one is inclined to believe that a pc material should have much worse transport — or in this case acceleration — properties than a crystalline one. In fact this is not the case. In the high resistive layers no space charge regions form along grain boundaries and nothing about additional inelastic scattering processes is known. The columnar growth of ZnS certainly does not stress the problem.

The fact that an MISIM structure was chosen by Inoguchi also appears now as an optimal incident: the charge-controlled breakdown of the S-layer is an absolutely stable one; in well-engineered devices it is safely undergone more than 10^{12} times within the lifetime of the device. What else has to be learned more about? Certainly the interface between an amorphous insulator and pc wide gap semiconductor is a "white area" of solid state physics but because of its importance in devices, the market of which will soon reach the 10^9 dollar/year limit, this will change.

Acknowledgements. I am indebted to G.O. Müller for innumerable discussions and I gratefully acknowledge the cooperation of many colleagues at the ZIE.

Literature

1. G. Destriau: J. Chem. Phys. **33**, 587 (1936)
2. H.F. Ivey: Advances in Electronics and Electron Physics Suppl. 1963
3. H.K. Henisch: Electroluminescence (Pergamon, New York 1962)
4. H. Hartmann, R. Mach, B. Selle: *Current Topics in Material Science*, ed. by E. Kaldis (North Holland Publ. Comp. Amsterdam, Vol.9, 1982)
5. P.J. Dean: J. Lum. **23**, 17 (1981)
6. Y. Fujita, J. Kuwata, M. Nishikawa, T. Tohda, T. Matsuoka, A. Abe, T. Nitta: Large scale ac thin film electroluminescent display panel. Japan Display **83**, 76 (1983)

7. S. Miura, K. Okamoto, S. Sato, S. Andoh, H. Ohnishi, Y. Hamakawa: Green color ACTFEL with sputtered $ZnS:TbF_3$. Japan Display **83**, 84 (1983)
8. H. Kozawaguchi, J. Ohwaki, B. Tsujiyama, K. Murase: Low voltage driven ac thin film electroluminescent devices. S.I.D. Int. Symp. 1982, 126
9. P.M. Alt, D.B. Dove, W.E. Howard: J. Appl. Phys. **53**, 5186 (1982)
10. S.K. Tiku: IEEE Trans. El. Dev.,ED **31**, 105 (1984)
11. L.G. Hale, R.D. Ketchpel, D.P. Ketchpel, I.S. Santha: High contrast thin film electroluminescent display device. Japan Display **83**? 80 (1983)
12. E.C. Freeman, D.H. Baird, J.R. Weaver: Use of coactivators in rare earth activated ZnS for electroluminescent displays. Japan Display **83**, 92 (1983)
13. R.E. Coovert, C.N. King, R.T. Tuege: Feasibility of a dual-color ACTFEL display. S.I.D. Int. Symp. 1982, 128
14. D. Theis: J. Lum. **23**, 191 (1981)
15. R. Menn, R. Tueta, A. Izrael, M. Braguier: IEEE Trans. Electron Devices, ED-**30**, 460 (1983)
16. P.J. Dean: Phys. Stat. Sol. (a) **81**, 625 (1984)
17. N.A. Vlasenko, Z.L. Denisova, V.S. Khomchenko: Phys. Stat. Sol. (a) **81**, 657 (1984)
18. H. Uede, Y. Kanatani, H. Kishishita, A. Fujimori, K. Okano: Thin film electroluminescence display units. S.I.D. Int. Symp. 1981, 28
19. T. Suntola: Performance of atomic layer epitaxy devices, S.I.D. Int. Symp. 1981, 20
20. D. Kahng: Appl. Phys. Lett. **13**, 210 (1968)
21. Y.S. Chen, D.C. Krupka: J. Appl. Phys. **43**, 4089 (1972)
22. E.J. Soxman, R.D. Ketchpel: Electroluminescent thin film research. JANAIR Report 72 09 03 (1972)
23. R. Mach, G.O. Müller: Phys. Stat. Sol. (a) **69**, 11 (1982)
24. G.O. Müller: Phys. Stat. Sol. (a) **81**, 597 (1984)
25. R. Mach, G.O. Müller: Phys. Stat. Sol. (a) **81**, 609 (1984)
26. W.E. Howard: ICL 1981, J. Lum. **23/24**, 835 (1981)
27. P.M. Alt: Thin film electroluminescent displays: device characteristics and performance. IBM Res. Report TC 9807 (343359) 1/9/83
28. A. Vecht: J. Crystal Growth **59**, 81 (1982)
29. A.G. Fischer: In *Lum. of Inorganic Solids*, ed. by P. Goldberg (Academic, New York 1966)
30. A. Vecht: J. Vac. Sci. Technol. **10**, 789 (1973)
31. P.J.F. Smith, N.J. Werring: Progress in d.c. electroluminescent displays and systems. Eurodisplay München **149** (1982)
32. R. Mach, W. Gericke, H. Treptow, W. Ludwig: Phys. Stat. Sol. (a) **49**, 667 (1978)
33. F.J. Bryant, A. Krier: Phys. Stat. Sol. (a) **81**, 681 (1984)
34. M.I. Abdalla, J.A. Thomas, A. Brenae, J.P. Noblanc: IEEE Trans. Electron Devices **28**, 694 (1981)
35. H. Ohnishi, H. Yoshino, N. Sakumo, K. Jeyasu: Green-emitting thin film d.c. EL devices with low threshold voltage. Japan Display **83**, 88 (1983)
36. H. Ohnishi, N. Sakuma, K. Ilyasu, Y. Hamakawa: J. Electrochem. Soc. **130**, 2115 (1983)
37. K. Okamoto, Y. Nasu, M. Okuyama, Y. Hamakawa: Low thresholdvoltage a.c. thin film electroluminescence devices 12th Conf. Solid State Devices Tokyo 1980, 123
38. T. Suntola, J. Antson, A. Pakkala, S. Lindfors: Atomic layer epitaxy for producing EL thin films S.I.D. Int. Symp. **80**, 108 (1980)
39. L.E. Tannas: Thin film electroluminescent emitter S.I.D. Int. Symp. 1981, 22
40. M. Yoshida, Y. Kahihara, T. Yamashita, K. Taniguchi, T. Inoguchi: Jap. J. Appl. Phys. **17**, 127 (1978)
41. Y. Oishi, T. Kato, Y. Hamakawa: Tunable color EL device Japan Display **83**, 570 (1983)

42. L.G. Hale, B. Garcia, R.D. Ketchpel, T.C. Lim: IEDM Digest 1980, 719
43. S. Miura, K. Okamoto, S. Sato, S. Andoh, H. Ohnishi, Y. Hamakawa: Green color ACTFEL with sputtered $ZnS:TbF_3$ Japan Display **83**, 84 (1983)
44. K. Okamoto, Y. Hamakawa: Appl. Phys. Lett. **35**, 508 (1979)
45. H. Kobayashi, S. Tanaka, T. Kunou, M. Shiiki, H. Sasakura: Various Tb compound luminescent centers in ZnS thin film electroluminescent devices. Japan Display **83**, 592 (1983)
46. T. Inoguchi, S. Mito: *Phosphor Films*, Topics in Appl. Phys. Vol.17, ed. by J.I. Pankove New York 1977
47. W.E. Howard: J. Lum **28**, 155 (1981)
48. K.W. Yang, S.J.T. Owen: Mechanism of the negative resistance characteristics in a.c. thin film electroluminescence devices. Int. Display Research Conf. 1982
49. W. Gericke, J.v. Kalben, R. Mach, G. Matzkeit, G.O. Müller: Memory devices using Al_2O_3 insulators. Intern. Workshop on Electroluminescence, Bad Stuer 1983
50. R. Mach, G.O. Müller, G. Schulz: Phys. Stat. Sol. (a) **81**, 723 (1984)
51. R. Mach, G.O. Müller, G. Schulz, J.v. Kalben, W. Gericke: Phys. Stat. Sol. (a) **81**, 733 (1984)
52. G.O. Müller, R. Mach: Phys. Stat. Sol. (a) **77**, K179 (1983)
53. R. Mach, G.O. Müller, W. Gericke, J.v. Kalben: Phys. Stat. Sol. (a) **75**, 187 (1983)
54. H. Sasakura, H. Kobayashi, S. Tanaka, J. Mita, T. Tanaka, H. Nakayama: J. Appl. Phys. **52**, 6901 (1981)
55. J.M. Hurd, C.N. King: J. Electr. Materials **8**, 879 (1979)
56. O. Goede, D.D. Thong: Phys. Stat. Sol. (b) **124**, 000 (1984)
57. W. Busse, H.E. Gumlich, B. Meissner, D. Theis: J. Lum. **12/13**, 693 (1976)
58. D. Theis, H. Oppolzer, G. Ebbinghaus, S. Schild: J. Cryst. Growth **63**, 47 (1983)
59. D. Theis: Phys. Stat. Sol. (a) **81**, 647 (1984)
60. V.P. Tanninen, M. Oikkonen, Y. Toumi: Thin Solid Films **109**, 283 (1983)
61. H. Venghaus, D. Theis, H. Oppolzer, S. Schild: J. Appl. Phys. **53**, 4146 (1982)
62. W. Gericke, J.v. Kalben, H. Reinsperger, B. Selle, L. Redlich, V. Gutan: Phys. Stat. Sol. (a) **81**, 715 (1984)
63. A. Preisinger, H.K. Pulker: Jap. J. Appl. Phys. Supp. 2, Part 1, 769 (1974)
64. K. Okamoto: High brightness low threshold-voltage thin film ac electroluminescence divices. Thesis Osaka University 1981
65. R.O. Törnqvist, J. Antson, J. Skarp, V.P. Tanninen: IEEE Trans. Electron. Dev. ED-**30**, 468 (1983)
66. V. Marello, L. Samuelson, A. Onton, W. Reuter: J. Appl. Phys. **52**, 3590 (1981)
67. H.S. Reehal, J.M. Gallego, C.B. Edwards: Appl. Phys. Lett. **40**, 258 (1982)

The Electrical Properties of Oxides Under Conditions of Oxidation, Reduction and Catalysis

S. Strässler, A. Reis, and D. Wieser
Cerberus AG, CH-8708 Männedorf, Switzerland

Many oxides show interesting electrical and chemical properties which are being explored in technological applications. Some examples are: their use as field dependent conductors (varistors), as gas sensors (humidity, oxygen, gases), and as catalysts. In our lecture we will analyze some problems connected with the basic mechanisms of such devices. We will include new concepts to elucidate the role of chemisorbed oxygen in the gas sensing mechanism.

I) Introduction

Oxides which are chemically inert and which exhibit semiconducting characteristics, have been found useful in many technological applications. In the present investigation we will be concerned mainly with the problem of how the ambient influences the electrical conductivity of metal oxides. This phenomenon is of particular interest for the understanding of gas sensors, varistors and also of catalysts. We approach the problem from the point of view of rate equations for the various defects in the bulk, the interface and the surface region. On the surface we consider adsorbed, ionosorbed and chemisorbed atoms. In section II, III, and IV we review the situation of bulk equilibria at very high temperatures. In section V, VI and VII we study how the rate equations determine the defect distribution at an interface. In section VIII, IX and X we give a very simplified picture of the catalytic processes on a surface and their influence on the electrical conductivity. The presentation will be mainly of didactic character.

II) Defect Models

The defect chemistry of metal oxides has been described in a number of reviews and articles [1]. Examples for defects are:

oxygen vacancies	:	V_O^x, V_O^+, V_O^{++} ,
metal vacancies	:	V_M^x, V_M^-, V_M^{--} ,
metal interstitials	:	M_i^x, M_i^+, M_i^{++} ,
oxygen interstitials	:	O_i^x, O_i^-, O_i^{--} ,

Here x refers to the neutral species and ± to the singly and ±± to the doubly ionized species. The concentrations of these defects are denoted by [...]. A possible way to test the defect models, is to measure the electrical conductivity σ of the oxides as a function of temperature and oxygen partial pressure P_{O_2} [2,3]. Recently MAHAN [4] analysing HAGEMARK's data [3], proposed that the first two defects are the most probable ones in ZnO. Let us discuss this case in more detail.

In equilibrium, the chemical balance between the different species in the solid and the gasphase is governed by to the following reactions:

Solid-gas,

$$ZnO(b) \leftrightarrow Zn(g) + \frac{1}{2} O_2(g) \quad , \tag{2.1}$$

$$Zn(b) \leftrightarrow Zn(g) + V_{Zn}^{x} \quad , \tag{2.2}$$

$$O(b) \leftrightarrow \frac{1}{2} O_2(g) + V_O^{x} \quad . \tag{2.3}$$

Electronic changes,

$$0 \leftrightarrow e + h \quad , \tag{2.4}$$

$$V_{Zn} \leftrightarrow V_{Zn}^{-} + h \quad , \tag{2.5}$$

$$V_{Zn}^{-} \leftrightarrow V_{Zn}^{--} + h \quad , \tag{2.6}$$

$$V_O^{x} \leftrightarrow V_O^{+} + e \quad , \tag{2.7}$$

$$V_O^{+} \leftrightarrow V_O^{++} + e \quad . \tag{2.8}$$

Here e refers to electrons, h to holes, g to the gas phase and b to the bulk. According to the laws of thermodynamics in equilibrium the electrochemical potentials for Zn and O of the gas phase must be equal to those of the solid phase:

$$\mu_{Zn}(g) = \mu_{Zn}(b) \quad , \tag{2.9}$$

$$\mu_{O_2}(g) = 2\mu_O(b) \quad . \tag{2.10}$$

The concentrations of the ionized species and of the electrons and holes,

$$n = [e] \quad , \tag{2.11}$$

$$p = [h] \quad , \tag{2.12}$$

follow from Fermi statistics. The probability f for the occupation of an electronic state E is given by

$$f = \frac{1}{e^{(E-\mu)/kT}+1} \quad , \tag{2.13}$$

where T is the temperature, k the Boltzmann constant and μ the chemical potential.

The mass action laws, corresponding to the reactions (2.1)-(2.8) follow from (2.9)-(2.13):

$$K(ZnO) = P_{Zn} \cdot P_{O_2}^{1/2} \quad , \tag{2.14}$$

$$K(V_{Zn}^{x}) = P_{Zn} \cdot [V_{Zn}^{x}] \quad , \tag{2.15}$$

$$K(V_{O}^{x}) = P_{O_2}^{1/2} \cdot [V_{O}^{x}] \quad , \tag{2.16}$$

$$K_i = n \cdot p \quad , \tag{2.17}$$

$$K(V_{Zn}^{-}) = [V_{Zn}^{-}] \cdot p/[V_{Zn}^{x}] \quad , \tag{2.18}$$

$$K(V_{Zn}^{--}) = [V_{Zn}^{--}) \cdot p/[V_{Zn}^{-}] \quad , \tag{2.19}$$

$$K(V_{O}^{+}) = [V_{O}^{+}] \cdot n/[V_{O}^{x}] \quad , \tag{2.20}$$

$$K(V_{O}^{++}) = [V_{O}^{++}] \cdot n/[V_{O}^{+}] \quad . \tag{2.21}$$

The equilibrium constants K(...) can be determined experimentally by measuring the various concentrations. The values obtained however will depend on the defect model assumed. For example, in the case of ZnO, there exist 3 different sets of values: KROEGER [1], HAGEMARK [3] and MAHAN [4]. Kröger assumes $V_O^{\mp}$, $V_{Zn}^{\mp}$ as the main defects at high temperatures. Hagemark assumes a Zn interstitial as the intrinsic donor. Mahan's values for the equilibrium constants are:

$$K(ZnO) = 1.5 \times 10^{10} \; e^{-4.89\beta} \; atm^{3/2} \quad , \tag{2.22}$$

$$K(V_{Zn}^{x}) = 1.33 \times 10^{30} \; e^{-6.75\beta} \; atm \; cm^{-3} \quad , \tag{2.23}$$

$$K(V_{O}^{x}) = 1.56 \times 10^{29} \; e^{-5.57\beta} \; atm^{1/2} \; cm^{-3} \quad , \tag{2.24}$$

$$K_i = N_c N_v \, e^{-\beta E_G} \; cm^{-6} \quad . \tag{2.25}$$

Here the electronic gap E_G is,

$$E_G = 3.34 - T/1250 \; eV \quad . \tag{2.26}$$

The effective density of states for electrons and holes is given by,.

$$N_c = 7.94 \times 10^{14} \; T^{3/2} \; cm^{-3} \quad , \quad \text{and} \tag{2.27}$$

$$N_v = 1.16 \times 10^{15} \; T^{3/2} \; cm^{-3} \quad . \tag{2.28}$$

The other constants are given by

$$K(V_{Zn}^{-}) = 2N_v \, e^{-0.8\beta} \; cm^{-3} \quad , \tag{2.29}$$

$$K(V_{Zn}^{--}) = 1/2N_v \, e^{-2.8\beta} \; cm^{-3} \quad , \tag{2.30}$$

$$K(V_{O}^{+}) = 2N_c \, e^{-0.05\beta} \; cm^{-3} \quad , \tag{2.31}$$

$$K(V_{O}^{++}) = 1/2N_c \, e^{-0.15\beta} \; cm^{-3} \quad . \tag{2.32}$$

In these equations $\beta=1/kT$ in units of eV. Obviously it is not easy to calculate these constants from first principles. Otherwise no controversy about the interpretation would occur. The situation is much simpler for the prefactors than for the activation energy. An example for such a calculation is given in the following section.

III) Theoretical prefactors

As an example, consider the equilibrium for the exchange,

$$M(b) \leftrightarrow M(g) + V_M^x \quad . \tag{3.1}$$

The equilibrium condition is,

$$\mu(b) = \mu(g) \quad . \tag{3.2}$$

The chemical potential is obtained via,

$$\mu = \left(\frac{\partial F}{\partial N}\right)_{V,T} \quad , \tag{3.3}$$

$$F = -kT \log Z \quad . \tag{3.4}$$

Here F is the free energy and Z the partition sum. For the gas phase,

$$Z = \frac{1}{N!} \left[\frac{V}{(2\pi h)^3} \cdot \int d^3p \; e^{-\frac{p^2}{2mk}}\right]^N \quad . \tag{3.5}$$

The evaluation of (3.3) gives,

$$\mu = kT \log P/P_c \quad , \tag{3.6}$$

where the characteristic pressure P_c is given by

$$P_c = \frac{kT}{\Omega_c} \quad , \tag{3.7}$$

and the characteristic volume Ω_c is given by

$$\Omega_c = \left(\frac{2\pi h^2}{mkT}\right)^{3/2} \quad . \tag{3.8}$$

For M = Zn,

$$m_{Zn} = 1.09 \cdot 10^{-22} \text{ gr} \quad . \tag{3.9}$$

Then

$$\Omega_c = \left(\frac{T_o}{T}\right)^{3/2} \; 2.18 \cdot 10^{-28} \text{ cm}^3 \quad . \tag{3.10}$$

where, $T_o = 1000^o$. For the solid phase (in an effective particle approximation)

$$Z_M = \left(\frac{1}{2\pi h}\right)^3 \int d^3p \int d^3r \exp\left(- \frac{p}{2mkT} - \frac{U(r)}{kT}\right) \quad . \tag{3.11}$$

The potential energy U(r) can be parameterized by

$$U = \frac{1}{2} m\omega_o^2 r^2 - \Delta \quad . \tag{3.12}$$

Here $\hbar\omega_o$ is the Debye energy and Δ is the energy for creating a vacancy. Within the ideal solution theory,

$$F = -kT\, N \log Z_M - kT \log \left(\frac{N}{N_S}\right) \quad , \tag{3.13}$$

where N_S is the total number of atoms at their proper sites. The evaluation of (3.11) and (3.13) gives,

$$\mu = -kT \log Z_M - kT \log \frac{c}{1-c} - \Delta \quad . \tag{3.14}$$

Here

$$Z_M = \left(\frac{kT}{\hbar\omega_o}\right)^{3/2} \quad , \quad \text{and} \tag{3.15}$$

$$c = 1 - \frac{N}{N_S} = [V_M^x] \cdot V_o \quad . \tag{3.16}$$

V_o is the volume of the unit cell of the crystal. In the limit $c \to 0$ it follows from Eqs. (3.1), (3.6) and (3.14) that,

$$P_M[V_M^x] = \frac{P_c}{V_o \cdot Z_M}\, e^{-\Delta/kT} \quad . \tag{3.17}$$

Comparison with (2.15) finally gives the following expression for the equilibrium constant,

$$K(V_M^x) = K^o(V_M^x)\, e^{-\Delta/kT} \quad , \tag{3.18}$$

where the prefactor becomes,

$$K^o(V_M^x) = \frac{P_c}{V_o \cdot Z_N} \quad . \tag{3.19}$$

For M=Zn expression (3.19) can be written as,

$$K^o(V_{Zn}^x) = \left(\frac{T_o}{T}\right)^{1/2} 0.7 \times 10^{30} \;\; \text{atm cm}^{-3} \quad . \tag{3.20}$$

The experimental value found by Hagemark is 1.33×10^{30}. The discrepancy is probably caused by the interpretation of ω_o and the temperature dependence of Δ.

For oxygen the same procedure gives two additional contributions due to the rotation and the vibration of the oxygen molecule,

$$Z_{rot} = \frac{4\pi}{(2\pi\hbar)^2} \int dM_x\, dM_y\, e^{-\frac{(M_x^2+M_y^2)}{2kT\cdot I}} \quad . \tag{3.21}$$

where I is the moment of inertia. In this case,

$$P_c = \frac{kT}{\Omega_c} \frac{1}{Z_{vib}} \frac{1}{Z_{rot}} \quad , \tag{3.22}$$

$$Z_{rot} = \frac{2I}{h^2} kT \quad , \tag{3.23}$$

$$Z_{vib} = \frac{kT}{h\omega_{vib}} \quad . \tag{3.24}$$

The corresponding equilibrium constant then becomes,

$$K^o(V_o^x) = \frac{P_c}{V_o \cdot Z_o} = (\frac{T_o}{T})^{3/4} \cdot 4.2 \times 10^{29} \text{ atm}^{1/2} \text{ cm}^{-3} \quad . \tag{3.25}$$

The experimental value MAHAN [4] obtained from analysing Hagemark's data is 1.56×10^{29}.

IV) Pressure Dependence of Electrical Conductivity

In this section we want to compare some predictions of the defect model of section II with electrical conductivity measurements. The conductivity is given by

$$\sigma = ne\mu_e + pe\mu_h \quad . \tag{4.1}$$

Here μ_e and μ_h are the mobilities for electrons and holes, respectively. The equations for the electron density n and the hole density p follow from charge neutrality and the mass action laws of section II.

The charge neutrality equation is,

$$n + [V_M^-] + 2[V_M^{--}] + [A^-] + 2[A^{--}] =$$

$$p + [V_O^+] + 2[V_O^{++}] + [D^+] + 2[D^{++}] \quad . \tag{4.2}$$

Here we have included donors [D] and acceptors [A]. In case of ZnO, i.e. Sb or Zn_i act as donors whereas eg. Li or Bi act as acceptors. The contribution of such donors to the conductivity at room temperature depends mainly on the donor ionization energy and the thermal treatment of the sample. The high temperature treatment determines the number of oxygen vacancies and the cooling procedure determines the number of frozen-in vacancies. The calculation of the donor concentration can be performed, if the temperature dependence of the equilibrium constants are known. Then all concentrations can be expressed in terms of n and P_{O_2}:

$$P = K_i/n \quad , \tag{4.3}$$

$$[V_{Zn}^-] = K(V_{Zn}^-) \cdot [V_{Zn}^x]/p \quad , \tag{4.4}$$

$$[V_{Zn}^x] = K(V_{Zn}^x) \cdot P_{O_2}^{1/2}/K_{ZnO} \quad , \tag{4.5}$$

$$[V_{Zn}^{--}] = K(V_{Zn}^{--}) \cdot [V_{Zn}^-]/p \quad , \tag{4.6}$$

$$[V_O^+] = K(V_O^+) \cdot [V_O^x]/n \quad , \tag{4.7}$$

$$[V_O^{++}] = K(V_O^{++})\cdot[V_O^{+}]/n \quad , \tag{4.8}$$

$$[V_O^{x}] = K(V_O^{x})/P_{O_2}^{1/2} \quad , \tag{4.9}$$

$$[V_{Zn}^{x}] = K(V_{Zn}^{x})/P_{Zn} \quad , \tag{4.10}$$

$$P_{Zn} = K(ZnO)/P_{O_2}^{1/2} \quad . \tag{4.11}$$

If for example, we take as donor a Zn interstitial, then the following reactions would have to be included:

$$Zn(g) \leftrightarrow Zn_i^{x} \quad , \tag{4.12}$$

$$Zn_i^{x} \leftrightarrow Zn_i^{+} + e \quad , \quad \text{or} \tag{4.13}$$

$$[Zn_i^{x}] = K(Zn_i^{x})/P_{Zn} \quad , \tag{4.14}$$

$$[Zn_i^{+}] = K(Zn_i^{+})/[Zn_i^{x}] \quad . \tag{4.15}$$

The values for $K(Zn_i^x)$ can be chosen from,

$$K(Zn_i^{x}) = \begin{cases} 2.23 \times 10^{15}\, e^{0.13\beta}\ \text{atm}^{-1}\,\text{cm}^{-3} \quad , & \text{(Kröger)} \\ 1.04 \times 10^{19}\, e^{0.68\beta}\ \text{atm}^{-1}\,\text{cm}^{-3} \quad , & \text{(Hagemark)} \end{cases} \tag{4.16}$$

Mahan's interpretation of Hagemark's experimental results is based on $Zn_i \rightarrow V_O$ and therefore $K(Zn_i^x) \rightarrow K(V_O^x)$. The main effect is, that this leads to a drastic reduction of the value for $K(V_O^x)$:

$$K(V_O^{x}) = \begin{cases} 3.87 \times 10^{30}\, e^{-4.29\beta}\ \text{atm}^{1/2}\,\text{cm}^{-3} \quad , & \text{(Kröger)} \\ 1.56 \times 10^{29}\, e^{-5.57\beta}\ \text{atm}^{1/2}\,\text{cm}^{-3} \quad . & \text{(Mahan)} \end{cases} \tag{4.17}$$

Recently EINZINGER [5] proposed that Mn in ZnO acts as a deep donor. In this case,

$$[D_{Mn}] = [D_{Mn}^{x}] + [D_{Mn}^{+}] \quad , \tag{4.18}$$

$$[D_{Mn}^{+}] = K(D_{Mn}^{+})\, [D_{Mn}^{x}]/n \quad . \tag{4.19}$$

Equation (4.2) for the determination of n is valid if the defects are in equilibrium with the ambient. For bulk samples one finds usually that above 1100°C equilibrium is reached in minutes. At lower temperatures, however, establishment of equilibrium may take many hours.

In order to calculate n at lower temperatures one takes

$$[V_O] = [V_O^{x}] + [V_O^{+}] + [V_O^{++}] \quad . \tag{4.20}$$

$$[V_{Zn}] = [V_{Zn}^{x}] + [V_{Zn}^{-}] + [V_{Zn}^{--}] \quad , \tag{4.21}$$

at the freeze-in temperature, and for temperatures below one considers only the reactions involving electronic rates. For interfaces and surfaces the freeze-in temperature can be much lower. This aspect becomes quite important when discussing gas sensors and varistors.

Let us now discuss a few consequences of these equations. Clearly there is some uncertainty with respect to models and numbers, so it is necessary to keep the discussion on a qualitative basis. Solving (4.2) numerically, assuming [D] = [A] = 0, we find that the number of oxygen vacancies can be estimated from the following approximation,

$$n \sim 2\,[V_0^{++}] \quad , \tag{4.22}$$

$$[V_0^{++}] = \frac{1}{2}\,[2 \cdot K(V_0^{++}) \cdot K(V_0^{x}) \cdot K(V_0^{+})]^{1/3}\, P_{O_2}^{-1/6} \quad . \tag{4.23}$$

At 1000°C and $P_{O_2} = 0.2$ atm one obtains

$$[V_0^{++}] = \begin{cases} 3.4 \times 10^{17} \ \text{cm}^{-3} \ , & \text{(Kröger)} \\ 1.4 \times 10^{15} \ \text{cm}^{-3} \ . & \text{(Mahan)} \end{cases} \tag{4.24}$$

Equation (4.22) leads to

$$\sigma \sim P_{O_2}^{\beta} \quad , \quad \beta = -1/6 \quad . \tag{4.25}$$

SAMSON and FONSTAD [6] found $\beta \sim 1/6.5$ for SnO above 1100°C. For ZnO RUDOLPH [2] found $\beta \sim 1/5.5$ for temperatures above 800°C.

If in (4.2) the oxygen interstitials dominate, the oxide becomes a hole conductor. Then the neutrality condition using the same approximation as in (4.9) becomes,

$$p \sim 2\,[O_i^{--}] \quad . \tag{4.26}$$

In this case,

$$\frac{1}{2}\, O_2(g) \leftrightarrow O_i^{x} \quad , \quad \text{or,} \tag{4.27}$$

$$K(O_i^{x}) = [O_i^{x}]/P_{O_2}^{1/2} \quad . \tag{4.28}$$

From

$$h + O_i^{-} \leftrightarrow O_i^{x} \quad , \tag{4.29}$$

$$h + O_i^{--} \leftrightarrow O_i^{-} \quad , \tag{4.30}$$

we have

$$[O_i^{--}] = K(O_i^{-})\, K(O_i^{--})/p^2 \quad . \tag{4.31}$$

Finally,

$$[O_i^{--}] = \frac{1}{2}[2\, K(O_i^{-}) K(O_i^{--}) K(O_i^{x})]^{1/3} \cdot P_{O_2}^{1/6} \quad . \tag{4.32}$$

This leads to,

$$\sigma \sim P_{O_2}^{\beta} \quad , \quad \beta = 1/6 \quad . \tag{4.33}$$

Such a behavior has been found for BaO, SrO, CaO, ThO_2, ZrO [2].

If interfaces or surfaces are involved, the behavior with respect to gas pressure may become much stronger. For example for sintered ZnO,

$$\sigma \sim P_{O_2}^{-\frac{1}{2}} \quad , \tag{4.34}$$

has been reported [7,8,9]. Recently [10] changes in the resistivity of ZnO varistors of the order of 10^5 have been observed as a function of oxygen pressure below 800°C.

Compared to the relatively simple situation of the bulk case as discussed in sections II, III and IV, the following complications arise in the case of interfaces and surfaces:

i) space charge effects,

ii) inhomogeneous distribution of defect concentrations,

iii) adsorption, chemisorption and catalytic reactions.

V) Double Schottky junction

In many polycrystalline semiconductors, double Schottky barriers form at the grain boundaries. In this case the electric conductivity is determined by the properties of the junctions. As has been demonstrated in Ref. [10], the junction can be influenced by the ambient at much lower temperatures than a homogeneous bulk sample. Obviously, changes in the concentrations of donors and acceptors occur in the vicinity of the interface. Recently some aspects of this problem have been studied in Ref. [4] and [5].

First we shall review the theory of double Schottky barriers without complications. This model is well known [11 - 14] and is included here as a zero order model to discuss various complications. The basic ingredient of the model is the assumption that there exists an excess negative charge eQ in the interface between two grains. This interface is assumed to be of small thickness. Poisson's equation for the electrostatic potential ϕ then becomes,

$$\frac{\partial^2\phi}{\partial x^2} = \frac{4\pi e^2}{\varepsilon}\,[-\,Q\delta(x) + N] \quad . \tag{5.1}$$

Here the positive charge density due to the ionized donors in the abrupt junction approximation is given by

$$N = \begin{cases} N_+ = \text{const} & \text{for} \quad \phi > 0 \\ 0 & \text{for} \quad \phi = 0 \quad . \end{cases} \tag{5.2}$$

The amount of positive charge in the depletion region,

$$-\,d < x < d \quad , \tag{5.3}$$

must be equal to the excess negative charge,

$$Q = 2N_+ d \quad . \tag{5.4}$$

Equation (5.1) has the solution,

$$\phi = \frac{1}{2} \frac{4\pi e^2}{\varepsilon} N_+ (x-d)^2 \quad . \tag{5.5}$$

The barrier height ϕ_B thus is given by

$$\phi_B = \frac{1}{2} \frac{4\pi e^2}{\varepsilon} \left(\frac{Q}{2}\right)^2 \frac{1}{N_+} \quad . \tag{5.6}$$

For a small junction voltage V, the current across the junction becomes,

$$j \sim V \exp(-(\phi_B - \mu)/kT) \quad . \tag{5.7}$$

Here μ is the chemical potential measured relative to the conduction band. The nonlinear behavior of the varistor can be used to determine the barrier ϕ_B.

VI) Inhomogeneity of vacancy-density near the interface

ZnO varistors are sintered at high temperature, then cooled slowly to room temperature. During the cooling process, the defect concentration becomes non uniform. Recently two such calculations have been performed for two different models in Ref. [4] and [5]. Both investigations came to the conclusion that the frozen-in oxygen vacancy concentration decreases near the interface. In this section we will show that for the double Schottky junction model, the frozen-in vacancy distribution can increase. At high temperatures where the oxygen vacancies are in equilibrium with the ambient, the positive charge density in Eq.(5.1) is given by

$$N_+ = 2[V_0^{++}] \quad , \tag{6.1}$$

where from section III, it follows that,

$$[V_0^{++}] = \frac{A(V_0^{++})}{n^2} P_{O_2}^{-\frac{1}{2}} \quad . \tag{6.2}$$

Here,

$$A(V_0^{++}) = K(V_0^{++}) K(V_0^{+}) K(V_0^{x}) \quad . \tag{6.3}$$

In the depletion region the electron density is given by

$$n = 2N_0 e^{-\phi/kT} \quad , \tag{6.4}$$

where N_0 is given by Eq.(4.22),

$$N_0 = [2A(V_0^{++})]^{1/3} P_{O_2}^{-1/6} \quad . \tag{6.5}$$

Equation (5.1) for the barrier becomes,

$$\frac{\partial^2 \phi}{\partial x^2} = \frac{4\pi e^2}{\varepsilon} [-Q\delta(x) + N_0 e^{+2\phi/kT}] \quad . \tag{6.6}$$

It can be written as,

$$\frac{1}{2}\left(\frac{\partial\phi}{\partial x}\right)^2 = \frac{4\pi e^2}{\varepsilon}\int_0^{\phi} q(\phi)d\phi \quad . \tag{6.7}$$

As a reference, we introduce the barrier height ϕ_o which would be established if the donor density would be frozen-in at the bulk value N_+^o,

$$\phi_o = \frac{1}{2}\,\frac{4\pi e^2}{\varepsilon}\left(\frac{Q}{2}\right)^2 \cdot \frac{1}{N_+^o} \quad . \tag{6.8}$$

Integration of (6.7) gives,

$$\phi_B = \frac{kT}{2}\log\left(\frac{2\phi_o}{kT} + 1\right) \quad . \tag{6.9}$$

The pressure dependence of ϕ_B follows from Eq.(6.5) and (6.8),

$$\phi_o = \frac{1}{2}\,\frac{4\pi e^2}{\varepsilon}\left(\frac{Q}{2}\right)^2 \frac{1}{[2A(V_0^{++})]^{1/3}} P_{O_2}^{1/6} \quad . \tag{6.10}$$

Equation (6.9) predicts a drastic reduction of the barrier due to the increase in the oxygen vacancy concentration:

$$\xi_B = [V_0^{++}]_{interface} / [V_0^{++}]_{bulk} = \frac{2\phi_o}{kT} + 1 \quad . \tag{6.11}$$

For a typical barrier height of $\phi_B \sim 0.4$ eV and a freeze temperature $T_F \sim 1000^oC$, (6.11) predicts an enhancement of 10^3. We note furthermore that the pressure dependence is weak. The spatial vacancy distribution can be calculated from (6.7). In units of the Debye length,

$$\lambda_D = \left(\frac{\varepsilon kT}{4\pi e^2 N_0}\right)^{1/2} \quad , \tag{6.12}$$

and for a normalized potential

$$\phi' = \phi/kT \quad , \tag{6.13}$$

(6.7) becomes,

$$\frac{d\phi'}{du} = (e^{2\phi'} + 1)^{1/2} \quad . \tag{6.14}$$

Close to the barrier region,

$$\phi' = \phi_B + \log\frac{1}{1+u\,\exp(\xi_B)} \quad . \tag{6.15}$$

The enhancement factor ξ then becomes,

$$\xi(u) = \xi_B\,\frac{1}{1+u\,\exp(\xi_B)} \quad . \tag{6.16}$$

Because typically λ_D is several 100 Å, we note that the region of high donor concentration is limited to a narrow region $\lesssim$ 100 Å close to the interface. The enhancement will be reduced if the temperature dependence of Q would be considered.

One should note that the dependence of ϕ_B on Q and P_{O_2} is quite weak. On the other hand, a strong dependence of the barrier height on the ambient was observed for temperatures much lower than a typical freeze-in temperature [10]. This problem will be studied in the next section.

VII) Pressure dependence of barrier height

Experimentally it is found [5] that the non-Ohmic property at the ZnO grain boundary appears by adding ions having an ionic radius larger than that of Zn. This suggest that the trapping state is caused by a deformation of the lattice. Such a stress can be released by vacancies. Possible reactions are

$$T^x + V_i^x \leftrightarrow T_v \quad , \tag{7.1}$$

or even

$$T^x + 2V_i^x \leftrightarrow T_v \quad , \tag{7.2}$$

if 2 vacancies are necessary to release the stress. Here T means an impurity and T_v refers to an impurity vacancy complex and V_i^x is a neutral oxygen vacancy in the interface region. For the ionization of the trap,

$$T^x + e \leftrightarrow T^- \quad . \tag{7.3}$$

The interface concentration thus becomes,

$$[T^-] = [T^x]\, e^{(\phi-\mu+E)/kT} \quad . \tag{7.4}$$

Here E is the ionization energy of the trap. For the following discussion, we assume that the temperature is below the freezing temperature of the bulk oxygen concentration. We furthermore approximate the inhomogeneous vacancy distribution $V_0^{++}(x)$ by some effective value N_{eff}. The barrier height is then given by

$$\phi_B = \frac{1}{2}\,\frac{4\pi e^2}{\varepsilon}\left(\frac{Q}{2}\right)^2 \frac{1}{N_{eff}} \quad . \tag{7.5}$$

The pressure dependence of N_{eff} is so weak, that the observed behavior must be explained via the dependence of the interface charge Q on the pressure. For the above model,

$$Q = [T^-] - [V_i^+] - 2[V_i^{++}] \quad . \tag{7.6}$$

The oxygen pressure enters the problem via

$$\frac{1}{2}\, O_2(g) + V_i^x \leftrightarrow O_i \quad , \quad \text{or} \tag{7.7}$$

$$[V_i^x] = K(V_i^x)/P_{O_2}^{1/2} \quad . \tag{7.8}$$

$[V_i^+]$ and $[V_i^{++}]$ have again a weak pressure dependence, so we conclude that the main effect is caused by the strong pressure dependence of (7.2),

$$[T_v] = K(T)\;[T^-][V_i^x]^2 \quad . \tag{7.9}$$

The total number of the species T is assumed to be fixed,

$$[T_t] = [T^x] + [T^-] + [T_v] \quad . \tag{7.10}$$

Let us first discuss the case where P_{O_2} is sufficiently large so that T_v is not important. In this case

$$Q = [T_t] \frac{1}{1+e^{(\phi_B-\mu-E)/kT}} \quad . \tag{7.11}$$

Two possibilities exist. If

$$\phi_B - \mu < E \quad , \tag{7.12}$$

the barrier is determined via,

$$\phi_B = \frac{1}{2}\frac{4\pi e}{\varepsilon}\frac{[T^+]^2}{N_{eff}} \quad . \tag{7.13}$$

If (7.12) is not satisfied the barrier is given by

$$\phi_B = E + \mu \quad . \tag{7.14}$$

EINZINGER [5] has recently proposed that it is (7.14) which determines the barrier height. Including $[T_v]$ in the equation for Q we have,

$$Q = [T_t] \frac{1}{1+e^{(\phi_B-\mu-E)/kT}\cdot(1+P_c/P_{O_2})} \quad , \text{ where,} \tag{7.15}$$

$$P_c = K(T)K^2(V_i^x) \quad . \tag{7.16}$$

Equation (7.15) defines the following relationship between the pressure and the conductivity

$$\sigma \text{ prop} \begin{cases} \text{const}, & P_{O_2} \lesssim P_c \\ P_{O_2}^{-1} \quad , & P_{O_2} \gtrsim P_c \quad . \end{cases} \tag{7.17}$$

Because the experiments have been performed at room temperature after annealing at $\sim 810^{\circ}$C the pressure dependence is given by

$$\sigma \text{ prop } P_{O_2}^{-\beta} \quad , \quad \beta \sim 3 \tag{7.18}$$

This is in good agreement with the observed value.

VIII Ionosorption and electrical conductivity

In this third part of our investigation, we assume that all bulk defects are frozen-in. In addition we assume that the surface dominates the electrical conductivity. This can be satisfied with 3 different arrangements:

i) accumulation layer,
ii) depletion layer which extends throughout a thin oxide film,
iii) depletion layer in the necks of a granular material (this is similar to the double Schottky junction).

It follows from Bardeen's [15] paper published in 1947 that the electrical conductivity of semiconducting materials may be sensitively affected by the interaction with the gas atmosphere surrounding the semiconductor. He proposed a model in which acceptor levels on the surface may trap or release electrons.

The general problem of the formation of space charge layers and the occupation of surface states has been treated in a large number of papers [16-20]. Here we will be concerned with the depletion layer case, which may become relevant for n-type metal oxides. Let us first review some of the basic equations which are very similar to the case studied in section V. Because in the surface problem we have only a single depletion region, the equation for the barrier height (7.5) becomes,

$$\phi_B = \frac{1}{2} \frac{4\pi e^2}{\varepsilon} \frac{Q^2}{N_{eff}} \quad . \tag{8.1}$$

for the case iii) or,

$$\phi_B \to 0 \quad , \tag{8.2}$$

for the case ii) where the thickness of the film or the dimensions of the grains become much smaller than the depletion width. Formulas for the general case have been derived in [20].

Obviously, the case (8.2) is the easiest way to investigate the influence of the ambient on the conductivity. Because of the logarithmic dependence of ϕ_B on Q via the equilibrium equation, the conclusions we obtain from (8.2) remain approximately valid even if (8.1) should be considered [20].

The ionosorption of a gas R on the surface is described by

$$R(g) \leftrightarrow R \quad , \text{ or} \tag{8.3}$$

$$[R] = K(R) \cdot P_R \quad . \tag{8.4}$$

From section III we have

$$K(R) = \frac{P_c}{F_o} \left(\frac{kT}{h\omega_o}\right) e^{-\Delta(R)/kT} \quad , \tag{8.5}$$

where F_o is the area per adsorption site, ω_o is the vibration frequency and $\Delta(R)$ the adsorption energy. Because P_c is of order 10^8 atm and $(kT/h\omega_o)^3 \sim 10$, we obtain a concentration,

$$c = [R]F_0 \ll 1 \quad , \tag{8.6}$$

unless $\Delta(R)/kT$ is extremely large. For ionosorption,

$$R + e \leftrightarrow R^- \quad , \text{ or} \tag{8.7}$$

$$[R^-] = K(R^-)[R][e] \quad . \tag{8.8}$$

Charge neutrality requires that in the compensated state,

$$[R^-] = dN_{eff} = [N^+] \quad , \tag{8.9}$$

where d is the thickness of the film. In the case iii), (8.9) has to be replaced by

$$[R^-] = N_{eff}\, d_D \tag{8.10}$$

where the width of the depletion region is given by

$$d_D = (\phi_B \frac{2\varepsilon}{4\pi N_{eff} \cdot e^2})^{1/2} \quad . \tag{8.11}$$

Combining (8.4), (8.10) and (8.11) the gas dependence of the conductivity becomes

$$\sigma \sim P_R^{-1} \quad . \tag{8.12}$$

IX) The role of oxygen, power laws and relaxation

If the formulas derived in the preceeding section are applied to ionosorption of oxygen, the theory predicts,

$$\sigma \sim P_{O_2}^{-1} \quad . \tag{9.1}$$

Such a behavior is not observed. Experiments usually give

$$\sigma \sim P_{O_2}^{\beta} \quad , \tag{9.2}$$

where β varies between $-\frac{1}{2}$ and zero, [2], [3], [6]-[9], [21].

One way to resolve the discrepancy [19], is to assume that the surface is sufficiently inhomogeneous and gives rise to a Freundlich isotherm. Equation (8.4) then is replaced by

$$[R] = K_F(R) \cdot P_R^{\gamma} \quad , \quad 0 < \gamma < 1 \quad . \tag{9.3}$$

The corresponding pressure dependence of the conductivity is,

$$\sigma \sim P_{O_2}^{-\gamma} \quad . \tag{9.4}$$

Another way to change the powerlaw, is to assume, that oxygen is being chemisorbed. The reaction route is then,

$$O_2(g) \leftrightarrow O_2 \quad , \tag{9.5}$$

$$O_2 \leftrightarrow O_2^- \quad , \tag{9.6}$$

$$O_2^- + e \leftrightarrow 2O^- \quad . \tag{9.7}$$

The presence of O^- has been stated in Refs. [23] and [24]. From (9.5) - (9.7), the following relations are derived:

$$[O_2] = K(O_2)P_{O_2} \quad , \tag{9.8}$$

$$[O_2^-] = K(O_2^-)[O_2][e] \quad , \tag{9.9}$$

$$[O^-]^2 = K(O^-)[O_2^-][e] \quad . \tag{9.10}$$

Elimination of $[O_2^-]$ gives,

$$[e] = [O^-]K(O_2)K(O_2^-)K(O^-)P_{O_2}^{-1/2} \quad . \tag{9.11}$$

If we assume that chemisorption dominates,

$$[O^-] \gg [O_2^-] \quad , \tag{9.13}$$

then the neutrality condition in the compensated state is,

$$[O^-] \sim [N^+] \quad . \tag{9.14}$$

The conductivity then becomes

$$\sigma \sim P_{O_2}^{-1/2} \quad . \tag{9.15}$$

There is a more direct way to investigate if O^- or O_2^- plays the dominant role. We have measured the change of the electrical conductivity of SnO_2 [27] as a function of time after exposure to a low oxygen pressure.

We find quite generally that the conductivity behaves as

$$\frac{d\sigma}{dt} \sim - \sigma^2 \quad . \tag{9.16}$$

In order to analyse the time dependence of the conductivity (9.5) - (9.7) have to be evaluated for a non equilibrium situation,

$$\frac{d[O_2^-]}{dt} = k(O_2^-)[O_2][e] - k'(O_2^-)[O_2^-] + k(O^-)[O_2^-][e] - k'(O^-)[O_2^-]^2 \quad ,$$

$$\frac{d[O^-]}{dt} = 2k(O^-)[O_2^-][e] - k'(O^-)[O^-]^2 \quad . \tag{9.17}$$

The initial condition is

$$[e] = [e]_o \quad , \tag{9.18}$$

where e_o refers to the electron density after the metal oxide has been exposed to a reducing atmosphere. The rate constants k, k' are related to the equilibrium constant K through,

$$K = k/k' \quad . \tag{9.19}$$

From the neutrality condition follows that,

$$\frac{d[e]}{dt} = - \frac{d[O^-]}{dt} - \frac{d[O_2^-]}{dt} \quad . \tag{9.20}$$

We find that the above set of equations leads to the experimental result (9.16) only if,

(i) $[O_2^-] \ll [O^-]$, or (9.21)

(ii) $k(O_2^-) \gg k(O^-)$. (9.22)

These results suggest that for SnO_2, the chemisorbed species of oxygen plays the dominant role.

X) Calatytic effects and electrical conductivity

The chemical activity of molecular oxygen in oxidation reactions is rather small. This can be understood on the basis of the high dissociation energy of oxygen,

$$O_2 \rightarrow 2O - 118\ \text{kcal/mol} \quad . \qquad (10.1)$$

Therefore in the absence of a catalyst e.g. homogeneous oxidation of hydrogen begins only at 400°C - 500°C,

$$H_2 + \frac{1}{2} O_2 \rightarrow H_2O + 57.8\ \text{kcal/mol} \quad . \qquad (10.2)$$

Another example is,

$$CO + \frac{1}{2} O_2 \rightarrow CO_2 + 67.6\ \text{kcal/mol} \quad . \qquad (10.3)$$

The reaction can be significantly enhanced in the presence of a calalyst:

$$R + \frac{1}{2} O_2 \xrightarrow{cat} RO_2 \quad . \qquad (10.4)$$

Already in 1817 Davy observed the oxidation of hydrogen over Pt. In this section we want to investigate how the reaction (10.4) affects the electrical conductivity of a metal oxide. The conductance change has been studied with a large number of oxides [22]. The general result is,

$$\sigma \sim P_R^{\beta} \quad , \qquad (10.5)$$

where β is found in the range,

$$0.3 < \beta < 0.8 \quad . \qquad (10.6)$$

Eq.(10.5) is valid only if P_R is higher than some critical pressure P_c. In the so-called associative [25, 26] model one assumes that the catalytic reaction proceeds directly between the adsorbed species R and the chemisorbed oxygen O^-,

$$R + O^- \rightarrow RO + e \quad , \qquad (10.7)$$

$$RO \rightarrow RO(g) \quad . \qquad (10.8)$$

In the so called redox model [25, 26] the following reaction steps are assumed,

$$R + O_s \rightarrow RO + V_s^+ + e \quad , \qquad (10.9)$$

$$O^- + V_s^+ \rightarrow O_s \quad . \qquad (10.10)$$

Here the index s refers to surface species. In both cases, the initial rate for the increase of the electron concentration in the conduction band becomes

$$\frac{d[e]}{dt} = k_{cat} \cdot [R] \quad . \qquad (10.11)$$

In our investigations on SnO_2 we have studied the rate of the electrical conductivity, after exposure of the oxide to a reducing environment. For both H_2 and CO we find that,

$$\frac{d\sigma}{dt} \sim P_R \quad . \qquad (10.12)$$

Comparison of (10.11) and (10.12) shows that the adsorption is described by a Henry law.

If after exposure to a reducing gas the sensor is again exposed to air or oxygen, we find that the conductivity initially decays to its original value according to

$$\frac{d\sigma}{dt} \sim - \sigma^2 \cdot P_{O_2}^{1/2} \quad . \tag{10.13}$$

But in this case, the time constant is much slower than that discussed in section IX. From a detailed analysis [27] we find, that the dependence of $d\sigma/dt$ on both σ and P_{O_2} can be explained by the redox mechanism for the catalytic reaction under the condition that surface reduction dominates, The steady state conductivity is determined by

$$\frac{d[V_s^+]}{dt} = k(V_s^+)[V_s^+]^2 - k_{cat} P_R = 0 \quad . \tag{10.14}$$

The electron concentration follows either from the steady state condition for the compensation or from the dominance of surface vacancies. In both cases the conductivity is given by

$$\sigma \sim P_R^{1/2} \quad . \tag{10.15}$$

Deviation from (10.15) are explained in Ref.[27] on the basis of a more detailed model for the catalytic process.

References:

1. F.A. Kröger: The Chemistry of Imperfect Crystals (North-Holland, Amsterdam, 1974), Nonstoichiometric Oxides, edited by O.Toft Sorensen, (Academic Press 1981)
2. J. Rudolph, Z.f. Naturforschung 14a, 727 (59)
3. K.I. Hagemark, J. Solid State Chem. 16, 293 (1976)
4. G.D. Mahan, J. Appl. Phys. 54, July (1983)
5. R. Einzinger, in Grain Boundaries in Semiconductors, edited by H.J.Leamy, G.E. Pike, and C.H. Seager (North-Holland, New York, 1982), p.343
6. S. Samson and C.G. Fonstad, J. Appl. Phys., 44, 4418 (73)
7. P. Amigues and S.J. Teichner, Discussion. Faraday Soc. 41, 362 (1966)
8. K. Ihokura, New Materials & New Processes 1, 43 (1981)
9. A. Jones, T.A. Jones, B. Mann and J.G. Firth, Sensors and Actuators 5, 75 (1984)
10. E. Sonder, M.M. Austin and D.L. Kinser, J. Appl. Phys. 54, 3566 (1983)
11. J.D. Levine, Crit. Rev. Solid State Sci. 5, 597 (1975)
12. J. Bernasconi, H.P. Klein, B. Knecht, and S. Strässler, J. Elec. Mat. 5, 473 (1976)
13. J. Bernasconi, S. Strässler, B. Knecht, H. Klein, and A. Menth, Sol. St. comm., 21, 867 (1977)
14. G.D. Mahan, L.M. Levinson, and H.R. Philipp, J. Appl. Phys. 50, 2799 (1979)
15. J. Bardeen, Phys. Rev. 71, 171 (1947)
16. H.J. Krusemeyer and O.G. Thomas, J. Phys. Chem. Solids 4, 78 (1958)
17. T.A. Goodwin and P. Mark, Progress in Surface Science 1, 1 (1972)
18. C.D. Kohl, Appl. Phys. A30, 127 (1983)
19. P.K. Clifford and D.T. Tuma, Sensors and Actuators 3, 233 and 255 (1983)
20. S. Strässler and A. Reis, Sensors and Actuators 4, 465 (1983)

21. N. Murakami, K. Tanaka, K. Sasaki and K. Ihokura, Ref. (22), p. 187
22. Proceedings of the International Meeting on Chemical Sensors, Fukuoka, Japan, T. Seiyama (editor), Elsevier, Amsterdam 1983
23. H. Chon and J. Pajares, J. Catalysis 14, 257 (1969)
24. Shin-Chia Chang, in Ref (22), p. 78
25. J. v.d.Berg, A.J. v.Dillen, J. v.d.Meijden, J.W. Geus in J.P. Bonelle et.al. (eds), Surface Properties and Catalysis by Non-Metals, 493-532, D. Reidel Publishing Co., 1983
26. G.I.Golodets, Studies in Surface Sciences and Catalysis 15, Elsevier 1983
27. A. Reis, S. Strässler and D. Wieser, Proc. 6th General Conference of EPS, Prague 1984

Evolution of Physical Models for ZnO-Varistors

Richard Einzinger

Siemens Research Laboratories, Otto-Hahn-Ring 6
D-8000 München 83, Fed. Rep. of Germany

The highly nonlinear conduction mechanism in ZnO based ceramics has attracted growing interest during the past decade. With increasing experimental experience model approaches have drifted from mechanisms related to massive non-zinc oxide intergranular layers like space charge limited current or Nordheim-Fowler tunneling to grain boundary effects where ZnO itself plays the dominant role. Firstly several attempts were made with surface state density distributions of foreign atoms. To allow for the measured high nonlinearity coefficient α in the range of $\alpha = 70$ or even more, additional effects have been taken into account arising from the generation of minority carriers - voltage dependent shortening of the spatial distance of energy states which are involved in quantum mechanical tunneling, or the barrier height control by capture of holes at surface states. In fact recent experimental work gives strong evidence for the presence of minority carriers above threshold voltage.

I. Introduction

Zinc oxide is a widely applied material throughout science and technology - from an antitoxic ingredient in medical liniments to a catalyst in the production of rubber or a piezoelectric substrate material for surface wave generation. In the late nineteen sixties a new application appeared - the zinc oxide varistor. The acronym varistor stands for variable resistor - a component which changes its resistivity as a function of the applied voltage. In combination with several other additive metal oxides like bismuth-, cobalt-, antimony-, manganese- and chromium oxide and organic binder substances, zinc oxide powder is compacted, for example, in the form of discs and subsequently fired at temperatures above 1200°C to give a ceramic body with excellent varistor properties. The figure of merit is the nonlinearity coefficient α (sometimes $\beta = 1/\alpha$) in the empirical power law

$$I = KV^{\alpha} \tag{1}$$

where K is a material constant. The α-value indicates the number of decades in current increase if the applied voltage increases by one decade. For standard varistor materials like SiC or TiO_2, the maximum α-values range between 2 and 7, whereas modern ZnO varistors get to values of 70 or even more. As shown in Fig. 1, this results in an almost rectangularly shaped symmetrical I-V-curve similar to a series connection of back-to-back Zener diodes.

Therefore ZnO varistors are favourably used as transient voltage suppression devices, with a high degree of adaptation to different technical demands allowed by the ceramics fabrication process. Thus the geometry of today's ZnO varistor devices ranges from small discs of a few millimeters diameter for electronic circuitry up to huge lightning arresters of several meters height used in energy distribution systems.

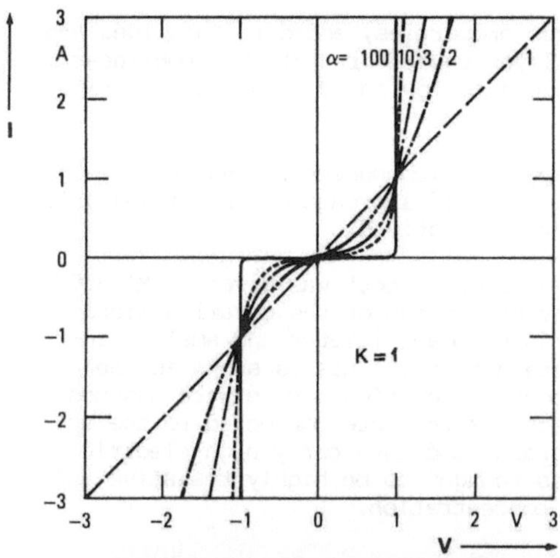

Fig. 1. Principle I-V-curves of varistors with different non-linearity coefficients

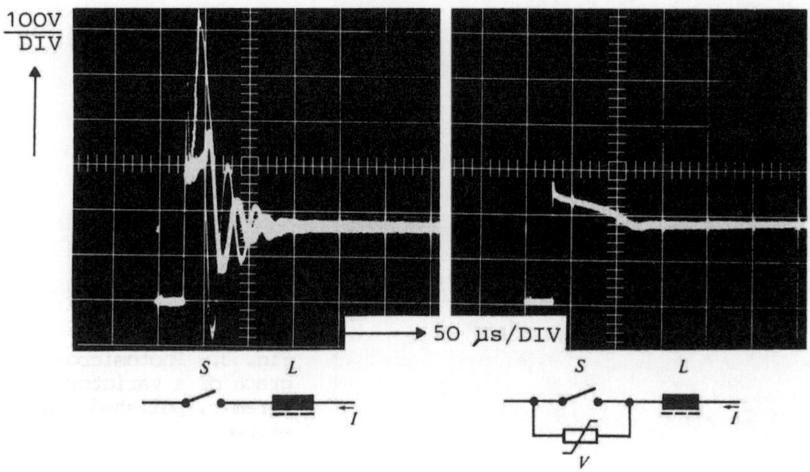

Fig. 2. Voltage vs time plot of transient (a) unprotected (b) ZnO varistor-protected contact

The application list covers the total area of electrotechnology and electronics. The main task is protection from destructive voltage transients which are either generated by the systems themselves, e.g. by switching of inductive loads, or by external events like lightning flashes. Figure 2 gives an example of transient voltage suppression for a telephone relay contact.

II. Development of Experimental Experience and Related Interpretations during the Last Decade

Nonlinear electrical conductivity of ZnO ceramics with a single additive, e.g. SiO_2, SnO_2, TiO_2, Bi_2O_3 had been investigated long before [1, 2]. Even then the varistor mechanism was supposed to be a grain boundary (G.B.) phenomenon. The relatively high dielectric constant, compared to pure ZnO ceramics which has no varistor properties, added to this idea and was investigated for use in boundary layer capacitors [3]. The nonlinear exponent α of binary systems nevertheless is low and does not exceed SiC-values.

Only after MATSUOKA [4] had developed multicomponent ZnO varistor ceramics with α-values up to 50, could a rapid increase of industrial interest, followed by scientific engagement, be noticed.

The first interpretation of the "varistor effect" was given by MATSUOKA [4]. It started from microscopical investigations of the ceramics microstructure, where ZnO grains of about 10 µm mean diameter and smaller intermediate layers of bismuth oxide were detected. This is shown in the photomicrograph of Fig. 3. Heterogeneous segregation layers were assumed to be the origin of the varistor effect, as they are connected to the applied voltage via low-resistive ZnO grains and thus carry high electric fields. The intergranular material was thought to be highly resistive and populated by electron traps in large concentration.

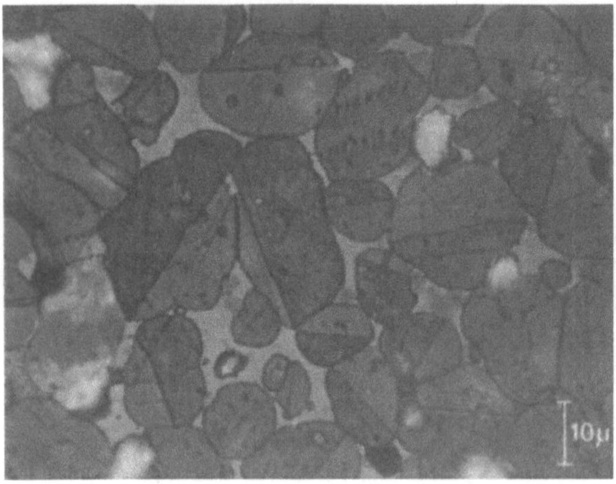

Fig. 3. Photomicrograph of a varistor ceramic, polished and etched

The breakdown mechanism was discussed in terms of the well-known space charge limited current phenomenon (SCLC)[5], where electrons which are injected at moderate rates are mostly trapped and do not contribute to the net current. At a certain injection level the traps get filled, thus the current increases steeply with applied voltage. Thus the SCLC I-V-response largely resembles a varistor I-V-curve. The threshold voltage, called "trap filled limit" (TFL), is given by

$$V_{TFL} = \frac{q}{\varepsilon\varepsilon_0} N_t d^2 \qquad (2)$$

where q is the elementary charge, $\varepsilon\varepsilon_0$ the permittivity, d material thickness, N_t trap density. V_{TFL} exhibits a square relationship on the material thickness d and a linear dependence on trap density N_t. This will be of importance later. From measurements of the threshold electrical field of macroscopic ceramics E_{th} = 100 V/mm and microscopic observation of d ~ 1 µm a threshold voltage for single intergranular layers of V_{th} ~ 1 V was estimated [4].

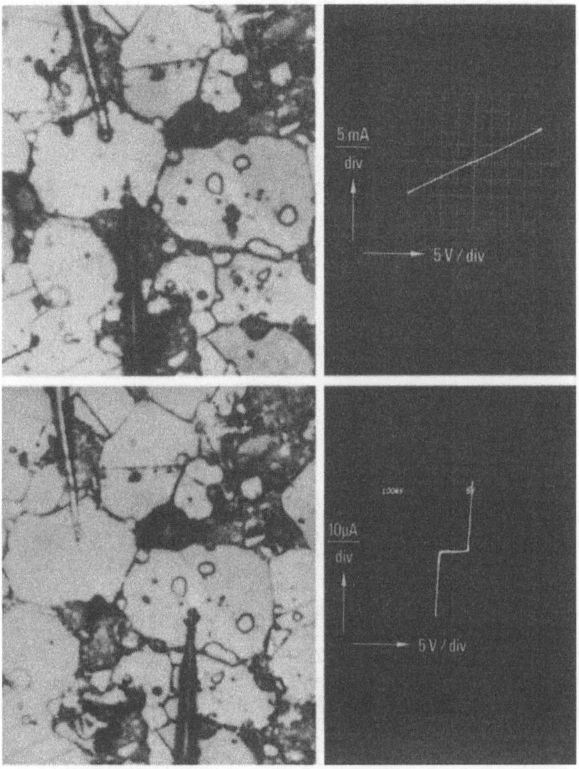

Fig. 4. Microcontact measurement on single grain (above) and grain pair (below)

At that time the present author tried to verify experimentally the above model approach. The aim was to measure threshold voltage directly in microscopic dimensions, i.e. on single intergranular layers. The measurement technique involved fine, electrochemically-etched tungsten wires with tip diameters of a few microns, which were micromanipulated on the polished ceramics surface in order to give contact to interesting material sections. These "microcontact measurements" at first confirmed that the varistor effect is located in the G.B. region [6]. Fig. 4 shows a specimen where the microprobes are welded onto a single ZnO grain by a capacitor discharge. The grain proves to be linearly resistive. 4-probe measurement indicated a specific resistance of the order $1\ \Omega cm$. By the displacement of one probe to the neighbouring ZnO grain the varistor effect shows up, with a threshold voltage of 3.4 V which is three times the previous value given by MATSUOKA. Measurements on hundreds of grain pairs in ceramics with different compositions exhibited that the threshold voltage is almost a constant, as shown in Fig. 5.

This experience did not correspond to the above model of SCLC, if a constant thickness of intergranular layers and a constant trap density had not to be assumed. Other contacting techniques like evaporation or sputtering of metal spots confirmed the narrow 3.4-V-distribution. Similar results were also reported by other workers [7, 8, 9].

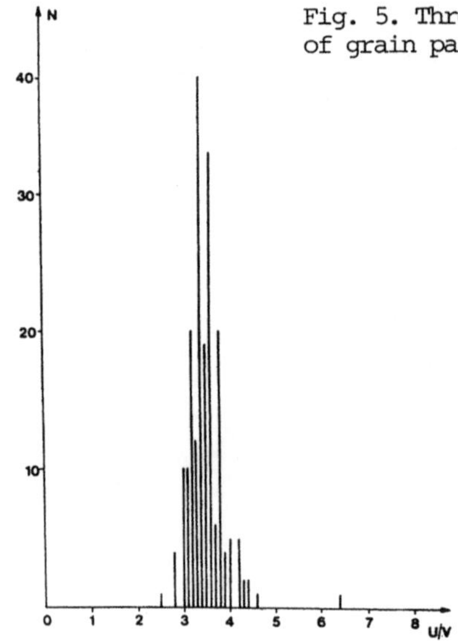

Fig. 5. Threshold voltage distribution of large number of grain pairs $N_{total} = 200$

Massive intergranular layers (d > 1 µm) as the origin of the highly nonlinear conductivity were finally ruled out by microcontact measurements on specimens of thinned ceramics with a defined $ZnO-Bi_2O_3-ZnO$ test area. They showed nonlinear coefficients not larger than $\alpha = 3$ up to voltages higher than the 3.4 V threshold by a factor 20. Above irreversible breakdown occurred.

A second model approach was given by LEVINSON AND PHILIPP [10]. They proposed much thinner non-ZnO-intergranular layers of the order of 100 Å thickness. They were discussed in terms of the Nordheim-Fowler tunneling mechanism. Here the interface layer makes up a roughly rectangular potential barrier in the energy band diagram, which is too high for charge carriers to be surmounted thermionically at small applied voltage and too thick to be conquered by quantum mechanical tunneling. At higher applied voltages the barrier is squeezed to a triangular shape, whereby the tunneling distance narrows and substantial tunnel current with highly nonlinear I-V-relationship sets in.

A disadvantage of this approach was, as in the previous one, that the measured narrow threshold voltage distribution implied a constant interface layer thickness which is improbable for a ceramics sintering mechanism.

The existence of varistor active massive intergranular layers was also doubted by an electron microscope work of CLARKE [11]. By lattice fringe imaging it was found that most of the G.B.s of real varistors are devoid of any other material than zinc oxide itself. This could be seen from lattice fringes of the grains reaching up to the G.B. in an undisturbed manner. This is demonstrated in Fig. 6.

The measured threshold voltage of single G.B.s of 3.4 V and its stability in different ceramic compositions where only ZnO is the common ingredient drew the attention to possible varistor mechanisms based on properties of

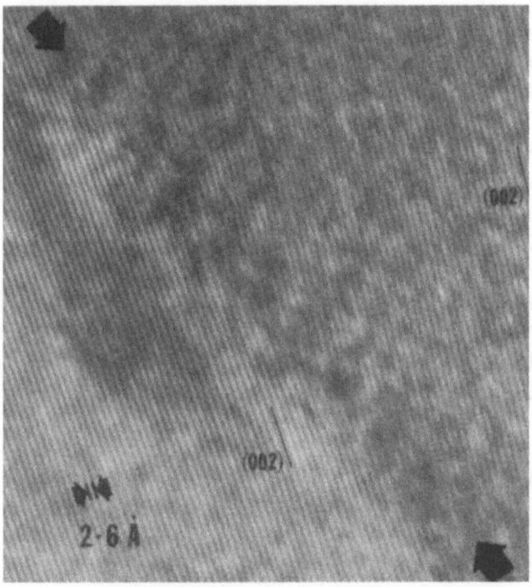

Fig. 6. Lattice fringe image of a G.B. segment in a commercial ZnO varistor material. The boundary plane is indicated by the arrows. The (002) lattice fringes in the adjacent grains continue up to the boundary, which is devoid of an intergranular film. From CLARKE [11]

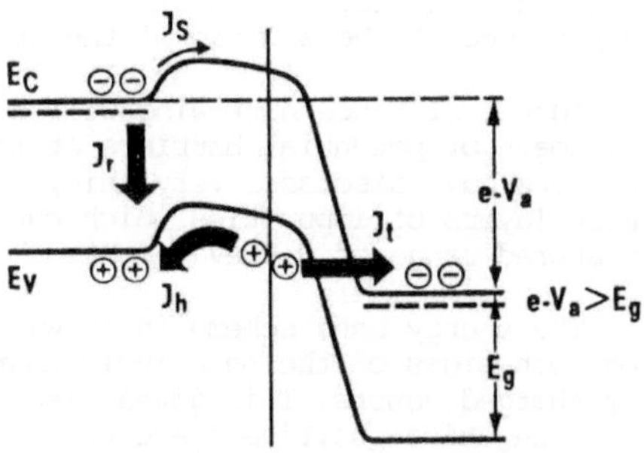

Fig. 7. Band-to-band tunnelling. J_t tunnel current, J_r indicates recombination mechanism, J_s thermionic over-barrier current. From [12]

the zinc oxide itself. The almost equivalent gap voltage 3.2 V of ZnO added to this idea, which was formulated by the present author in terms of a band-to-band tunneling mechanism [12].
Fig. 7 gives the principle outline. The form of the potential barrier differs from the previous models, as here no surface states are assumed to be responsible, but rather a compensation of donors within the grains near the G.B.

Thus, roughly, an npn double diode-like energy band scheme applies. With bias voltage V_a the decrease of electron potential energy on the positively biased side permits a non-zero tunnel probability if valence band states of the negatively biased grain correspond in energy to conduction band states of the positively biased grain. This holds for

$$eV_a > E_g - \Phi(V_a) \tag{3}$$

The I-V-response which was calculated for the band-band tunnel effect by a formulae given by SZE [13]

$$J_t = \frac{\sqrt{2m_n^*}\, q^3 F V_a}{4\pi^2 \hbar^2 E_g^{1/2}} \exp\left(-\frac{4\sqrt{2m_n^*}\, E_g^{3/2}}{3q\hbar F}\right)$$

$$F = \frac{V_a + V_g}{d_i}, \quad E_g = qV_g, \quad m_n^* = 0,27 \tag{4}$$

d_i = tunnel distance

clearly indicated that this approach can not explain α-values higher than α = 25 if an additional effect is not taken into account which had already been introduced by the Nordheim-Fowler mechanism: a voltage dependent narrowing of the tunnel distance d_i.
If this effect was a realistic input, I-V-curves of real varistor components could be realized as well as the small temperature dependence $Tk_V = dV/VdT$ above threshold, which here is given by the temperature dependence of the ZnO gap of the order of 10^{-4}/K.

Tunnel distance narrowing will be treated below. At present the establishment of potential barriers at the ZnO-ZnO G.B. is important. Several workers have discussed very thin, ideally monoatomic, thus hardly detectable layers of impurities which can result in potential barriers in the measured range of 0,6 eV height.

The energy band scheme is shown in Fig. 8, where electrons are depleted on both sides of the negatively charged G.B. and thereby uncover positively charged donors. This gives rise to a spacial distribution ρ(x) of positive-negative-positive space charge. Insertion into poisson's equation (for one dimension)

$$\frac{\delta^2 \Phi(x)}{\delta x^2} = - \frac{\rho(x)}{\varepsilon\varepsilon_0} \tag{5}$$

and double integration ends at the parabolic potential barrier of Fig. 8. The voltage dependence of the barrier height and thereby the I-V-relationship is given by [14]

$$\Phi(V) = S^2 \left(1 - \frac{V}{4\Phi_0} \frac{1}{S^2}\right)^2 \Phi_0 \tag{6}$$

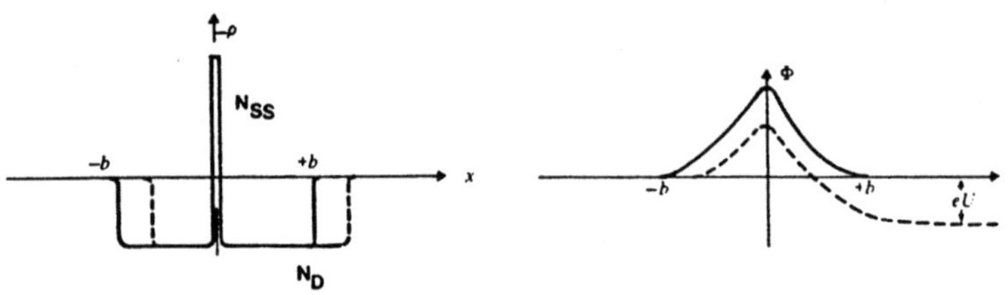

Fig. 8. Space charge (left) and potential barrier (right) for grain-grain contact with charged surface states

In eq. (6) $S = Q_{SS}/Q_{0SS}$ holds for the surface charge ratio at non zero and zero bias, Φ_0 is the potential barrier at zero bias. If S = 1, Φ(V) tends towards zero at $V = 4\Phi_0$, i.e. 2.4 V, with α-values up to 20 in the 1 V to 2 V range. This is in clear discrepancy to the measured 3.4 V-threshold and the weak nonlinearity 1 < α < 5 of the prebreakdown region.

Therefore several attempts were made to find surface state density relations on energy $N_{SS}(E)$ which could provide an S(V)-dependence able to back up the potential barrier Φ(V) up to threshold voltage. At V_{th} a rapid decrease of Φ(V) should then allow for high α-values.

Simple analytic $N_{SS}(E)$-relationships like constant, exponentially rising, exponentially decreasing and monoenergetic functions were tested [15, 16, 17]. None of them was successful in the above sense.

Possibly very specific $N_{SS}(E)$-functions could be fitted. These surely would be true only for specific impurities. In contradiction, as already mentioned, ZnO varistors are produced with very different additives.

Measurements on varistor ceramics were performed in terms of I-V-deconvolution for $N_{SS}(E)$, as well as capacitance measurements [18]. For different samples different distributions were analyzed. Here a principle problem with measurements on macroscopic varistor ceramics arises. Several chemical phases are present besides cobalt-doped ZnO which can basically contribute to the macroscopic electrical properties. This had been demonstrated by the present author in measurements on the microscopic scale by an ultrasonic cutting technique [19]. Islands of grain-grain junctions were prepared by separating the test area from its surrounding ceramics. In Fig. 9 the specimen includes a small triangle of bismuth oxide at the lower end of the G.B. After I-V-curve measurement this small heterogeneous inclusion was split off. The subsequent I-V-measurement on the purified junction clearly revealed that the prebreakdown behaviour of the junction had been masked by the bismuth oxide bypass. The same holds for capacitance properties, as shown in Fig. 10.

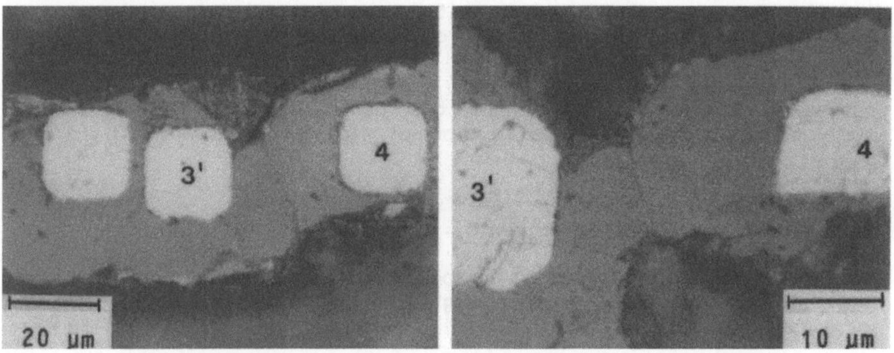

Fig. 9. "Island" specimen with Bi_2O_3-contaminated G.B. (left) and purified G.B. (right)

Therefore, the interpretation of measurements on macroscopic ceramics can be misleading if they are related to the true varistor junctions.

The contribution of bismuth oxides phases to the leakage current as well as to degradation phenomena is rather probable. The stabilisation of high-temperature metastabile β-, γ- and δ-Bi_2O_3-phases at room temperature by dopants as applied in varistors like Sb, Cr, Mn, Co is well known in literature [20]. X-ray analysis in component ceramics indicated the presence of β-, γ- and δ-phases [21]. Electrical measurements in the Bi_2O_3-phase system are known for higher temperatures [22]. An extrapolation down to room temperature indicates that the δ-phase largely contributes to the electrical conductivity, even in small compositional amount [23]. Phase transformations from less conductive to more conductive phases triggered by heavy duty action with hot channel formation could result in degradation phenomena with growing leakage current.

These phenomena had to be contended with in real varistors. Though technical countermeasures have been developed, the physical background is poorly

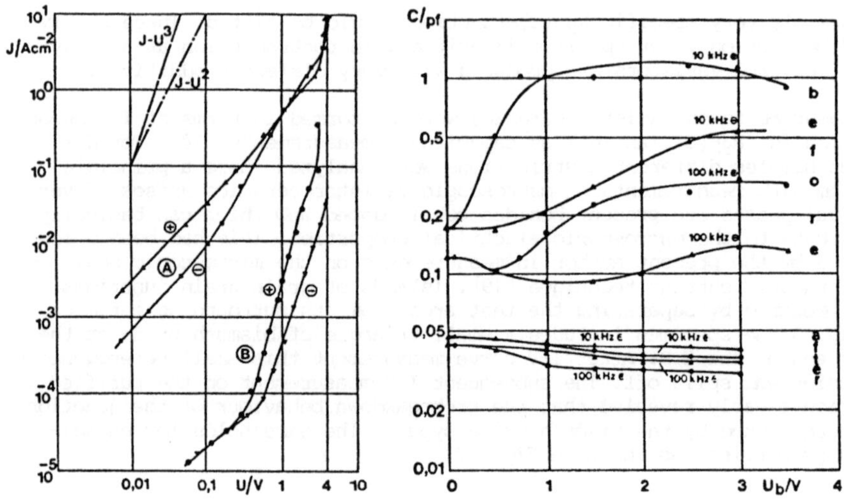

Fig. 10. I-V-curve (left) and capacitance vs bias (right) before (A) and after (B) elimination of Bi_2O_3 from G.B.

understood. To date, degradation has been discussed mainly for the varistor active potential barriers in terms of zinc ion transport [23].

III. Recent Models and Experiments

More recent approaches tried to introduce additional mechanisms for the control of the potential barrier height, as $N_{SS}(E)$ modelling did not solve the problem.

III.1. First, MAHAN et al. [25] proposed a modified tunnel mechanism. As sketched in Fig. 11, the potential barrier is governed by a charged oxide layer of about 30 Å thickness which is composed of "another oxide than zinc oxide". Above a certain applied voltage the hole chemical potential μ_H drops below the valence band edge of the negatively biased grain.

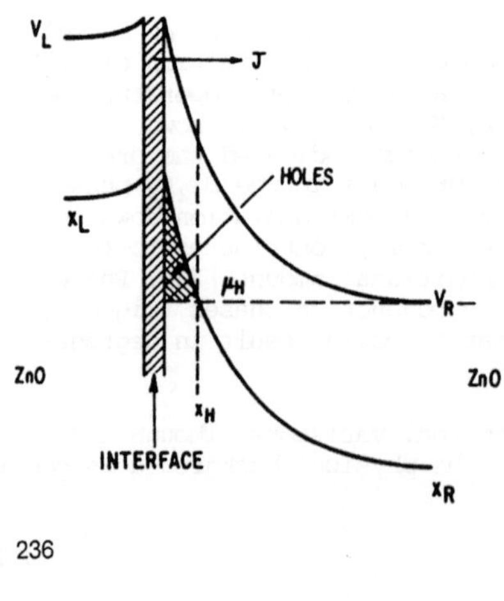

Fig. 11. MAHAN's hole model. Electrons tunneling from the interface are indicated by the arrow labelled J

Therefore holes appear at the G.B. This hole population is assumed to have a low recombination rate and introduces an additional positive space charge next to the negatively charged oxide layer. This is followed by a drastic deformation of the normally parabolic band edge decrease at the positively biased grain in terms of a steeper levelling off. Thereby electrons which are trapped in the interface oxide have a larger probability to tunnel out. With increasing bias the interface loses its potential barrier-supporting negative charge very progressively. Accordingly, the potential barrier decreases until the flat band condition is reached on the negative side.

This model of a tunneling electron-controlled barrier with voltage dependent tunnel length was extensively evaluated for I-V-response and is in good agreement with the measured threshold voltage 3.4 V, its temperature coefficient and high α-values. Nonetheless the assumed quasi static hole population has been doubted [26], as holes next to trapped electrons should undergo substantial recombination. Besides, varistor active oxide layers of non-ZnO material have not been detected experimentally.

III.2. A new input originated from PIKE [27]. It starts from the known potential barrier model with charged surface states at the G.B. As indicated in Fig.12, holes are created by impact ionisation of valence band states by conduction band electrons which have gained enough energy from the electric field.

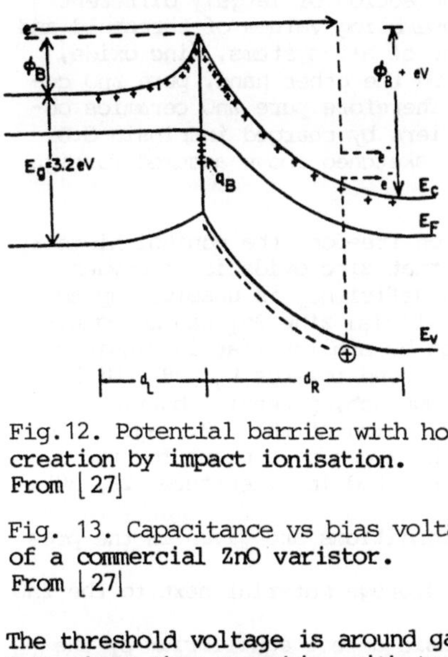

Fig.12. Potential barrier with hole creation by impact ionisation. From [27]

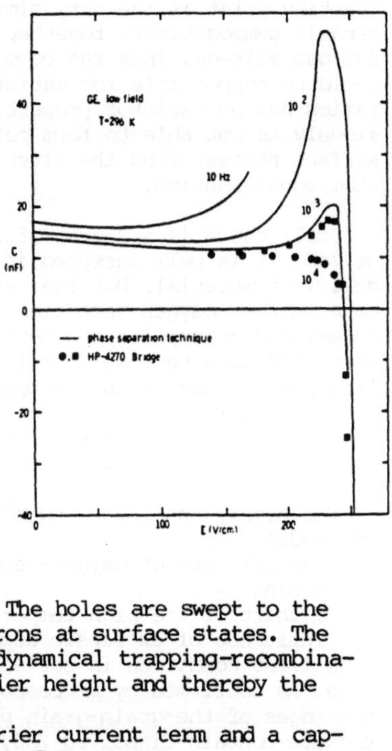

Fig. 13. Capacitance vs bias voltage of a commercial ZnO varistor. From [27]

The threshold voltage is around gap voltage. The holes are swept to the G.B. where they recombine with trapped electrons at surface states. The net charge at the G.B. states is given by a dynamical trapping-recombination balance. It controls the potential barrier height and thereby the varistor current-voltage relationship.
The varistor current is given by an over-barrier current term and a capture-reemission term for the charge balance at G.B. states.

$$J = J_{over} + J_{bal}$$

Both contributions undergo bias dependent, non-zero angle phase relationships which can explain a strange, experimentally evaluated frequency dependent capacitance - voltage relationship. It is shown in Fig. 13.

Near threshold voltage, the capacitance rises with applied bias voltage. This is due to the increase of the capacitance relevant +90 ° -out-of-phase term in J_{bal}. At threshold, the capacitance drops and goes to negative values. According to PIKE's work, this can be explained by the change in polarity of charge carriers when minority carriers (holes) appear. Then the phase relationship changes to -90 ° out of phase, indicating negative capacity. Thus the appearance of negative capacitance is seen as a clue to the presence of minority carriers at and above threshold voltage. To the present author's knowledge an I-V-curve evaluation of PIKE's model has not yet been published. One difficulty involved may be the lack of experimental data on minority carriers in ZnO, as bulk p-type ZnO is unknown. A second point is the prebreakdown region, where again specific $N_{SS}(E)$-functions must be arbitrarily assumed, as in previous models, in order to back up the potential barrier until PIKE's mechanism of minority carrier controlled barrier sets in.

A confusing situation arises from the experiment of Fig. 10, where the capacitance increase with bias voltage must be attributed to the ZnO-Bi_2O_3-ZnO-structure and not to the true ZnO-ZnO varistor junction.

III.3. The present author's approach steps in from another side. The starting point is the very similar varistor action of largely different ceramic compositions, together with the resembling values of threshold and ZnO gap voltage. Thus the common ingredient of all systems, zinc oxide, could be responsible for varistor action. On the other hand, pure ZnO ceramics has no varistor properties at all. Therefore pure ZnO ceramics obviously is not able to form potential barriers by charged intrinsic G.B. surface states. Thus the literature models sketched above assumed foreign atom contamination.

Still there is a degree of interpretation freedom, the nonstoichiometry of ZnO. It is well accepted in literature that zinc oxide is an oxygen deficient material. But the nature of this deficiency is unsolved up to date. Either oxygen vacancies V_U, or interstitial zinc Zn_i can be equally assumed to explain the n-type conduction by similar flat intrinsic donors. ESR data have set the interpretation trend towards V_U [28]. With this presupposition in the present model approach, potential barriers are formed by a loss of donors because of reoxidation of the G.B. region. For ZnO single crystals, surface treatment with oxygen as a means for the establishment of nonlinear conduction is reported in literature [28, 30].

For highly nonlinear ZnO ceramics two conditions are given in the present model [31]:

i. the presence of oxygen-(and zinc) ion storage material next to the ZnO grains
ii. a shift of the high-temperature intrinsic defect equilibrium by the influence of extrinsic dopants.

The first condition is satisfied by the varistor ceramics microstructure. As confirmed by microscopy, bismuth oxide "pockets" are present at the edges of the grain-grain packing. X-ray analysis confirmed that these pockets contain dissolved ZnO, in accordance with the phase diagram Bi_2O_3-ZnO [32].

During sintering at high temperatures above 1200 °C, the zinc oxide lattice is in equilibrium with its surrounding atmosphere. The calculation of defect concentrations of pure ZnO, performed with a set of mass reaction equations [31,40] and KRÖGER's energy level diagram and mass reaction constants [33] shows that donor active oxygen vacancies overwhelm acceptor active zinc vacancies in concentration. During the cooling period, equilibrium defect concentrations want to decrease. As the defect mobility vanishes rapidly with falling temperature, the grain interior "freezes in" high-temperature relevant defect concentrations. At room temperature, a redistribution of charges must be considered because of the much easier establishment of electronic equilibrium. As shown in Fig. 14, n-type conductivity with electron concentrations of the order of 10^{17} cm^{-3} is achieved, if the ionic equilibrium is frozen above 1100 °C. The calculation is performed by semiconductor Fermi statistics on doubly ionizable donors and acceptors [34,35], where the charged species n, p, singly ionized $V_O{\cdot}$ and V_{Zn}' and doubly ionized $V_O{\cdot\cdot}$ and V_{Zn}'' obey the neutrality condition

$$n + [V_{Zn}'] + 2[V_{Zn}''] = p + [V_O{\cdot}] + 2[V_O{\cdot\cdot}] \qquad (7)$$

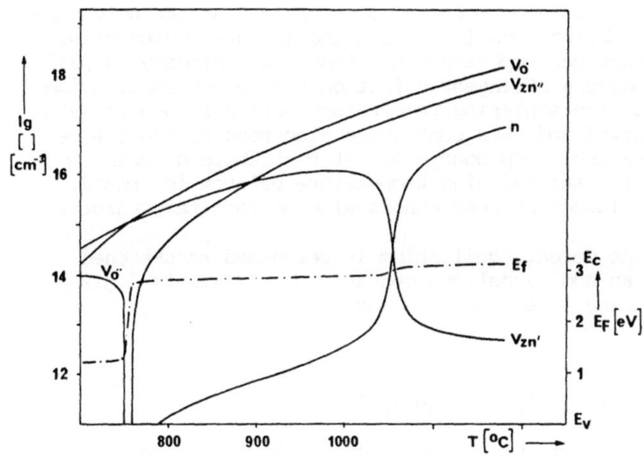

Fig. 14. Calculated concentrations of charged species and Fermi level in relation to "freezing temperature" of high-temperature equilibrium

The calculation involves the known "frozen-in" high temperature concentrations

$$[V_O] = [V_O^x] + [V_O{\cdot}] + [V_O{\cdot\cdot}] \qquad (8)$$

$$[V_{Zn}] = [V_{Zn}^*] + [V_{Zn}'] + [V_{Zn}''] \;, \qquad (9)$$

the known energy levels E_{d1} of V_O^x, E_{d2} of $V_O{\cdot}$, E_{a1} of V_{Zn}' and E_{a2} of V_{Zn}'', with the convention $E_v = 0$. The room temperature concentrations are then related to the unknown Fermi level E_f by

$$[V_O{\cdot}] = \frac{[V_O]\; 2\exp\,(E_{d1} - E_f)/KT)}{1 + \exp\,((E_{d1} + E_{d2} - 2E_f)/KT) + 2\exp\,((E_{d1} - E_f)/KT)} \qquad (10)$$

$$[V_O{\cdot\cdot}] = \frac{[V_O]\; \exp\,((E_{d1} + E_{d2} - 2E_f)/KT)}{1 + \exp\,((E_{d1} + E_{d2} - 2E_f)/KT) + 2\exp\,((E_{d1} - E_f)/KT)} \qquad (11)$$

$$[V_{Zn}'] = \frac{[V_{Zn}]\; 2\exp\,((E_f - E_{a1}/KT)}{1 + \exp\,((2E_f - E_{a1} - E_{a2})KT) + 2\exp\,((E_f - E_{a1})/KT)} \qquad (12)$$

$$[V_{Zn}''] = \frac{[V_{Zn}] \exp (2E_f - E_{a1} - E_{a2})/KT}{1 + \exp ((2E_f - E_{a1} - E_{a2})/KT) + 2\exp ((E_f - E_{a1})/KT)} \quad (13)$$

$$n = N_C \exp ((E_F - E_C)/KT) \quad (14)$$

$$p = N_V \exp ((E_V - E_F)/KT) \quad (15)$$

After solving eq. (7) for E_f the charged species are calculated from eq. (10) to eq. (15).

From Fig. 14 one sees that the conductivity decreases rapidly if the high-temperature defect equilibrium follows temperature below 1100 °C. G.B.'s are able to do this. Thus, pure ZnO should form potential barriers because of the defect concentration gradient from high values inside the grain to lowering values towards G.B.

The calculation of the potential barrier involves the establishment of the room temperature space charge distribution along the frozen-in $[V_0]$- and $[V_{Zn}]$-profile and its treatment in Poisson's equation which is two times integrated. As a result, for pure ZnO a potential barrier height below 0.1 eV is calculated, being much lower than the measured value of 0.6 eV of varistors. This fact was emphazised by MAHAN [36]. Already in [31], the present author proposed a stringent effect on the defect equilibrium by extrinsic donors: At high sintering temperature extrinsic donors lying deep in the gap are ionized and thus contribute electrons to the charge balance. At room temperature, deep donors do not contribute directly to the conductivity. But if a shifted high temperature balance is frozen, there is an indirect influence by deep donors also at room temperature.

The calculation of the defect equilibrium is performed as sketched above. It now involves an additional charged species D$^{\bullet}$ (singly ionizable donor) in eq. (7) according to a mass reaction

$$D^x \rightleftarrows D^{\bullet} + e$$

$$\frac{[D^{\bullet}]\, n}{[D^x]} = K_D = \frac{1}{2} N_C \exp (- E_D/KT) \exp \beta_D/K$$

and results in a depression of the native donor V_0. This can be seen in terms of a "condensation" of electrons at appropriate levels as a reaction to increased "electron pressure". As a consequence, an increase of $[V_{Zn}]$ is triggered.

This shift of the defect equilibrium is increased with decreasing temperature. Fig. 15 gives the concentrations $[V_0^{\bullet}]$ and $[V_{Zn}']$ which contribute most to $[V_0]$ and $[V_{Zn}]$, in relation to the extrinsic donor level E_D for three different temperatures.

As a consequence, defect equilibria in the G.B. range, which can follow decreasing temperature, are progressively shifted during the cooling process. Thus the concentration profile, which establishes according to the boundary conditions of grain interior and G.B. defect concentrations, is drastically changed. Fig.16 shows profiles of V_0 and V_{Zn} for undoped and extrinsically doped ZnO with $N_D = 2 \times 10^{18} cm^{-3}$, $E_D = 0.7 eV$, $dT/dt = 0.5$ K/min.

For $N_D = 0$, donor active V_0 are dominant throughout and only a small potential barrier results as mentioned above. For the case of extrinsic doping, V_0 and V_{Zn} profiles intersect near the G.B. with acceptor-active V_{Zn} dominating in the G.B. range.

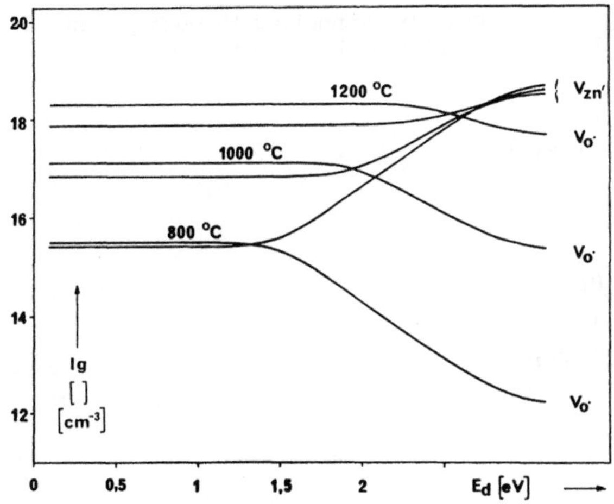

Fig. 15. Calculated high temperature defect equilibrium, singly ionized V_{Zn}' and $V_O\cdot$ concentrations for three different temperatures in relation to extrinsic donor level E_d

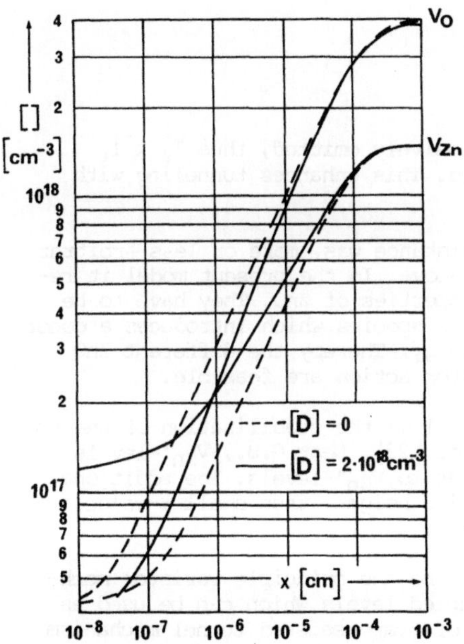

Fig. 16. Calculated concentration profiles of zinc- and oxygen vacancies for undoped and extrinsically doped ZnO near G.B.

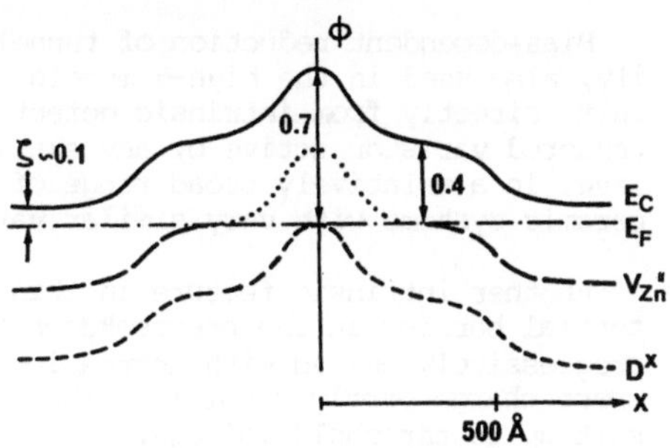

Fig. 17. Potential barrier for extrinsically donor doped ZnO. $\lfloor D \rfloor = 2 \times 10^{18}$ cm^{-3}, cooling velocity $dT/dt = 0.5$ K/min

The calculation of the potential barrier is performed as above and yields an interesting result. As shown in Fig. 17 the barrier includes preshoulders on both sides of the central peak.

They arise from a local pinning of the Fermi level at the V_{Zn}''-level which is 0,4 eV below E_C. The central peak height is given by $E_D - \zeta = 0,7$ eV $- 0,1$ eV $= 0,6$ eV.

With bias voltage $eV < E_g - \Phi(V)$, the preshoulder at the positively biased side contributes fully to the tunnel distance l_1 which is too large for substantial tunnel current. This is shown in Fig. 18. With increased

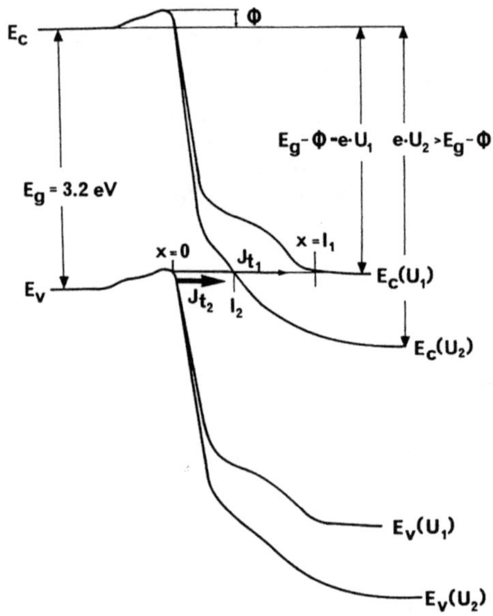

Fig. 18. Tunneling through potential barrier at bias voltage U_1 and U_2. As $l_2 < l_1$, $J_{t_2} >> J_{t_1}$

bias voltage, the preshoulder is progressively omitted, thus $l_2 < l_1$ undergoes a voltage-dependent shortening. This enhances tunneling with high α-values.

Bias-dependent reduction of tunnel distance was, more or less arbitrarily, also used in the high-α models of above. In the present model it results directly from intrinsic defect properties of ZnO. They have to be rendered varistor active by any extrinsic species which introduces a donor level in a relatively broad range of energy. Thereby the different ZnO ceramic systems with very similar varistor action are feasible.

Another intrinsic feature in this model is the stabilisation of the potential barrier in the prebreakdown range [23]. Near G.B., V_{Zn}'-levels progressively unload with increasing bias to V_{Zn}''-levels. The additional space charge partly stabilizes the barrier height, thus small α-values result up to threshold voltage.

At V_{th}, band tunneling is supposed to be the principle varistor mechanism. As the gap contains several occupied levels which can be used as "steps" by tunneling electrons, a sort of trap-assisted tunnel mechanism is to be considered.

At present, this approach has not yet been evaluated for I-V-curves. Recently, there have been doubts on the validity of KRÖGER's mass reaction constants. In [36] the reinterpretation of experimental data yields a new set of constants whereby the room temperature resistivity of pure ZnO rises by a factor 30 compared to measured values. The situation which comes from extrinsic doping has not been considered in [36].

As in the above sketched recent models, the present approach includes the generation of minority carriers above threshold voltage. Besides the appearance of negative capacitance in PIKE's work, optically active hole-electron recombination could also be a strong evidence for minority carriers.

III.4. Recent Experiments

GREUTER performed experiments where electrically biased varistor ceramics was inspected for light output. In fact, light emission was observed [37]. The photon energy was next to gap energy in one type of emission. A subgap emission was additionally detected in the range of 1.5 eV. It is interpreted as a recombination process on a gap state, possibly on the Co^{2+}-level.

SCHWING [35] performed ZnO bicrystal experiments, where two single crystals were sandwiched with an intermediate layer of bismuth oxide plus other dopants. These model varistors have threshold voltages around 3.12 V in the range of maximum α. Highest α-values were above 80. Sometimes a doubled V_{th} of 6 V appeared. It is interpreted as a series connection of two complete micro varistors, which are formed by a recrystallisation of two epitaxial ZnO layers on the crystal surfaces during cooling. This idea was supported by additional experiments where one single crystal was dipped into Zn-rich bismuth oxide melt and then retracted. This specimen contained one complete microvaristor. The forming mechanism is explained in terms of a reoxidation process of the ZnO surface, similar to the present authors approach.

BAUMGARTNER [38, 39] investigated ZnO epitaxial layers in ceramic sandwich specimen. This work first confirmed that epitaxial growth occurs on the G.B.'s from Zn-rich intermediate liquid phases. In contrast to SCHWING's experience, electrical measurements by microprobes showed no varistor effect. The grain-epitaxial layer contact was highly conductive and ohmic, as well as the epitaxial layer itself. The reason for this discrepancy is unsolved to date.

<u>Literature cited:</u>

1 M.S. Kosman and E.G. Pettsold, Uchenye Zapisky Leningr. Gosudarst. Pedagog. 207, 191 (1961)
2 Kh. S. Valeyev and M.D. Mashkovich, J. Techn. Phys. (USSR) 27, 1649 (1957)
3 R.A. Delaney and Kaiser, H.D., IEEE Trans. on Parts, Mat. and Packaging, Vol. PMP2 No. 1/2 (1966
4 M. Matsuoka, Jpn. J. Appl. Phys. 10 (6), 736 (1971)
5 R.H. Bube, Photoconductivity of Solids, Wiley and Sons, New York (1960)
6 R. Einzinger, Ber. Dt. Ker. Ges. 52 (7), 244 (1975]
7 J.T.C. van Kemenade and R.K. Eijnthoven, Ber. Dt. Ker. Ges. 55 (6), 330 (1978)
8 G.D. Mahan, L.M. Levinson and H.R. Philipp, Appl. Phys. Lett. 33 (9], 830 (1978)
9 D.L. Krivanek and P. Williams, Appl. Phys. Lett. 34 (11), 805 (1979)
10 L.M. Levinson and H.R. Philipp, GE Rept. 75, CRD 175 (1975)
11 Dr.R. Clarke, J. Appl. Phys. 49, 2407 (1978)
12 R. Einzinger, Appl. Surf. Sci. 3, 390-408 (1979)
13 S.M. Sze, Physics of semiconductor devices, Wiley, New York (1969)
14 G.D. Mahan, L.M. Levinson, H.R. Philipp, J. Appl. Phys. 50 (4], 2799 (1979)
15 J. Bernasconi, H.P. Klein and Strässler, J. Electron. Mat. 5 (5) (1976)
16 J.D. Levine, CRC Crit. Rev. Sol. St. Sci. 5, 597 (1975)
17 G.E. Pike and C.H. Seager, J. Appl. Phys. 50 (5), 3414-3422 (1979)
18 P.L. Hower, T.K. Gupta, J. Appl. Phys. 50 (7], 4847 (1979)

19 R. Einzinger, Adv. in Ceramics, Vol. 1, 359-374 (1981)
20 G. Gattow and D. Schütze, Z. anorg. Chem. 328, 44 (1964)
21 M. Inada, Jpn. J. Phys. 18 (8), 1439 (1979)
22 H.A. Harwig and A.G. Gerards, J. sol. st. Chem. 26, 265 (1978)
23 R. Einzinger, Dissertation TU München, 168, (1982)
24 T.K. Gupta, W.G. Carlson and B.O. Hall, Mat. Res. Soc. Symposia Proc. 5, New York, 393-399 (1982)
25 G.D. Mahan, L.M. Levinson and H.R. Philipp, Appl. Phys. Lett. 33 (9), 830-832 (1978)
26 G.E. Pike, Symposium on Grain Boundaries, Mat. Res. Soc. Boston (1981)
27 G.E. Pike, Mat. Res. Soc. Symposia, Proc. 5, 369-379 (1982) New York
28 B. Schallenberger and A. Hausmann, Z. Phys. B44, 143 (1981)
29 G. Heiland, Z. Phys. 148, 15 (1957)
30 A. Shimizu et al, Jpn J. Appl. Phys. 17 (8), 1435 (1978)
31 R. Einzinger, Mat. Res. Soc. Symposia Proc. 5, 343-355 (1982)
32 G.M. Safranow, Russ. J. Anorg. Chem. 16 (3), 460 (1971)
33 F.A. Kröger, The Chemistry of Imperfect Crystals, Amsterdam (1964)
34 R. Wernicke, Dissertation RWTH Aachen, 21 (1975)
35 U. Schwing, Dissertation Univ. Karlsruhe, 110 (1984)
36 G.D. Mahan, J. Appl. Phys. 54 (7), 3825-3832 (1983)
37 F. Greuter, Contribution to Symposium on Polykristalline Halbleiter, MPI Stuttgart (Febr. 1984)
38 I. Baumgartner and R. Einzinger, Proc. 5th Int. Round Table Conf on Sintering, Portoroz, Yugoslavia, 367-371 (1981)
39 I. Baumgartner, Dissertation Univ. München, 142 (1983)
40 W. Heywang, Amorphe und Polykristalline Halbleiter, Band 8, Springer-Verlag, 201-237 (1984)

Index of Contributors

MIX
Papier aus verantwortungsvollen Quellen
Paper from responsible sources
FSC® C105338

If you have any concerns about our products,
you can contact us on
ProductSafety@springernature.com

In case Publisher is established outside the EU,
the EU authorized representative is:
Springer Nature Customer Service Center GmbH
Europaplatz 3, 69115 Heidelberg, Germany

Printed by Libri Plureos GmbH
in Hamburg, Germany